THE LOW SURFACE BRIGHTNESS UNIVERSE
IAU Colloquium 171

COVER ILLUSTRATION:
The picture on the front cover is of the giant Low Surface Brightness Galaxy Malin 3. Inset is M51 shown on the same scale. The image of Malin 3 has been contrast enhanced.

A SERIES OF BOOKS ON RECENT DEVELOPMENTS IN ASTRONOMY AND ASTROPHYSICS

Managing Editor, D. Harold McNamara
Production Manager, Enid L. Livingston

A.S.P. CONFERENCE SERIES PUBLICATIONS COMMITTEE

Sallie Baliunas, Chair
Carol Ambruster
Catharine Garmany
Mark S. Giampapa
Kenneth Janes

© Copyright 1999 Astronomical Society of the Pacific
390 Ashton Avenue, San Francisco, California 94112

All rights reserved

Printed by BookCrafters, Inc.

First published 1999

Library of Congress Catalog Card Number: 99-62757
ISBN 1-886733-92-9

Please contact proper address for information on:

PUBLISHING:
Managing Editor
PO Box 24463
211 KMB
Brigham Young University
Provo, UT 84602-4463 USA

Phone: 801-378-2298
Fax: 801-378-4049
E-mail: pasp@astro.byu.edu

ORDERING BOOKS:
Astronomical Society of the Pacific
CONFERENCE SERIES
390 Ashton Avenue
San Francisco, CA 94112 - 1722 USA

Phone: 415-337-2624
Fax: 415-337-5205
E-mail: catalog@aspsky.org
Web Site: http://www.aspsky.org

A SERIES OF BOOKS ON RECENT DEVELOPMENTS IN ASTRONOMY AND ASTROPHYSICS

Vol. 1-Progress and Opportunities in Southern Hemisphere Optical Astronomy: CTIO 25th Anniversary Symposium
ed. V. M. Blanco and M. M. Phillips　　　　　　　　　　　　ISBN 0-937707-18-X

Vol. 2-Proceedings of a Workshop on Optical Surveys for Quasars
ed. P. S. Osmer, A. C. Porter, R. F. Green, and C. B. Foltz　　　ISBN 0-937707-19-8

Vol. 3-Fiber Optics in Astronomy
ed. S. C. Barden　　　　　　　　　　　　　　　　　　　　ISBN 0-937707-20-1

Vol. 4-The Extragalactic Distance Scale: Proceedings of the ASP 100th Anniversary Symposium
ed. S. van den Bergh and C. J. Pritchet　　　　　　　　　　　ISBN 0-937707-21-X

Vol. 5-The Minnesota Lectures on Clusters of Galaxies and Large-Scale Structure
ed. J. M. Dickey　　　　　　　　　　　　　　　　　　　　ISBN 0-937707-22-8

Vol. 6-Synthesis Imaging in Radio Astronomy: A Collection of Lectures from the Third NRAO Synthesis Imaging Summer School
ed. R. A. Perley, F. R. Schwab, and A. H. Bridle　　　　　　　ISBN 0-937707-23-6

Vol. 7-Properties of Hot Luminous Stars: Boulder-Munich Workshop
ed. C. D. Garmany　　　　　　　　　　　　　　　　　　　ISBN 0-937707-24-4

Vol. 8-CCDs in Astronomy
ed. G. H. Jacoby　　　　　　　　　　　　　　　　　　　　ISBN 0-937707-25-2

Vol. 9-Cool Stars, Stellar Systems, and the Sun. Sixth Cambridge Workshop
ed. G. Wallerstein　　　　　　　　　　　　　　　　　　　ISBN 0-937707-27-9

Vol. 10-Evolution of the Universe of Galaxies: Edwin Hubble Centennial Symposium
ed. R. G. Kron　　　　　　　　　　　　　　　　　　　　ISBN 0-937707-28-7

Vol. 11-Confrontation Between Stellar Pulsation and Evolution
ed. C. Cacciari and G. Clementini　　　　　　　　　　　　　ISBN 0-937707-30-9

Vol. 12-The Evolution of the Interstellar Medium
ed. L. Blitz　　　　　　　　　　　　　　　　　　　　　　ISBN 0-937707-31-7

Vol. 13-The Formation and Evolution of Star Clusters
ed. K. Janes　　　　　　　　　　　　　　　　　　　　　　ISBN 0-937707-32-5

Vol. 14-Astrophysics with Infrared Arrays
ed. R. Elston　　　　　　　　　　　　　　　　　　　　　ISBN 0-937707-33-3

Vol. 15-Large-Scale Structures and Peculiar Motions in the Universe
ed. D. W. Latham and L. A. N. da Costa　　　　　　　　　　ISBN 0-937707-34-1

Vol. 16-Proceedings of the 3rd Haystack Observatory Conference on Atoms, Ions and Molecules: New Results in Spectral Line Astrophysics
ed. A. D. Haschick and P. T. P. Ho　　　　　　　　　　　　ISBN 0-937707-35-X

Vol. 17-Light Pollution, Radio Interference, and Space Debris
ed. D. L. Crawford　　　　　　　　　　　　　　　　　　　ISBN 0-937707-36-8

Vol. 18-The Interpretation of Modern Synthesis Observations of Spiral Galaxies
ed. N. Duric and P. C. Crane　　　　　　　　　　　　　　ISBN 0-937707-37-6

Vol. 19-Radio Interferometry: Theory, Techniques, and Applications, IAU Colloquium 131
ed. T. J. Cornwell and R. A. Perley　　　　　　　　　　　　ISBN 0-937707-38-4

Vol. 20-Frontiers of Stellar Evolution: 50th Anniversary McDonald Observatory (1939-1989)
ed. D. L. Lambert　　　　　　　　　　　　　　　　　　　ISBN 0-937707-39-2

Vol. 21-The Space Distribution of Quasars
ed. D. Crampton   ISBN 0-937707-40-6

Vol. 22-Nonisotropic and Variable Outflows from Stars
ed. L. Drissen, C. Leitherer, and A. Nota   ISBN 0-937707-41-4

Vol. 23-Astronomical CCD Observing and Reduction Techniques
ed. S. B. Howell   ISBN 0-937707-42-4

Vol. 24-Cosmology and Large-Scale Structure in the Universe
ed. R. R. de Carvalho   ISBN 0-937707-43-0

Vol. 25-Astronomical Data Analysis Software and Systems I
ed. D. M. Worrall, C. Biemesderfer, and J. Barnes   ISBN 0-937707-44-9

Vol. 26-Cool Stars, Stellar Systems, and the Sun, Seventh Cambridge Workshop
ed. M. S. Giampapa and J. A. Bookbinder   ISBN 0-937707-45-7

Vol. 27-The Solar Cycle: Proceedings of the National Solar Observatory/Sacramento Peak
12th Summer Workshop
ed. K. L. Harvey   ISBN 0-937707-46-5

Vol. 28-Automated Telescopes for Photometry and Imaging
ed. S. J. Adelman, R. J. Dukes, Jr., and C. J. Adelman   ISBN 0-937707-47-3

Vol. 29-Viña Del Mar Workshop on Catacysmic Variable Stars
ed. N. Vogt   ISBN 0-937707-48-1

Vol. 30-Variable Stars and Galaxies
ed. B. Warner   ISBN 0-937707-49-X

Vol. 31-Relationships Between Active Galactic Nuclei and Starburst Galaxies
ed. A. V. Filippenko   ISBN 0-937707-50-3

Vol. 32-Complementary Approaches to Double and Multiple Star Research, IAU Collouquium
135
ed. H. A. McAlister and W. I. Hartkopf   ISBN 0-937707-51-1

Vol. 33-Research Amateur Astronomy
ed. S. J. Edberg   ISBN 0-937707-52-X

Vol. 34-Robotic Telescopes in the 1990s
ed. A. V. Filippenko   ISBN 0-937707-53-8

Vol. 35-Massive Stars: Their Lives in the Interstellar Medium
ed. J. P. Cassinelli and E. B. Churchwell   ISBN 0-937707-54-6

Vol. 36-Planets Around Pulsars
ed. J. A. Phillips, S. E. Thorsett, and S. R. Kulkarni   ISBN 0-937707-55-4

Vol. 37-Fiber Optics in Astronomy II
ed. P. M. Gray   ISBN 0-937707-56-2

Vol. 38-New Frontiers in Binary Star Research: Pacific Rim Colloquium
ed. K. C. Leung and I.-S. Nha   ISBN 0-937707-57-0

Vol. 39-The Minnesota Lectures on the Structure and Dynamics of the Milky Way
ed. Roberta M. Humphreys   ISBN 0-937707-58-9

Vol. 40-Inside the Stars, IAU Colloquium 137
ed. Werner W. Weiss and Annie Baglin   ISBN 0-937707-59-7

Vol. 41-Astronomical Infrared Spectroscopy: Future Observational Directions
ed. Sun Kwok   ISBN 0-937707-60-0

Vol. 42-GONG 1992: Seismic Investigation of the Sun and Stars
ed. Timothy M. Brown   ISBN 0-937707-61-9

Vol. 43-Sky Surveys: Protostars to Protogalaxies
ed. B. T. Soifer ISBN 0-937707-62-7

Vol. 44-Peculiar Versus Normal Phenomena in A-Type and Related Stars, IAU Colloquium 138
ed. M. M. Dworetsky, F. Castelli, and R. Faraggiana ISBN 0-937707-63-5

Vol. 45-Luminous High-Latitude Stars
ed. D. D. Sasselov ISBN 0-937707-64-3

Vol. 46-The Magnetic and Velocity Fields of Solar Active Regions, IAU Colloquium 141
ed. H. Zirin, G. Ai, and H. Wang ISBN 0-937707-65-1

Vol. 47-Third Decennial US-USSR Conference on SETI
ed. G. Seth Shostak ISBN 0-937707-66-X

Vol. 48-The Globular Cluster-Galaxy Connection
ed. Graeme H. Smith and Jean P. Brodie ISBN 0-937707-67-8

Vol. 49-Galaxy Evolution: The Milky Way Perspective
ed. Steven R. Majewski ISBN 0-937707-68-6

Vol. 50-Structure and Dynamics of Globular Clusters
ed. S. G. Djorgovski and G. Meylan ISBN 0-937707-69-4

Vol. 51-Observational Cosmology
ed. G. Chincarini, A. Iovino, T. Maccacaro, and D. Maccagni ISBN 0-937707-70-8

Vol. 52-Astronomical Data Analysis Software and Systems II
ed. R. J. Hanisch, R. J. V. Brissenden, and Jeannette Barnes ISBN 0-937707-71-6

Vol. 53-Blue Stragglers
ed. Rex A. Saffer ISBN 0-937707-72-4

Vol. 54-The First Stromlo Symposium: The Physics of Active Galaxies
ed. Geoffrey V. Bicknell, Michael A. Dopita, and Peter J. Quinn ISBN 0-937707-73-2

Vol. 55-Optical Astronomy from the Earth and Moon
ed. Diane M. Pyper and Ronald J. Angione ISBN 0-937707-74-0

Vol. 56-Interacting Binary Stars
ed. Allen W. Shafter ISBN 0-937707-75-9

Vol. 57-Stellar and Circumstellar Astrophysics
ed. George Wallerstein and Alberto Noriega-Crespo ISBN 0-937707-76-7

Vol. 58-The First Symposium on the Infrared Cirrus and Diffuse Interstellar Clouds
ed. Roc M. Cutri and William B. Latter ISBN 0-937707-77-5

Vol. 59-Astronomy with Millimeter and Submillimeter Wave Interferometry, IAU Colloquium 140
ed. M. Ishiguro and Wm. J. Welch ISBN 0-937707-78-3

Vol. 60-The MK Process at 50 Years: A Powerful Tool for Astrophysical Insight: A Workshop of the Vatican Observatory
ed. C. J. Corbally, R. O. Gray, and R. F. Garrison ISBN 0-937707-79-1

Vol. 61-Astronomical Data Analysis Software and Systems III
ed. Dennis R. Crabtree, R. J. Hanisch, and Jeannette Barnes ISBN 0-937707-80-5

Vol. 62-The Nature and Evolutionary Status of Herbig Ae / Be Stars
ed. P. S. Thé, M. R. Pérez, and E. P. J. van den Heuvel ISBN 0-937707-81-3

Vol. 63-Seventy-Five Years of Hirayama Asteroid Families: The role of Collisions in the Solar System History
ed. Y. Kozai, R. P. Binzel, and T. Hirayama ISBN 0-937707-82-1

Vol. 64-Cool Stars, Stellar Systems, and the Sun, Eighth Cambridge Workshop
ed. Jean-Pierre Caillault ISBN 0-937707-83-X

Vol. 65-Clouds, Cores, and Low Mass Stars
ed. Dan P. Clemens and Richard Barvainis   ISBN 0-937707-84-8

Vol. 66- Physics of the Gaseous and Stellar Disks of the Galaxy
ed. Ivan R. King   ISBN 0-937707-85-6

Vol. 67-Unveiling Large-Scale Structures Behind the Milky Way
ed. C. Balkowski and R. C. Kraan-Korteweg   ISBN 0-937707-86-4

Vol. 68-Solar Active Region Evolution: Comparing Models with Observations
ed. K. S. Balasubramaniam and George W. Simon   ISBN 0-937707-87-2

Vol. 69-Reverberation Mapping of the Broad-Line Region in Active Galactic Nuclei
ed. P. M. Gondhalekar, K. Horne, and B. M. Peterson   ISBN 0-937707-88-0

Vol. 70-Groups of Galaxies
ed. Otto G. Richter and Kirk Borne   ISBN 0-937707-89-9

Vol. 71-Tridimensional Optical Spectroscopic Methods in Astrophysics, IAU Colloquium 149
ed. G. Comte and M. Marcelin   ISBN 0-937707-90-2

Vol. 72-Millisecond Pulsars: A Decade of Surprise.
ed. A. A. Fruchter, M. Tavani, and D. C. Backer   ISBN 0-937707-91-0

Vol. 73-Airborne Astronomy Symposium on the Galactic Ecosystem: From Gas to Stars to Dust
ed. M. R. Haas, J. A. Davidson, and E. F. Erickson   ISBN 0-937707-92-9

Vol. 74-Progress in the Search for Extraterrestrial Life: 1993 Bioastronomy Symposium
ed. G. Seth Shostak   ISBN 0-937707-93-7

Vol. 75-Multi-Feed Systems for Radio Telescopes
ed. D. T. Emerson and J. M. Payne   ISBN 0-937707-94-5

Vol. 76-GONG '94: Helio- and Astero-Seismology from the Earth and Space
ed. Roger K. Ulrich, Edward J. Rhodes, Jr., and Werner Däppen ISBN 0-937707-95-3

Vol. 77-Astronomical Data Analysis Software and System IV
ed. R. A. Shaw, H. E. Payne, and J. J. E. Hayes   ISBN 0-937707-96-1

Vol. 78-Astrophysical Applications of Powerful New Databases: Joint Discussion No. 16 of the 22nd General Assembly of the IAU
ed. S. J. Adelman and W. L. Wiese   ISBN 0-937707-97-X

Vol. 79-Robotic Telescopes: Current Capabilities, Present Developments, and Future Prospects for Automated Astronomy
ed. Gregory W. Henry and Joel A. Eaton   ISBN 0-937707-98-8

Vol. 80-The Physics of the Interstellar Medium and Intergalactic Medium
ed. A. Ferrara, C. F. McKee, C. Heiles, and P. R. Shapiro   ISBN 0-937707-99-6

Vol. 81-Laboratory and Astronomical High Resolution Spectra
ed. A. J. Sauval, R. Blomme, and N. Grevesse   ISBN 1-886733-01-5

Vol. 82-Very Long Baseline Interferometry and the VLBA
ed. J. A. Zensus, P. J. Diamond, and P. J. Napier   ISBN 1-886733-02-3

Vol. 83-Astrophysical Applications of Stellar Pulsation. IAU Colloquium 155
ed. R. S. Stobie and P. A. Whitelock   ISBN 1-886733-03-1

Vol. 84-The Future Utilisation of Schmidt Telescopes, IAU Colloquium 148
ed. Jessica Chapman, Russell Cannon, Sandra Harrison, and Bambang Hidayat   ISBN 1-886733-05-8

Vol. 85-Cape Workshop on Magnetic Cataclysmic Variables
ed. D. A. H. Buckley and B. Warner   ISBN 1-886733-06-6

Vol. 86-Fresh Views of Elliptical Galaxies
ed. Alberto Buzzoni, Alvio Renzini, and Alfonso Serrano   ISBN 1-886733-07-4

Vol. 87-New Observing Modes for the Next Century
ed. Todd Boroson, John Davies, and Ian Robson   ISBN 1-886733-08-2

Vol. 88- Clusters, Lensing, and the Future of the Universe
ed. Virginia Trimble and Andreas Reisenegger   ISBN 1-886733-09-0

Vol. 89-Astronomy Education: Current Developments, Future Coordination
ed.John R. Percy   ISBN 1-886733-10-4

Vol. 90-The Origins, Evolution, and Destinies of Binary Stars in Clusters
ed. E. F. Milone and J. C. Mermilliod   ISBN 1-886733-11-2

Vol. 91-Barred Galaxies, IAU Colloquium 157
ed. R. Buta, D. A. Crocker, and B. G. Elmegreen   ISBN 1-886733-12-0

Vol. 92-Formation of the Galactic Halo–Inside and Out
ed. H. L. Morrison and A. Sarajedini   ISBN 1-886733-13-9

Vol. 93-Radio Emission from the Stars and the Sun
ed. A. R. Taylor and J. M. Paredes   ISBN 1-886733-14-7

Vol. 94-Mapping, Measuring, and Modelling the Universe
ed. Peter Coles, Vincent Martinez, and Maria-Jesus Pons-Borderia   ISBN 1-886733-15-5

Vol. 95-Solar Drivers of Interplanetary and Terrestrial Disturbances: Proceedings of 16th
International Workshop, National Solar Observatory/Sacramento Peak
ed. K. S. Balasubramaniam, S. L. Keil, and R. N. Smartt   ISBN 1-886733-16-3

Vol. 96- Hydrogen-Deficient Stars
ed. C. S. Jeffery and U. Heber   ISBN 1-886733-17-1

Vol. 97-Polarimetry of the Interstellar Medium
ed.W. G. Roberge and D. C. B. Whittet   ISBN 1-886733-18-X

Vol. 98-From Stars to Galaxies: The Impact of Stellar Physics on Galaxy Evolution
ed. Claus Leitherer, Uta Fritze-von Alvensleben, and John Huchra   ISBN 1-886733-19-8

Vol. 99-Cosmic Abundances: Proceedings of the 6th Annual October Astrophysics Conference
ed. Stephen S. Holt and Geroge Sonneborn   ISBN 1-886733-20-1

Vol. 100-Energy Transport in Radio Galaxies and Quasars
ed. P. E. Hardee, A. H. Bridle, and J. A. Zensus   ISBN 1-886733-21-X

Vol. 101-Astronomical Data Analysis Software and Systems V
ed. George H. Jacoby and Jeannette Barnes   ISSN   1080-7926

Vol. 102-The Galactic Center, 4th ESO/CTIO Workshop
ed. Roland Gredel   ISBN 1-886733-22-8

Vol. 103-The Physics of Liners in View of Recent Observations
ed. M. Eracleous, A. Koratkar, C. Leitherer, and L. Ho   ISBN 1-886733-23-6

Vol. 104-Physics, Chemistry, and Dynamics of Interplanetary Dust, IAU Colloquium 150
ed. Bo A. S. Gustafson and Martha S. Hanner   ISBN 1-886733-24-4

Vol. 105-Pulsars: Problems and Progress, IAU Colloquium 160
ed. M. Bailes, S. Johnston, and M. A. Walker   ISBN 1-886733-25-2

Vol. 106-Minnesota Lectures on Extragalactic Neutral Hydrogen
ed. Evan D. Skillman   ISBN 1-886733-26-0

Vol. 107-Completing the Inventory of the Solar System: A Symposium held in conjuunction with
the 106th Annual Meeting of the ASP
ed. Terrence W. Rettig and Joseph M. Hahn   ISBN 1-886733-27-9

Vol. 108-M. A. S. S. Model Atmospheres and Spectrum Synthesis: 5th Vienna Workshop
ed. S. J. Adelman, F. Kupka, and W. W. Weiss ISBN 1-886733-28-7

Vol. 109-Cool Stars, Stellar Systems, and the Sun, Ninth Cambridge Workshop
ed. Roberto Pallavicini and Andrea K. Dupree ISBN 1-886733-29-5

Vol. 110-Blazar Continuum Variability
ed. H. R. Miller, J. R. Webb, and J. C. Noble ISBN 1-886733-30-9

Vol. 111-Magnetic Reconnection in the Solar Atmosphere: Proceedings of a Yohkoh Conference
ed. R. D. Bentley and J. T. Mariska ISBN 1-886733-31-7

Vol. 112-The History of the Milky Way and Its Satellite System
ed. A. Burkert, D. H. Hartmann, and S. R. Majewski ISBN 1-886733-32-5

Vol. 113-Emission Lines in Active Galaxies: New Methods and Techniques, IAU Colloquium 159
ed. B. M. Peterson, F. Z. Cheng, and A. S. Wilson ISBN 1-886733-33-3

Vol. 114-Young Galaxies and QSO Absorption-Line Systems
ed. Sueli M. Viegas, Ruth Gruenwald, and Reinaldo R. de Carvalho ISBN 1-886733-34-1

Vol. 115-Galactic and Cluster Cooling Flows
ed. Noam Soker ISBN 1-886733-35-X

Vol. 116-The Second Stromlo Symposium: The Nature of Elliptical Galaxies
ed. M. Arnaboldi, G. S. Da Costa, and P. Saha ISBN 1-886733-36-8

Vol. 117- Dark and Visible Matter in Galaxies
ed. Massimo Persic and Paolo Salucci ISBN 1-886733-37-6

Vol. 118-First Advances in Solar Physics Euroconference: Advances in the Physics of Sunspots
ed. B. Schmieder, J. C. del Toro Iniesta, and M. Vázquez ISBN 1-886733-38-4

Vol. 119-Planets Beyond the Solar System and the Next Generation of Space Missions
ed. David R. Soderblom ISBN 1-886733-39-2

Vol. 120-Luminous Blue Variables: Massive Stars in Transition
ed. Antonella Nota and Henny J. G. L. M. Lamers ISBN 1-886733-40-6

Vol. 121-Accretion Phenomena and Related Outflows, IAU Colloquium 163
ed. D. T. Wickramasinghe, G. V. Bicknell and L. Ferrario ISBN 1-886733-41-4

Vol. 122-From Stardust to Planetesimals: Symposium held as part of the 108th Annual Meeting of the ASP
ed. Yvonne J. Pendleton and A. G. G. M. Tielens ISBN 1-886733-42-2

Vol. 123-The 12th 'Kingston Meeting': Computational Astrophysics
ed. David A. Clarke and Michael J. West ISBN 1-886733-43-0

Vol. 124-Diffuse Infrared Radiation and the IRTS
ed. Haruyuki Okuda, Toshio Matsumoto, and Thomas L. Roellig ISBN 1-886733-44-9

Vol. 125- Astronomical Data Analysis Software and Systems VI
ed. Gareth Hunt and H. E. Payne ISBN 1-886733-45-7

Vol. 126-From Quantum Fluctuations to Cosmological Structures
ed. D. Vallis-Gabaud, M. A. Hendry, P. Molaro, and K. Chamcham ISBN 1-886733-46-5

Vol. 127-Proper Motions and Galactic Astronomy
ed. Roberta M. Humphreys ISBN 1-886733-47-3

Vol. 128- Mass Ejection from AGN (Active Galactic Nuclei)
ed. N. Arav, I. Shlosman, and R. J. Weymann　　　　　　　　ISBN 1-886733-48-1

Vol. 129-The George Gamow Symposium
ed. E. Harper, W. C. Parke, and G. D. Anderson　　　　　　　ISBN 1-886733-49-X

Vol. 130-The Third Pacfic Rim Conference on Recent Development on Binary Star Research
ed. Kam-Ching Leung　　　　　　　　　　　　　　　　　　ISBN 1-886733-50-3

Vol. 131-Boulder-Munich II: Properties of Hot, Luminous Stars
ed. Ian D. Howarth　　　　　　　　　　　　　　　　　　　ISBN 1-886733-51-1

Vol. 132-Star Formation with the Infrared Space Observatory (ISO)
ed. João L. Yun and René Liseau　　　　　　　　　　　　　ISBN 1-886733-53-X

Vol. 133-Science with the NGST
ed. Eric P. Smith and Anuradha Koratkar　　　　　　　　　　ISBN 1-886733-53-8

Vol. 134-Brown Dwarfs and Extrasolar Planets
ed. Rafael Rebolo, Eduardo L. Martin,
and Maria Rosa Zapatero Osorio　　　　　　　　　　　　　ISBN 1-886733-54-6

Vol. 135-A Half Century of Stellar Pulsation Interpretations: A Tribute to Arthur N. Cox
ed. P. A Bradley and J. A. Guzik　　　　　　　　　　　　　ISBN 1-886733-55-4

Vol. 136- Galactic Halos: A UC Santa Cruz Workshop
ed. Dennis Zaritsky　　　　　　　　　　　　　　　　　　　ISBN 1-886733-56-2

Vol. 137-Wild Stars in the Old West: Proceedings of the 13th North American Workshop
on Cataclysmic Variables and Related Objects
ed. S. Howell, E.Kuulkers, and C. Woodward　　　　　　　　ISBN 1-886733-57-0

Vol. 138-1997 Pacific Rim Conference on Stellar Astrophysics
ed. Kwing L. Chan, K. S. Cheng, and Harinder P. Singh　　　　ISBN 1-886733-58-9

Vol. 139-Preserving the Astronomical Windows, Proceedings of Joint Discussion
No. 5 of the 23rd General Assembly of the IAU
ed. Syuzo Isobe and Tomohiro Hirayama　　　　　　　　　　ISBN 1-886733-59-7

Vol. 140-Synoptic Solar Physics – 18th NSO/Sacramento Peak Summer Workshop
ed. K. S. Balasubramaniam, J. W. Harvey, and D. M. Rabin　　ISBN 1-886733-60-0

Vol. 141-Astrophysics from Antarctica
ed. Giles Novak and Randall H. Landsberg　　　　　　　　　ISBN 1-886733-61-9

Vol. 142-The Stellar Initial Mass Function, 38th Herstmonceux Conference
ed. Gerry Gilmore and Debbie Howell　　　　　　　　　　　ISBN 1-886733-62-7

Vol. 143-The Scientific Impact of the Goddard High Resolution Spectrograph
ed. John C. Brandt, Thomas B. Ake III, and Carolyn Collins Petersen　ISBN 1-886733-63-5

Vol. 144- Radio Emission from Galactic and Extragalactic Compact Sources,
IAU Colloquium 164
ed. J. Anton Zensus, G. B. Taylor, and J. M. Wrobel　　　　　ISBN 1-886733-64-3

Vol. 145-Astronomical Data Analysis Software and Systems VII
ed. Rudolf Albrecht, Richard N. Hook, and Howard A. Bushouse　ISBN 1-886733-65-1

Vol. 146-The Young Universe: Galaxy Formation and Evolution at
Intermediate and High Redshift
ed. S. D'Odorico, A. Fontana, and E. Giallongo　　　　　　　ISBN 1-886733-66-X

Vol. 147-Abundance Profiles: Diagnostic Tools for Galaxy History
ed. Daniel Friedli, Mike Edmunds, Carmelle Robert,
and Laurent Drissen　　　　　　　　　　　　　　　　　　ISBN 1-886733-67-8

Vol. 148-Origins
ed. Charles E. Woodward, J. Michael Shull,
and Harley A. Thronson, Jr.   ISBN 1-886733-68-6

Vol. 149-Solar System Formation and Evolution
ed. D. Lazzaro, R. Vieira Martins, S. Ferraz-Mello,
J. Fernández, and C. Beaugé   ISBN 1-886733-69-4

Vol. 150-New Perspectives on Solar Prominences, IAU Colloquium 167
ed. David Webb, David Rust, and Brigitte Schmieder   ISBN 1-886733-70-8

Vol. 151-Cosmic Microwave Background and Large Scale Structure of the Universe
ed. Yong-Ik Byun and Kin-Wang Ng   ISBN 1-886733-71-6

Vol. 152-Fiber Optics in Astronomy III
ed. S. Arribas, E. Mediavilla, and F. Watson   ISBN 1-886733-72-4

Vol. 153-Library and Information Services in Astronomy III, (LISA III)
ed. Uta Grothkopf, Heinz Andernach, Sarah Stevens-Rayburn,
and Monique Gomez   ISBN 1-886733-73-2

Vol. 154-Cool Stars, Stellar Systems, and the Sun, Tenth Cambridge Workshop
ed. Robert A. Donahue and Jay A. Bookbinder   ISBN 1-886733-74-0

Vol. 155-Second Advances in Solar Physics Euroconference:
Three-Dimensional Structure of Solar Active Regions
ed. Costas E. Alissandrakis and Brigitte Schmieder   ISBN 1-886733-75-9

Vol. 156-Highly Redshifted Radio Lines
ed. C. L. Carilli, S. J. E. Radford, K. M. Menten and
G. I. Langston   ISBN 1-886733-76-7

Vol. 157-Annapolis Workshop on Magnetic Cataclysmic Variables
ed. Coel Hellier and Koji Mukai   ISBN 1-886733-77-5

Vol. 158-Solar and Stellar Activity: Similarities and Differences
ed. C. J. Butler and J. G. Doyle   ISBN 1-886733-78-3

Vol. 159-BL Lac Phenomenon
ed. Leo O. Takalo and Aimo Sillanpää   ISBN 1-886733-79-1

Vol. 160-Astrophysical Discs, An EC Summer School
ed. J. A. Sellwood and Jeremy Goodman   ISBN 1-886733-80-5

Vol. 161-High Energy Processes in Accreting Black Holes
ed. Juri Poutanen and Roland Svensson   ISBN 1-886733-81-3

Vol. 162-Quasars and Cosmology
ed. Gary Ferland and Jack Baldwin   ISBN 1-886733-83-X

Vol. 163-Star Formation in Early Type Galaxies
ed. Jordi Cepa and Patricia Carral   ISBN 1-886733-84-8

Vol. 164-Ultraviolet–Optical Space Astronomy Beyond HST
ed. Jon A. Morse, J. Michael Shull and Anne L. Kinney   ISBN 1-886733-85-6

Vol. 165-The Third Stromlo Symposium: The Galactic Halo
ed. Brad K. Gibson, Tim S. Axelrod, and Mary E. Putman   ISBN 1-886733-86-4

Vol. 166-Stromlo Workshop on High-Velocity Clouds
ed. Brad K. Gibson and Mary E. Putman   ISBN 1-886733-87-2

Vol. 167-Harmonizing Cosmic Distance Scales in a Post-HIPPARCOS Era
ed. Daniel Egret and André Heck   ISBN 1-886733-88-0

Vol. 168-New Perspectives on the Interstellar Medium
ed. A. R. Taylor, T. L. Landecker, and G. Joncas   ISBN 1-886733-89-9

Vol. 169-11th European Workshop on White Dwarfs
ed. J.-E. Solheim and E. G. Meištas   ISBN 1-886733-91-0

Vol. 170-The Low Surface Brightness Universe, IAU Colloquium 171
ed. J. I. Davies, C. Impey, and S. Phillipps   ISBN 1-886733-92-9

Book orders or inquiries concerning these volumes should be directed to the:

Astronomical Society of the Pacific Conference Series
390 Ashton Avenue
San Francisco, CA 94112-1722 USA

Phone: 415-337-2126   E-mail: catalog@aspsky.org
Fax: 415-337-5205   Web Site: http://www.aspsky.org

ASTRONOMICAL SOCIETY OF THE PACIFIC
CONFERENCE SERIES

Volume 170

THE LOW SURFACE BRIGHTNESS UNIVERSE
IAU Colloquium 171

Proceedings of an IAU Colloquium held at
Cardiff, Wales
5-10 July, 1998

Edited by
J. I. Davies, C. Impey, and S. Phillipps

Contents

Preface   xx

Observational selection effects

Historical Introduction
K. C. Freeman   3

Is there a Low Surface Brightness Universe ?
M. J. Disney   9

Optical Galaxy Selection
S. McGaugh   19

Selection effects at 21cm
F. H. Briggs   27

Constraints on the unseen galaxy population from the Lyα forest
K. M. Lanzetta, Hsiao-Wen Chen, J. K. Webb and X. Barcons   35

Luminosity and surface brightness distributions

Low luminosity galaxies in large surveys
J. P. Huchra   45

The space density of spiral galaxies as a function of their luminosity, surface brightness and scalesize
R. S. de Jong and C. Lacey   52

The faint end of the galaxy luminosity function in rich clusters
R. M. Smith, S. Phillipps, S. P. Driver and W. J. Couch   60

Optical and near-IR field luminosity functions
J. Loveday   68

Low surface brightness galaxies in deep surveys
H. C. Ferguson   76

Low surface brightness dwarf galaxies in the Bristol-Anglo-Australian Observatory Virgo cluster survey
J. B. Jones, S. Phillipps, J. M. Schwartzenberg and Q. A. Parker   84

A dichotomy between HSB and LSB galaxies
M. Verheijen and B. Tully   92

Low Surface Brightness galaxy surveys

Multi-Wavelength Surveys for Galaxies Hidden by the Milky Way
R. C. Kraan-Korteweg — 103

A survey for low surface brightness dwarf galaxies around M31
T. E. Armandroff, J. E. Davies and G. H. Jacoby — 111

The Fornax spectroscopic survey - Low Surface Brightness galaxies in Fornax
M. J. Drinkwater, S. Phillipps and J. B. Jones — 120

Found: High Surface Brightness compact galaxies
M. J. Drinkwater, S. Phillipps, J. B. Jones, M. D. Gregg
Q. A. Parker, R. M. Smith, J. I. Davies and E. M. Sadler — 128

Counting the ghosts: optical field surveys for Low Surface Brightness galaxies
J. J. Dalcanton — 131

The radial extent of the Fornax cluster Low Surface Brightness galaxy population
J. I. Davies, A. Kambas, Z. Morshidi-Esslinger and R. Smith — 138

The discovery of red Low Surface brightness galaxies
K. O'Neil — 145

Dwarf galaxies in nearby groups
T. Bremnes, B. Binggeli and P. Prugniel — 154

Low Surface Brightness Dwarf Galaxies in Nearby Clusters
K. Chiboucas and M. Mateo — 157

Properties of Low Surface Brightness galaxies

The interstellar medium in Low Surface Brightness galaxies
W. J. G. de Blok — 161

Dwarf galaxies as Low Surface Brightness galaxies
E. D. Skillman — 169

Low Surface Brightness galaxies beyond $z=0.5$: existence, detection and properties
G. Bothun — 177

Environmental effects on the faint end of the luminosity function
S. Phillipps, J. B. Jones, R. M. Smith, W. J. Couch and S. P. Driver — 183

Morphology and stellar populations in the gas-rich, giant
Low Surface Brightness galaxies
P. Knezek   191

The extreme outer regions of disk galaxies: star formation and
metal abundances
A. Ferguson, R. Wyse and J. Gallagher   196

The Low Surface Brightness galaxy HIPASS1126-72
V. Kilborn, E. de Blok, L. Stavely-Smith and R. Webster   204

The structure of the super thin spiral galaxy UGC7321
L. D. Matthews, J. S. Gallagher and W. van Driel   207

HST WFC-2 imaging of four nearby Low Surface Brightness
galaxies
K. O'Neil   210

Kinematics of giant Low Surface Brightness galaxies
T. E. Pickering   214

What causes the HI holes in gas-rich Low Surface Brightness
dwarfs ?
K. L. Rhode, J. J. Salzer and D. J. Westpfahl   221

VLA HI imaging of the Low Surface Brightness dwarf
galaxy DDO47
F. Walter and E. Brinks   224

The evolution of Low Surface Brightness galaxies

The fate of Low Surface Brightness galaxies in clusters and the
origin of the diffuse intra-cluster light
B. Moore, G. Lake, J. Stadel and T. Quinn   229

The structure of the multi-phase ISM in Low Surface Brightness
galaxies
M. Spaans   237

The star formation histories of Low Surface Brighness galaxies
E. F. Bell, R. G. Bower, R. S. de Jong, B. J. Rauscher,
D. Barnaby, D. A. Harper, M. Hereld and R. F. Loewenstein   245

Gas rich Low Surface Brightness galaxies - progenitors of blue
compact dwarfs ?
J. J. Salzer and S. A. Norton   253

Morphological aspects of star formation in dwarf galaxies
N. Brosch, A. Heller and E. Almoznino — 261

The connection between dE and dI galaxies
B. W. Miller — 271

Star formation thresholds in Low Surface Brightness dwarf galaxies
L. van Zee — 274

Testing environmental influences on star formation with a sample of Low Surface Brightness dwarf galaxies in the Vigo cluster
A. Heller, E. Almoznino and N. Brosch — 282

HI and QSO absorption lines

The Parkes Multi-beam blind HI survey
R. L. Webster, V. Kilborn, J. C. O'Brien, L. Stavely-Smith, M. E. Putman and G. Banks — 291

HI in Karachentsev objects: properties of new nearby dwarf galaxies
W. K. Huchtmeier — 299

An HI survey of Low Surface Brightness galaxies selected from the APM survey
S. Côté, T. Broadhurst, J. Loveday and S. Kolind — 307

Using weak MgII lines to chart Low Surface Brightness galaxies
V. Le Brun and C. W. Churchill — 315

Simulations of Lyα absorption from Low Surface Brightness galaxies
S. W. Linder — 323

First results from the HI Parkes zone of avoidence survey
P. A. Henning, L. Stavely-Smith, R. C. Kraan-Korteweg and E. M. Sadler — 331

Results from the Dwingeloo obscured galaxies survey
A. J. Rivers, P. A. Henning, R. C. Kraan-Korteweg, O. Lahav and W. B. Burton — 334

HI properties of Low Surface Brightness dwarf and blue compact dwarf galaxies
C. L. Taylor, E. Brinks and E. D. Skillman — 337

Extra-galactic background light

The HST/LCO measurement of the optical extra-galactic
background light
R. A. Bernstein   341

Optical diffuse light in clusters of galaxies
R. Vílchez-Gómez   349

Diffuse ultraviolet background radiation
R. C. Henry   357

Detecting the Low Surface Brightness Universe: the extra-galactic
background light and LSB galaxies
P. Väisänen and E. V. Tollestrup   365

The baryonic mass content of the Universe

Cosmic baryon density from primordial nucleosynthesis and
other evidence
B. E. J. Pagel   375

Chemical constraints, baryonic mass and the chemical
evolution of Low Surface Brightness galaxies
M. G. Edmunds   383

Helium abundances in the most metal-deficient dwarf galaxies
Y. I. Izotov   390

What we don't know about the Universe
C. Impey   393

Preface

In July 1998, nearly a hundred astronomers gathered in Wales to consider the Low Surface Brightness (LSB) Universe. IAU Colloquium 171 took place at the University of Cardiff and it marked the 60th birthday of Mike Disney, the Cardiff astronomer who, following those other well known dissidents Fritz Zwicky and Halton Arp, was a pioneer in recognizing that the observed population of galaxies is highly influenced by surface brightness selection. It is now clear that many galaxies are much more diffuse or unevolved than the Milky Way. There are interesting consequences to the truism that our galaxy catalogs only contain galaxies that we can detect.

The study of the LSB universe is an ongoing battle between technology and available photons - after all, Messier's objects were LSB by the standards of the 18th century. Over the past ten years or so, the locus of the debate has shifted from the existence of LSB galaxies to their cosmological significance. Most of the contributions at the Colloquium reflected the traditional optical study of galaxies. However, much evidence was also presented on the gas content of galaxies using the 21cm line of neutral hydrogen. Participants also heard about attempts to learn about the LSB Universe using quasar absorption lines, which are sensitive to extremely small amounts of neutral gas.

IAU Colloquium 171 was a bracing mix of new data and challenging ideas. The combination of optical and radio astronomers, looking at high and low redshift targets, proved to be a success and spurred many discussions outside the formal sessions. Indeed, from humble beginnings, the current breadth and depth of the subject was emphasised by the existence of a whole community dedicated to its study.

As the meeting progressed, controversial statements and unanswered questions were written on a blackboard at the front of the lecture theater. By the end of the week, nearly 50 had been listed. Even so, there was little real controversy at the meeting itself - who is there to argue with at a meeting of the already converted!

Participants from hot summer climates were also grateful for the blustery Welsh weather, which was best experienced during a clifftop walk on the third afternoon. Another conference highlight was the dinner in the great hall of Caerphilly Castle, built in the 13th century and one of the major fortresses of Europe. As Mike Disney reminded us after the dinner, astronomers are indeed lucky to be paid to travel the world and ponder the exotica of the Universe. The meeting also coincided with the final stages of the soccer World Cup, providing - especially for our French and Dutch colleagues - an excuse to visit the local hostelries to watch the matches on television and celebrate or drown sorrows!

We thank the Department of Physics at the University of Cardiff for hosting the colloquium and for financial support. Thanks are also due to Rodney Smith for his hard work chairing the Local Organising Committee, to our other session organisers and chairmen Greg Bothun, Elias Brinks, Mike Disney, Ken Freeman, Thijs van der Hulst and Yuri Izotov and to the Cardiff Conference Office, especially Su Hayward-Lewis, our conference co-ordinator.

Jon Davies, Chris Impey, Steve Phillipps

Searching for LSB - I

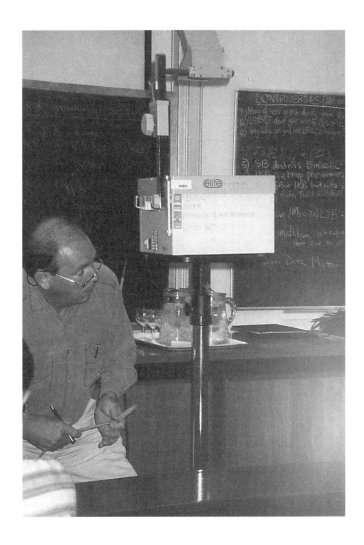

It's not under here !

Historical Introduction

K.C. Freeman

Mount Stromlo and Siding Spring Observatories, Private Bag, Weston Creek PO, ACT 2611, Australia

Abstract. Our ideas about the surface brightness distribution of galaxies has changed greatly since 1970. I contrast the view at that time with our present view of the subject, and then briefly discuss some topics in the studies of Low Surface Brightness Galaxies (LSBG) that I think are particularly interesting: the Tully-Fisher relation, LSBG as systems of high angular momentum, LSBG in clusters of galaxies, and the potential impact of the HIPASS survey.

1. Introduction

Around 1970 I became interested in the exponential disk with surface brightness distribution $I(R) = I_\circ \exp(-\alpha R)$, assuming that it corresponded to an exponential surface density. I worked out the rotation curve for a self-gravitating exponential disk and was then curious about the range of values for the two parameters for the surface brightness distribution. I took *the* 36 disk galaxies with available surface photometry at the time. The data were mostly photographic, and the sample was inhomogeneous. Of the 36 galaxies, 28 seemed to have very similar blue central surface brightnesses $\mu_B(0)$ around 21.6 mag arcsec^{-2} after inclination correction for the path length through the galaxies. This was surprising, because much of the photometry was fairly rough. The sample also included a single low surface brightness galaxy, the Local Group member IC1613, with central surface brightness $\mu_B(0) = 23.7$.

The conclusions of this work (Freeman 1970) are sometimes overstated. For example (from a recent paper)

> In 1970 Freeman found that all spiral galaxies have a narrow distribution of central surface brightness centered around $\mu_B(0) = 21.65\pm0.3$ mag arcsec^{-2}.

This was all before dark matter in galaxies was well established. The apparently uniform central surface brightness for many of the galaxies, if interpreted as a uniform central surface density for the self-gravitating disks, led directly to a relationship between the angular momentum J and mass M of the disks of the form $J \propto M^{7/4}$. This seemed interesting for understanding the formation of disk galaxies, and several explanations for this relationship soon followed. This was the observational situation in 1970. There was no attempt at this stage to correct for volume/visibility effects.

The current situation is very different. Thanks to the work by Disney, Davies and Phillipps here in Cardiff, and to many who made the major photographic and CCD surveys over the last ~ 15 years, we now realise the importance of visibility effects and the existence of disk galaxies in significant numbers with much lower surface brightness. The present view of the true distribution of central surface brightnesses is summarised by Bothun et al. (1997: Fig. 1); the volume-corrected galaxy density per magnitude of surface brightness appears roughly flat from $\mu_B(0) = 21.5$ down to at least 24, although there remain differences of opinion about how to do this volume correction correctly. In any case, it seems clear that a significant fraction of disk galaxies have $\mu_B(0) > 23$ mag arcsec^{-2}.

Why is this important ? The true distribution of surface brightnesses for disk galaxies affects our understanding of their formation processes. It is relevant to the distribution of the baryon to dark matter ratio for individual galaxies, and to the total baryon and dark matter contribution of galaxies in the universe.

2. The Present View of Properties of LSBG

(For references, please see the major reviews by Impey and Bothun (1997: IB) and Bothun et al. (1997: BIM).

- disks are mostly exponential.

- LSB does *not* necessarily mean low luminosity and low mass: see IB (Fig. 2) and BIM (Fig. 3). Giants with LSB disks exist, although they are rare and are usually detected optically by their high surface brightness (HSB) inner regions. Tables 1 and 2 give observational parameters and derived masses and M/L ratios for four giant LSB galaxies (see also Freeman, 1997). Note their large scale lengths and their large M/L values. In Table 1, V_o is the systemic velocity in km s^{-1}, W_o is the corrected HI velocity width in km s^{-1} and h is the scale length in kpc. The quantities in Table 2 are in solar units.

- Galaxies cover most of the μ_o - h plane for $\mu_o > 20$: the only apparently unpopulated region is for the high surface brightness systems with large scale lengths (see BIM Fig. 4).

- Colors and surface brightness are not related: fading is not the dominant reason for the LSB (see BIM Fig. 7).

- LSB disks are dark matter dominated at almost all radii (*e.g.* de Blok and McGaugh 1997). LSB halos tend to be less dense and have larger core radii; this may just be the consequence of the weaker baryonic compression of the halo by the lower surface brightness disk as it forms. The dark/luminous mass ratio is typically > 20. This apparently high ratio may not be so extreme. The dark/luminous mass ratio is reliably known for only one HSBG (the Milky Way), for which it is again > 20 (*e.g.* Freeman 1996).

Table 1: Dynamical and Structural Parameters for LSB Giants

	V_o	$B(0)_c$	W_o	M_B	h
Malin 1	24750	26	455	-21.0	55
F568-6	13830	23.4	674	-21.2	16
1226+0105	23655	23.3	405	-21.6	12
NGC 5084	1550		700	-20.8	

Table 2: Masses and M/L Ratios for LSB Giants

	M_{tot}	M_{tot}/L	M_{HI}
Malin 1	2×10^{12}	55	1×10^{11}
F568-6	9×10^{11}	20	2×10^{10}
1226+0105	2×10^{12}	30	2×10^{10}
NGC 5084	1.3×10^{12}	45	9×10^{9}

Table 3: Zero Surface Brightness Disks

	DDO 154	NGC 2915
Stellar Mass	$5 \times 10^7 M_\odot$	$3 \times 10^8 M_\odot$
HI Mass	2.7×10^8	1.3×10^9
Total Mass	3.8×10^9	2.7×10^{10}
M_{total}/M_{baryon}	12	17

3. The Tully-Fisher Relation

Zwaan et al. (1995) showed that most LSBG and HSBG follow the same Tully-Fisher relation. There are a few galaxies that do not follow this common Tully-Fisher law. Examples include the two gas-rich and dark matter dominated galaxies DDO 154 and NGC 2915 (Meurer et al. 1996): in these two galaxies, most of the baryons are in the extended HI disk component. DDO 154 and NGC 2915 lie about two magnitudes fainter than the usual M_B - (velocity width) relation, but rise to it if the dominant HI component is notionally converted to stars with an M/L ratio ≈ 1. Their properties are summarised in Table 3. Their stellar mass is about an order of magnitude less than their HI mass, which in turn is another order of magnitude less than the "total" dynamical mass. The last row of the table shows that the ratio of total mass to baryonic mass is again similar to that for the Milky Way (~ 20). The observation that these HI-dominated galaxies DDO 154 and NGC 2915 lie on the Tully-Fisher relation when their HI mass is notionally transformed to light is a further indication that the dark/baryon mass ratio is roughly constant from galaxy to galaxy, even for these extreme systems. We can regard these very extended HI disks as zero surface brightness (ZSB) disks.

IB consider the baryon contribution from LSB galaxies. The apparently flat number distribution of surface densities suggests that the baryon density Ω_b lies in the range $0.014 < \Omega_b < 0.025$. The likely nucleosynthesis bounds on Ω_b are $0.02 < \Omega_b < 0.03$. The volume-corrected estimate of the Ω_b contribution from luminous matter is about 0.003 (Persic and Salucci 1992). So it is possible that a substantial fraction of the baryons could lie in LSBG. However Briggs (1997) finds that LSBG with $M > 10^7 M_\odot$ contribute $< 10\%$ to the local HI density, and probably also little to the total mass and luminosity density.

4. LSBG as systems of high angular momentum

The angular momentum of aggregating systems in the early universe is probably tidally acquired. Simulations show that the parameter $\lambda = J|E|^{1/2}G^{-1}M^{-5/2}$ for halos lies in the range 0 to 0.15, with mean value 0.05 (J is the angular momentum, E the binding energy and M the mass). Assume that gas and dark matter are initially well mixed with the same specific angular momentum J/M, and are truncated at a radius r_t. The gas dissipates to form an equilibrium exponential disk with scale length h in the potential of the halo. Then the collapse factor $r_t/h = \sqrt{2}/\lambda \approx 30$ (Fall and Efstathiou 1980). For example, for the Milky Way, $h \approx 4$ kpc so $r_t \approx 120$ kpc which is consistent with the M31 timing estimates. High-λ galaxies have longer scale lengths and lower surface densities. For example, increasing λ from 0.05 to 0.15 increases h by a factor 3 in the mean and reduces the surface density by about a factor 10. This is consistent with the larger scale lengths observed for LBSG by Zwaan et al. (1995).

Dalcanton et al. (1997) and Jiminez et al. (1998) made more detailed models of this kind. Dalcanton et al. adopt a probability distribution of λ from simulations, take a Schechter-like distribution of masses, include baryonic compression and predict the distribution of disk galaxies in the (μ_o, h) plane. Comparing their expected distribution with the observed distribution suggests that many galaxies remain to be discovered in some regions of this plane.

5. LSBG in Clusters of Galaxies

The cluster environment is often regarded as hostile to the formation and survival of LSBG. However the harrassment process (Moore et al. 1996) may contribute to the production of LSBG in clusters. Although LSBG in clusters are interesting for understanding the baryon content of clusters, they may tell us more about the dynamics of the cluster environment than about the nature of the LSBG phenomenon. We know from planetary nebula studies that a significant fraction of the stellar mass of the Virgo cluster lies in the intracluster medium (e.g. Mendez et al. 1997, Feldmeier et al. 1998). These intracluster stars are probably harrassment debris. The bimodality of the K-band distribution of central surface brightness for disk galaxies in the UMa cluster, discovered by Tully and Verheijen (1997), may be another possible dynamical effect of the cluster environment, although the long crossing time ($\sim 0.5 H_o^{-1}$) may argue against this interpretation. It would be interesting to know if the UMa cluster has an intracluster stellar debris population like that of the Virgo cluster.

6. HIPASS and LSBG

The HIPASS survey (see Webster's paper in this volume) covers about 27,000 square degrees of sky and will identify gas-rich galaxies independent of their optical properties. It will give an optically unbiased view of the distribution of galaxies in the (surface brightness - luminosity) plane. The HIPASS survey has its own selection effects. For galaxies that overfill the Parkes beam (*i.e.* galaxies with diameter in kpc $> 4.4\times$ distance in Mpc), the surface density limit is $N_{HI} > 5 \times 10^{17} \Delta V^{1/2}$ cm^{-2}, where ΔV is the velocity width of the HI profile in km s^{-1}. For smaller galaxies, the HI mass limit is $M_{HI} > 5 \times 10^4 D^2 \Delta V^{1/2} M_\odot$, where D is the distance in Mpc. For example, for $D = 10$ Mpc and $\Delta V = 100$ km s^{-1}, $M_{HI} > 5 \times 10^7 M_\odot$. Objects like NGC 2915 can be detected out to 50 Mpc. Galaxies with $\Delta V = 150$ km s^{-1} that fill the Parkes beam can be detected down to surface density levels of 6×10^{18} cm^{-2}.

Despite the opportunities that HIPASS offers, we should remember the results of Zwaan *et al.* (1997) from their study of the Arecibo strip: 61 detections in 65 square degrees down to a limiting $N_{HI} \approx 10^{18}$ cm^{-2}. They found that:

- The HI mass function for galaxies down to $M_{HI} = 10^7 M_\odot$ is similar to the HI mass function for optical galaxies.

- There does not appear to be a large class of gas-rich dwarfs or LSBG that were not previously detected optically. The HI content is dominated by high-mass galaxies with $M \sim 10^9 - 10^{10} M_\odot$.

- The lower limit to the average HI surface density is about $10^{19.7}$ cm^{-2}, despite the sensitivity of the survey to much lower surface densities. Is this limit due to ionization of lower column density systems or is it more fundamental ?

References

Bothun, G., Impey, C., McGaugh, S. 1997, PASP, 109, 745
Briggs, F. 1997, ApJ, 484, 618
Dalcanton, J., Spergel, D., Summers, F. 1997, ApJ, 482, 659
de Blok, W.J.G., McGaugh, S. 1997, MNRAS, 290, 533
Fall, S.M., Efstathiou, G. 1980, MNRAS, 193, 189
Feldmeier, J., Ciardullo, R., Jacoby, G. 1998, ApJ, 503, 109
Freeman, K.C. 1970, ApJ, 160, 811
Freeman, K.C. 1996, in "Unsolved Problems of the Milky Way" (IAU Symposium 169), ed L. Blitz & P. Teuben (Kluwer), 645
Freeman, K.C. 1997, Proc.Astron.Soc.Austr., 14, 4
Impey, C., Bothun, G. 1997, ARA&A, 35, 267

Jiminez, R., Padoan, P., Matteuci, F., Heavens, A. 1998, MNRAS, 299, 123
Meurer, G.M., Carignan, C., Beaulieu, S.F., Freeman, K.C. 1996, AJ, 111, 1551
Mendez, R.H., Guerrero, M.A¿, Freeman, K.C., Arnaboldi, M., Kudritzki, R.P., Hopp, U., Capaccioli, M., Ford, H. 1997, ApJ, 491, L23
Moore, B., Katz, N., Lake, G., Dressler, A., Oemler, A. 1996, Nature, 379, 613
Persic, M., Salucci, P. 1992, MNRAS, 258, 14P
Tully, B., Verheijen, M. 1997, ApJ, 484, 145
Zwaan, M.A., van der Hulst, J.M., de Blok, W.J.G., McGaugh, S.S. 1995, MNRAS, 273, 35P
Zwaan, M.A., Briggs, F.H., Sprayberry, D., Sorar, E. 1997, ApJ, 490, 173

Is there a Low Surface Brightness Universe ?

M. J. Disney

Physics and Astronomy, Cardiff University, PO BOX 913, Cardiff. CF2 3YB, UK. e-mail:mjd@astro.cf.ac.uk

Abstract.

I have been asked to review some of the arguments for and against the existence of a significant population of Hidden Galaxies, and to concentrate on the distant past and the possible future of this subject, leaving more recent observations to others. In particular I ask how should we correct for selection effects and how are we to resolve contradictory evidence?

1. INTRODUCTION

Our subject is built upon the following conjecture: *"Most of the galaxies, even in our neighbourhood, remain to be discovered because galaxies are extremely hard to detect through our atmosphere and against our sky."*

Some of the conjectured missing galaxies will be so dim (i.e. be of such low surface brightness) as to be lost completely below our parochial sky; of others, which I call "Crouching Giants", we shall see only their bright cores - causing us to misconstrue them as dwarfs. Many genuine nearby dwarfs and high SB galaxies will be hidden among a sea of apparently more numerous background giants while others, "Masquerades" as I call them, will appear so bright and compact as to be indistinguishable, at first glance, from stars.

Examples of all these types of "hidden galaxy" certainly exist; the controversy enters in assigning a significance to them relative to the well-known and long studied population of "Normal galaxies". Some dismiss them as rare oddities while others believe them to contain the majority of galactic light - or mass. Lacking certain knowledge it is a good question to ask which of the two extreme points of view will make for the better working hypothesis. I contend that, for the galaxy explorer at least, it is more fruitful to assume that large hidden populations remain to be found and to seek new observing techniques to detect and count them. At the same time we all recognise that certain (very difficult) observations, such as detection of the Extra-Galactic Background Light, must set upper bounds to our wildest imaginings. For now these upper bounds are rather uncertain (but improving) so that it is still not unreasonable to contend that hidden galaxies might contain several times more light in total than the known population. If eventually they prove to emit less light they may nevertheless contain the preponderance of mass, in the form of dark matter, while we already know for sure that their total cross-section - important for the absorption of background radiation (e.g. Quasar absorption line systems), should comfortably exceed that of conventional galaxies (e.g. Linder, 1998).

"The Conjecture", as I shall call it, is heuristically a strong one, in the sense that it challenges us all to refute it using every observational and theoretical tool to hand. Indeed, it has been decisively refuted several times already - only to rise again and again from its grave. At this meeting we shall certainly hear at least one more new and apparently decisive refutation. Are we not therefore, as honest scientists, bound to abandon it for good? Popperian logic urges "yes", while historical precedent answers "no". How are we to reconcile these contradictory reactions to the same pieces of evidence when only one of them, surely, can be sound? The answer, as I shall argue in Section 7, is to be found in the Bayesian approach to evidence.

My own interest in the subject was aroused in 1969 while making my first observations of galaxies, using the brand new 90-inch at the University of Arizona. Spectroscopic observations had to be guided by watching the object on the slit with the naked eye and I spent many a cold night so staring at NGC ellipticals. It was natural to wonder if there was anyone out there staring back at the Milky Way. I estimated that from the core of an elliptical, where the local sky would necessarily be very bright, our own galaxy, and indeed most of the galaxies in the Universe, would be quite invisible. My sympathy for elliptical based astronomers was cut short when I realised that we might be blinded by the same effect ourselves. At about the same time Ken Freeman (1970) noted, without explanation, that the best studied disk galaxies all have the same surface brightness (SB) while, before us both, Arp (1965) had pointed out that any object, to appear diffuse, must be confined to a certain (rather broad) band in the apparent size/apparent magnitude plane. But I have to thank some typically sceptical remarks from Ron Ekers, during a seminar of mine at Groningen, for forcing me five years later to put my own ideas (Disney, 1976) into a quantitatively defensible form.

2. VISIBILITY THEORY - AND ITS PITFALLS

Our objective must be to enumerate the true number-density of galaxies $N_T(L_T, \Sigma)$ $dL_T.d\Sigma$ in the total-luminosity (L_T)/ Surface-brightness (Σ) plane - which I call the Bivariate Brightness Distribution (BBD). To correct the observed number $N_0(L_T, \Sigma)$ for selection effects, one needs to know the "Visibility" $V(L_T, \Sigma)$ - which denotes the relative volumes within which different galaxies can be detected using the observational procedure in question. Clearly $N_T \propto N_0/V$.

It is easy to show that V is a separable function of L_T and Σ (Disney & Phillipps, 1983) i.e.

$$V(L_T, \Sigma) = L_T^{3/2} \Lambda(\Sigma) \qquad (1)$$

so that one can FIX L_T and examine V as a function $\Lambda(\Sigma)$ of the SB alone. My first idea (1976) was that $\Lambda(\Sigma) \propto \Theta_{ap}^3$ - where Θ_{ap} is the apparent or isophotal size of a galaxy seen against our sky. Sure enough, and most excitingly, $\Lambda(\Sigma)$ turned out to be a steeply humped function centred, in the case of both ellipticals and spirals, at the very values of SB which turn up most commonly

Is there a LSB Universe

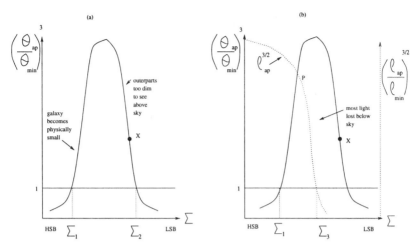

Figure 1. The Visibility Volume $\Lambda(\Sigma)$ of a galaxy as a function of SB Σ in schematic form. Assume the galaxy is of FIXED total luminosity L_T and lies at a fixed distance d_0. Its apparent (isophotal) angular size Θ_{ap} is a humped function of Σ and so is $d_{max}/d_0 = \Theta_{ap}/\Theta_{min}$, where d_{max} is the maximum distance to which the galaxy could be seen as larger than some minimum catalogue limit Θ_{min}. Figure (a) shows $\Lambda(\Sigma) \propto (d_{max}/d_0)^3 \sim (\Theta_{ap}/\Theta_{min})^3$. Any such galaxy with $\Sigma_1 > \Sigma > \Sigma_2$, including the dim object X would be visible out to d_0 or further. But what if there is also an apparent (isophotal) luminosity limit l_{min} (implicit or explicit)? Now $d_{max}/d_0 = (l_{ap}/l_{min})^{1/2}$ and $(l_{ap}/l_{min})^{3/2}$ is the dotted line in (b). The modified Visibility volume is now $\Lambda'(\Sigma)$, always the lesser of the two lines. $\Lambda'(\Sigma)$ is very sharply peaked at P (where $\Sigma_{CAT} \equiv l_{min}/\Theta_{min}^2$) and only galaxies in the much narrower range $\Sigma_1 > \Sigma > \Sigma_3$ would be visible at d_0, and X, although large enough, is selected against because it is apparently far too faint.

in catalogues (Fig 1). There is a peak because HSBGs *are* physically too small, while LSBGs, *appear* too small, with only their centres visible. Such coincidences impressed me then, and many authors continue to use the same SB Visibility function $\Lambda(\Sigma)$ today (e.g. Impey & Bothun 1997, in their excellent review).

But $\Lambda(\Sigma)$ is far from the whole story. To begin with, it yields a peak which is too broad to fit the very narrow SB distributions observed. The full-blown Visibility Function $\Lambda'(\Sigma)$ (Figure 1b) which includes limits on both apparent size (Θ_{lim}) AND apparent luminosity (l_{lim}), is both more accurate, and a great deal more interesting than the earlier $\Lambda(\Sigma)$ (Disney & Phillipps, 1983). Since $\Lambda'(\Sigma)$ is the lower envelope of two plunging and intersecting curves (Figure 1b) it is not continuous, and its peak is very sharp. More importantly it fits the data better and explains why dim galaxies like X are so hard to see (see caption). Samples need to be tested (V/V_{max}) for completeness in both angular size and apparent magnitude. (Note that CCD samples have median SBs no dimmer than plates because both Θ_{lim} and l_{lim} are smaller so $\Sigma_{CAT} \sim$ same.)

As one who has got it wrong myself, I can ruefully claim that Galaxy Visibility is a difficult subject, if only because it is three-dimensional. And it is difficult (for experienced astronomers in particular), to believe its dramatic consequences. Thus Allen and Shu (1979) projected onto the wrong (i.e. degenerate) plane and got almost everything wrong subsequently. Thus van der Kruit (1987) used the correct theory but normalised his two Visibility curves arbitrarily (which you mustn't) and failed to check (which you must) that his sample was complete in both Θ_{lim} and l_{lim} (which it wasn't). Thus McGaugh (1996) has introduced scalelength h in place of L_T then claimed that $\Lambda(\Sigma)$ is not peaked. He's wrong first because he doesn't hold L_T fixed while varying Σ, and second because $V(h, \Sigma) = L_T^{3/2}\Lambda(\Sigma) = (h^2\Sigma)^{3/2}\Lambda(\Sigma) = h^3 F(\Sigma)$, and although $F(\Sigma)$ may not be peaked, $\Lambda(\Sigma)$ still is. If very intelligent astronomers can make such mistakes, so can we all. I believe the majority of workers using Visibility today are still using it wrongly - in particular, they are seriously under-counting LSBGs like X by sticking to the simplistic $\Lambda(\Sigma)$, instead of using $\Lambda'(\Sigma)$.

3. LUMINOSITY-VISIBILITY AND LUMINOSITY FUNCTIONS

I have been asked to say a few words about this much simpler - but still contentious subject. Why do nearly all LF's (not only for optical galaxies) have a faint-end slope close to or just below $\alpha = 1.5$? The answer, I believe, is "because in almost all cases they are based on flux-limited surveys with not enough distances known". In other words the 1.5 is meaningless and proves only that space is roughly Euclidean out to the median distance of the survey. The proof is simple for if we assume that α is only a slowly varying function of L then the local slope of the Schechter function is:

$$\frac{dlogN(L)}{dlogL} \sim -\alpha - \frac{L}{L_\star} \qquad (2)$$

Because of the $L_T^{3/2}$ in (1) the *observed* number (per dL) in a flux-limited sample will peak sharply at $L_{max} = L_\star(3/2\text{-}\alpha)$ where the local slope must therefore be 1.5! If you force-fit a Schechter function to such flux-limited data then, in order to accommodate the exponential choke at the high luminosity end, you will find an α of slightly less than 1.5 for the low luminosity end.

For the above reason, it is not surprising to me that observations in clusters (e.g. Driver at al. 1994, Trentham 1997, Phillipps et al. 1998) find steeper LF's than ones found in the field. Cluster LF's may have their own problems, but they do not require the vast numbers of redshifts (usually lacking) that are needed to unambiguously establish the flat LF's often claimed for the field (e.g. Disney 1997, Marzke et al 1994).

4. ARE THERE HIGH SB GALAXIES ?

Allen and Shu (1979) put forward often quoted, but actually circular arguments against the existence of many HSBGs. At the same time our own analysis of the selection against HSBGs was totally inadequate (Disney 1976, Disney &

Phillipps 1983) because most HSBGs will be missed, not because they will masquerade as stars as we assumed, but because they will masquerade as background giant galaxies - a far more serious selection effect (Disney, 1998a).

HSBGs will only be uncovered in numbers by large spectroscopic or multiband photometric surveys, or with a large F-O-V space-camera. I am impressed with the early 2DF fibre work by Drinkwater & Gregg 1998) which finds unsuspected HSBGs at the few per cent level ($\sim$ 5 per square degree brighter than B_J=19.7) and this is likely to be an underestimate because galaxies containing absorbing dust cannot attain very high SBs in *the optical*. If you contract a disc galaxy in the radial direction, the effectiveness of its dust as an absorber increases as it is pushed nearer to the accompanying stars - and there will be an upper limit to that SB not much greater than the Freeman value (Jura 1980; Disney & Phillipps 1987; Disney, Davies & Phillipps 1989). However, the question of how much absorption there is in discs is a tricky one, and the early work on inclinations and IRAS observations is now regarded as inconclusive (see Davies & Burstein, 1995, for the debate) but should shortly be illuminated by the first submillimeter survey of bright galaxies being carried out by SCUBA (Dunne et al. 1998). The high sub-mm background derived from COBE suggests that half the starlight from galaxies is absorbed by dust (Dwek et al. 1998, Trewhella et al. 1997).

5. THE STRONGEST ARGUMENTS IN FAVOUR OF THE CONJECTURE

are currently, in my opinion:

(a) The large number of fortuitous numerical coincidences (about 8) otherwise required to explain the observations, if the Conjecture is false (Disney and Phillipps, 1987).

(b) A complete lack of correlation between Blue SB and Blue colour (B-V) - which makes no sense at all unless it is due to selection. Increased star formation should push both parameters up together (Disney & Phillipps 1985, Bothun, Impey & McGaugh 1997, O'Neil 1997).

(c) The close correspondence between calculated Visibility and median observed distance in a large sample of spirals (Davies et al. 1994).

(d) The large number of QSOALs containing metals. What else could they sensibly be - if not LSBGs and dwarfs (Phillipps et al. 1991, Linder, 1998) ?

(e) The lack of correlation observed between the apparent SBs of galaxies, and their galactic latitudes, even in latitudes where the foreground absorption is known to be significant (Davies et al. 1993)

(f) Equal numbers of galaxies found per 1 magnitude bin in SB between $\mu_0(B) = 21$ and 26.5 (Turner et al. 1993, McGaugh 1996).

6. THE STRONGEST CHALLENGES TO THE CONJECTURE

at present are:

(a) The failure to find optically invisible galaxies in deepish 21-cm pencil beams (Zwaan et al. 1997).

(b) The claim that all QSOALs can be accounted for by enormous gaseous halos surrounding giant galaxies (Lanzetta et al. 1995)

(c) The claim that fluctuations in the EBL are too low to accommodate many LSBGs (Vogeley, 1998).

7. REFUTATION AND THE BAYESIAN APPROACH TO EVIDENCE

How can honest minds reconcile such contradictory evidence? I believe the answer lies in the Bayesian approach to scientific reasoning which is now undergoing a renaissance. The Bayesian's belief in some hypothesis h, as it is affected by some evidence e, is given by P(h|e) "the probability of h given e" which, according to Bayes' theorem is given by:

$$P(h|e) = \frac{P(e|h).P(h)}{P(e|h).P(h) + P(e|\sim h).P(\sim h)} \quad (3)$$

where $\sim h \equiv$ "not h".

Such an approach is explicitly subjective because it relies on the assignation of an explicit prior belief or probability P(h) to the hypothesis under question. As such it was severely criticised by Fisher and other "classical" statisticians who proposed allegedly more objective inference tests in its place, and it went out of fashion. It is now enjoying a renaissance because the classical tests have been found to suffer from previously unacknowledged subjective ailments of their own - see for example Howson & Urbach's (1989) readable polemic on the subject.

Consider what conclusions two astronomers A (whose long previous experience lends him high confidence in the Conjecture) and B (who is pretty sceptical about it) draw from some new evidence, e.g. from Zwaan et al's (1997) recent failure to turn up invisible galaxies in an Arecibo strip survey.

If h = "The Conjecture" then:

A(*Believer*)		B(*Sceptic*)	
Prior: $P_A(h)=0.9$,	Thus $P(\sim h=0.1)$	$P_B(h)=0.2$,	Thus $P_B(\sim h)=0.8$
$P(e|h)=0.3^*$,	Thus $P(e|\sim h)=0.7$	$P(e|h)=0.2^*$,	Thus $P(e|\sim h)=0.8$
Bayes: $P_A(h|e)=$	$\dfrac{0.3 \times 0.9}{0.3 \times 0.9 + 0.7 \times 0.1}$	$P_B(h|e)=$	$\dfrac{0.2 \times 0.2}{0.2 \times 0.2 + 0.8 \times 0.8}$
=	0.8 (down from 0.9)	=	0.06 (down from 0.2)

(*There is also a lesser subjective element here, hence the slightly different estimates of P(e|h) by A and B).

Thus according to the Bayesian approach, both can maintain their different opinions with integrity, though A's confidence in the Conjecture is slightly eroded whilst B's scepticism is reinforced. And surely that is how it should be. Through their respective priors, each can factor into the equation all the other evidence which they believe bears on the matter. The fiction of a single refutation being enough to change an honest mind, particularly in an observational science, is seen for what it is.

8. BLIND 21-CM HI SURVEYS

The obvious way to get a completely dark sky has always been to use "redshift" to discriminate local brightness from distant dimness - and where better to do that than 21-cm? Unfortunately receiver noise - and other forms of system noise "illuminate" the 21-cm sky in their own fashion, so that any radio telescope, irrespective of size, has a column density limit (Disney & Banks 1997)

$$N_{HI}(cm^{-2}) \approx 10^{18} T_S \sqrt{\Delta V(kms^{-1})/t_{obs}(sec)} \qquad (4)$$

where T_S = system temperature, ΔV is the HI line-width, and t_{obs} the integration time. Generally speaking, one might expect LSBGs to have low N_{HI}'s as well (see de Blok et al. 1996), so that earlier 21-cm blind surveys set few interesting limits on the population of LSBGs - although claims were sometimes made to the contrary (e.g. Shostak 1977).

But technology is changing fast; lower noise amplifiers; faster and cheaper correlators; multi-beam foci on single dishes and wider bandwidth receiver systems on interferometers, mean that over the next few years extreme LSBGs (i.e. with $\mu_0 > 27$ Bμ) will become rather easy to pick up at 21-cm if they exist and if they have "normal" M_{HI}/L_B's in the range 0.3 to 3. The one fly in the ointment is the possibility that such low column density systems might be ionised by the intergalactic radiation field - as some have argued (Maloney, 1993, Dove & Schull, 1994).

What we need to do is look. A start has been made with HIPASS (HI Parkes All Sky Survey, see Staveley-Smith et al. 1996) which is covering the entire southern sky out to 12700 kms^{-1} with sufficient sensitivity to pick up some 10^4 galaxies independently of their optical properties, while Jodrell Bank is taking on the Northern sky. We are at the same time carrying out deeper surveys of selected regions so that we should soon know, one way or another, if large numbers of HI-rich LSBGs exist. The shallow HIPASS survey will only pick up optically invisible galaxies (on the DSS) if they have extreme M_{HI}/L_B's of > 25 in solar units but deeper surveys will be more interesting. In a 5600 seconds/point search of a $4° \times 8°$ patch, we find (Disney et al. 1998) 106 sources, of which 5 have no optical counterparts on the DSS.

9. OUTSIDE THE ATMOSPHERE

The sky will be darker and different, as was pointed out by O'Connell (1987). For instance, in the optical, Bryn Jones and I (1997) have looked at the SBs of galaxies in the HDF - to find the extraordinary result that they are much the same as those seen from the ground (see also Schade, 1995). Yet because SB $\propto (1+z)^{-4}$, they ought to be down by 3-5 magnitudes - depending on the uncertain k-corrections. I can think of only three alternate explanations for this extraordinary result - and it is extraordinary though barely remarked upon, all of which are implausible: selection bringing to light a rich population of intrinsically HSBGs not seen so far from the ground; an evolutionary conspiracy in which SB conveniently rises into the past so as to just cancel the cosmological effects; or the Universe isn't expanding, i.e. the redshift has failed the classical Tolman (1930) test.

To either side of the optical window the sky ought to be invitingly dark, but instruments planned to specifically exploit this opportunity to look for low SB structures seem sadly to be lacking from NASA's repertoire. For instance in the NUV, O'Connell (1987) showed that the contrast between a galaxy SED and the sky should be 2-3 magnitudes larger than it is from the ground-based optical, while the NIR sky seen from the HST with NICMOS is 200 times darker than it is from Earth. Surely there are some exciting opportunities here; maybe we can design the last camera on HST - the WFC-3, or the first on NGST to exploit these windows. Interested parties please e-mail me.

10. IS THE CONJECTURE REALLY TRUE?

Suppose that for every galaxy in the UGC catalogue ($\sim 10^4$ of them) there exists another Shadow Galaxy of the same luminosity and at the same distance - but with 10 times its radius. Such "Shadows" would have extremely low SBs ($\mu_0(B)$'s $> 27 B\mu$) and HI column densities ($N_{HI} \sim 10^{18}$ cm^{-2}). In order to be sure of finding only 25 members of such a clustered population, I estimate you would need to take 10^4 very deep CCD frames, or 10^4 even deeper 21-cm integrations. Since we haven't made such an effort, or anything approaching it, we cannot be sure that the Shadow Universe is not there. And if it is there, it is most likely to turn up serendipitously - as it did with the Crouching Giant Malin 1 (Bothun et al. 1987). But the most promising place to look is among QSOALs, where I believe it is turning up already. The Shadow population would have a total cross-section 2 orders of magnitude greater than the familiar one and would neatly account for many of the otherwise puzzling properties of QSOALs.

Having confessed to a Bayesian leaning I include in Table 1, for what it's worth, my own estimate of the odds against the truth of the Conjecture, as they have been changed by various pieces of evidence over the years.

11. THE FUTURE OF OUR SUBJECT

There is nothing wrong with ambition in science. We may all ask (see Hardy, 1984):

"*Here, on the level sand, Between the sea and land, What shall I build or write, Against the fall of night?*"

but ambition is sometimes carried to such lengths by aggressive individuals as to poison a whole subject area, drive out the competition and thus slow progress to a crawl. To maintain our own momentum, by preserving the enjoyment in our fascinating area, I hope we will therefore:

(a) Recognise there is such a lot more to do that we are unlikely to be "scooped". According to Harwit(1981) we have made no more than 10% of the key discoveries in astronomy.

(b) Remember that "Astronomy is not Physics" - indeed it is so incomplete that one refutation can never be enough. According to the Bayesian view there is usually room for more than one interpretation of the same evidence.

Epoch	Odds Against	Evidence
1975	$10^6 : 1$	Original prior belief
1976	$50 : 1$	First selection effects paper (Disney 1976)
1977	$10^3 : 1$	Failure to find "Icebergs" in off-beams (Shostak 1977)
1983	$10 : 1$	Visibility calculations done properly (Disney & Phillipps 1983)
1985	$5 : 1$	No correlation between SB & colour (Disney & Phillipps 1985)
1987	$3 : 1$	Many LSBGs found in Fornax (Phillipps et al. 1987)
1987	$2 : 1$	Malin 1 found (Bothun et al. 1987)
1990	$1 : 1$	Shostak's failure explained (unpub. see Disney & Banks 1997)
1993	$1 : 2$	No correlation found between SB and b_{II} (Davies et al. 1993)
1994	$1 : 4$	Distribution of ESO SBs fits Visibility (Davies et al. 1994)
1995	$1 : 2$	Huge halos invoked to explain QSOALs (Lanzetta et al. 1995)
1995	$1 : 3$	LSBGs in equatorial survey (Sprayberry et al. 1993)
1997	$1 : 4$	Blind HI survey of CenA group (Banks et al. 1998)
1997	$1 : 5$	LSBGs found in deep CCD survey (Dalcanton et al. 1997)
1997	$1 : 3$	No very dim galaxies in Arecibo strip (Zwaan et al. 1997)
1998	$1 : 10$	HSBGs found in redshift surveys (Drinkwater & Gregg 1998)
1998	$1 : 5$	Low SB fluctuations in HDF (Vogeley, 1998)

Table 1. Odds Against The Conjecture

(c) Never write anonymous referees reports either on papers or proposals. The referees job is surely to encourage and improve good science, not discourage or disprove bad, for in an open society bad science will sink to its own level quickly enough.

(d) Celebrate and reference each others work, and expect the same in return. In that connection, I want to sincerely thank Greg Bothun, Julianne Dalcanton and Chris Impey for being overgenerous to me.

Acknowledgments. I am deeply grateful to those close colleagues who have contributed so much to my own understanding and enjoyment of this subject, most particularly Jon Davies, Steve Phillipps, Alan Wright, Ron Ekers, Ed Kibblewhite, Peter Boyce, John Bryn Jones, Gareth Banks and Robert Minchin.

References

Allen, R.J. & Shu, F.H. 1979, ApJ, 227, 67
Arp, H. 1965, ApJ, 145, 402
Banks, G.D. 1998, "21-cm Blind Searches for galaxies", PhD Thesis, Cardiff
de Blok, W.J.G. et al., 1996, MNRAS, 283, 18
Bothun, G. Impey, C. & McGaugh, S. 1997, PASP, 109, 745
Bothun, G. et al., 1987, AJ, 94, 23
Dalcanton, J.J. et al. 1997, AJ, 114, 635
Davies, J.I. & Burstein, D. 1994, ed: "The Opacity of Spiral Discs", Kluwer
Davies, J.I. et al. 1994, MNRAS, 268, 984
Davies, J.I. et al. 1993, MNRAS, 260, 491
Disney, M.J. & Banks, G.D. 1997, Publ. Astron. Soc. Australia, 14, 69

Disney, M.J. & Phillipps, S. 1987, Procs "Theory and Observational Limits in Cosmology", ed: Stoeger, publ. Specola Vaticana, p.385
Disney, M.J. & Phillipps, S. 1985, MNRAS, 216, 53
Disney, M.J. 1998a, in prep
Disney, M.J. et al. 1998b, Publ. Astron. Soc. Australia, in press
Disney, M.J. & Phillipps, S. 1983, MNRAS, 205, 1253
Disney, M.J. 1976, Nature, 263, 573
Disney, M.J. 1997, Procs. IAU Symposium 179, ed. McLean, B.J., p11
Disney, M.J. Davies, J.I. & Phillipps, S. 1989, MNRAS, 239, 939
Driver, S.P. et al., 1994, MNRAS, 266, 155
Dove, J.B. & Schull, J.M. 1994, ApJ, 423, 196
Drinkwater, M.J. & Gregg, M.D. 1998, MNRAS, 296, L15
Dunne, L. & Eales, S.A. 1998, *in prep*
Freeman, K.C. 1970, ApJ, 160, 811
Hardy, G.H. 1984, "A Mathematicians Apology", C.U.P.
Harwitt, M. 1981, "Cosmic Discovery", Harvester Press
Howson, C. & Urbach, P. 1989, "Scientific Reasoning", Open Court, Chicago
Impey, C.D. & Bothun, G.D. 1997, Ann. Rev. Astr. Ap., 35, 267
Jones, J.B. & Disney, M.J. 1997, "Procs of The HST and the High redshift Universe", eds: Tanvir N.R., p151
Jura, M. 1980, ApJ, 238, 337
van der Kruit, P.C. 1987, AA, 173, 59
Lanzetta, K.M. et al. 1995a, ApJ, 442, 538
Linder, S.M. 1998, ApJ, 495, 637
Maloney, P. 1993, ApJ, 414, 41
McGaugh, S.S. 1996, MNRAS, 280, 337
O'Connell, R. 1987, AJ, 94, 867
O'Neil, K. 1997, PhD Thesis, University of Michigan
Phillipps, S. et al. 1987, MNRAS, 229, 505
Phillipps, S. et al. 1991, MNRAS, 242, 235
Phillipps, S. et al. 1998, ApJ, 493, L59
Schade, D. et al. 1995, ApJ, 451, L1
Shostak, G.S. 1977, AA, 54, 919
Sprayberry, D. et al. 1995, AJ, 109, 558
Staveley-Smith, L. et al. 1996, Publ. Astron. Soc. Australia, 13, 243
Tolman, R.C. 1930, Proc. N.A.S., 16, 511
Trentham, N. 1997, MNRAS, 286, 133
Trewhella, M. et al. 1997, MNRAS, 288, 397
Turner, J.A. et al., 1993, MNRAS, 261, 39
Vogeley, M.S. 1998, astro-ph/9711209
Zwaan, M. et al. 1997, AJ, 490, 173

The Low Surface Brightness Universe, IAU Col. 171
ASP Conference Series, Vol. 170, 1999
J. I. Davies, C. Impey and S. Phillipps, eds.

Optical Galaxy Selection

Stacy McGaugh

Department of Astronomy, University of Maryland, USA

Abstract. Our view of the properties of galaxies is strongly affected by the way in which we survey for them. I discuss some aspects of selection effects and methods to compensate for them. One result is an estimate of the surface brightness distribution. I believe this is progress, but considerable uncertainty remains.

1. Limits and Selection Effects

A basic goal of galaxy research is to identify and characterize the galaxy populations which inhabit the universe. Astronomers undertake large surveys to discover galaxies, but this is only part of the battle. In order to transform from the observed, apparent distributions of galaxy properties and recover the desired intrinsic distributions, one must understand and correct for the selection characteristics of galaxy surveys.

An essential fact about galaxies, as they appear on survey plates, is that they are not point sources. Some of the consequences of this simple fact were first quantified by Disney (1976), who pointed out that the intrinsic distribution could be very different from the apparent one. At the least, it is necessary to generalize the formal procedures developed for point sources (Disney & Phillipps 1983; McGaugh et al. 1995).

The appearance of galaxies requires many parameters to approximate: they have multiple components (bulge and disk); they have spiral arms and dust lanes; their morphology can be a strong function of wavelength depending on their recent star formation history; etc. So, we simplify. Here I will pretend that a tolerable approximation can be had with axially averaged radial surface brightness profiles. This reduces the dimensionality of the problem immensely. The surface brightness profiles of galaxies obtained in this way are tolerably described as a combination of bulge and disk (4 parameters) or with a generalized (Sérsic 1969) profile (3 parameters). For brevity, I will reduce this further to 2 parameters by limiting the discussion to galaxies dominated by exponential disks:

$$\mu(r) = \mu_0 + 1.086\frac{r}{h}. \quad (1)$$

This reduces complex galaxies to a characteristic central surface brightness μ_0 and scale size h. In some cases this is a fair approximation.

Galaxies inhabit a large range in surface brightness and scale length (Figure 1). Real limits appear at high surface brightness and luminosity. Here, objects are prominent and are easily selected by surveys. The limit at L^* represents

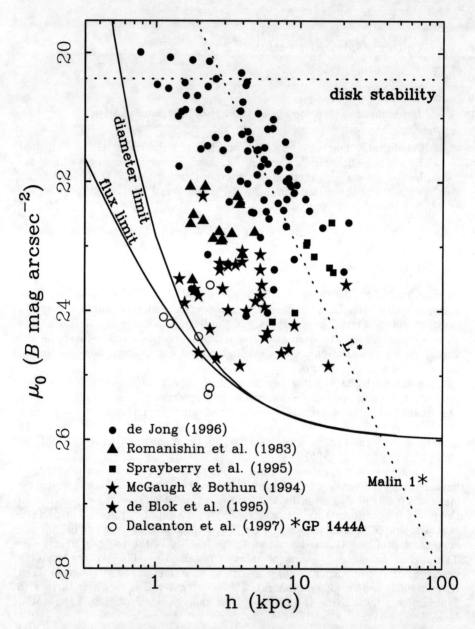

Figure 1. The distribution of disk galaxies in the central surface brightness-scale length plane. Disks exist at all (μ_0, h) up to maxima in μ_0 and L. Lower bounds are imposed by observational selection limits. Illustrative selection lines are drawn for $\mu_\ell = 26$, typical of good photographic plates.

the well known turndown in galaxy numbers at this point. The limit at high μ_0 may represent a physical restriction based on the requirement that disks be stable (Milgrom 1989; McGaugh & de Blok 1998), though a massive bulge or dark halo could in principle stabilize even higher surface brightness disks. At the dim and small end, there is no clearly defined physical limit: our knowledge here is circumscribed by selection effects. Galaxies which are intrinsically small or low surface brightness are hard to find, regardless of how common they may be. Examples of very low surface brightness galaxies are known to exist, implying that the sparsely populated space in Figure 1 is just waiting to be filled.

In addition to a survey diameter or flux (magnitude) limit, there must (at the least) be a second survey parameter quantifying the limiting isophote μ_ℓ at which these are measured (Disney 1976). These are shown in Figure 1 by extracting from equation 1 the radius or flux of a galaxy at the limiting isophote:

$$r > r_\ell = 0.92h(\mu_\ell - \mu_0) \qquad (2)$$

for diameter selection and

$$m < m_\ell = \mu_0 - 5\log h - 2.5\log[f(\mu_0, \mu_\ell)] + C \qquad (3)$$

for selection by apparent magnitude. See McGaugh et al. (1995) for the definition of $f(x)$.

The problem becomes obvious: to reach small scale lengths is difficult; to find galaxies dimmer than the isophotal limit is impossible. However, the situation is even worse than illustrated in Figure 1, where it is assumed that galaxies can be detected right down to the formal survey limit. Even maintaining this fiction, one is really battling the volume sampling function $V(\mu_0, h)$ (Disney & Phillipps 1983; McGaugh et al. 1995) illustrated in Figure 2. This is much more restrictive than the lines in Figure 1 imply.

2. Volume Sampling and Intrinsic Properties

The volume sampled by any given survey is a strong function of both the survey parameters (μ_ℓ, r_ℓ or m_ℓ) and the intrinsic properties of galaxies (μ_0, h). A progressively larger volume is sampled for intrinsically larger and higher surface brightness galaxies. Such galaxies will always dominate the apparent numbers of objects in catalogs regardless of the intrinsic bivariate distribution $\Phi(\mu_0, h)$ (McGaugh 1996).

It is possible, in principle, to correct for $V(\mu_0, h)$ and recover the intrinsic distribution from the observed one. A good example is given by de Jong (1996 and these proceedings). Here I restrict myself to the projection of the bivariate distribution $\Phi(\mu_0, h)$ along the surface brightness axis, which I call the surface brightness distribution $\Phi(\mu_0)$ (Figure 3). I derive this by *assuming* μ_0 and h are uncorrelated (see Figure 1). While almost certainly not true in detail, this procedure does not produce a result which is grossly at odds with the results of the more complete analysis of de Jong.

The surface brightness distribution is broad, being *roughly* flat below the Freeman (1970) value. Low surface brightness galaxies are numerous, with approximately equal numbers of galaxies in each bin of central surface brightness.

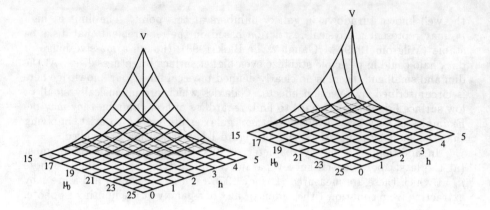

Figure 2. The volume sampled by (a) diameter and (b) flux limited surveys as a function of surface brightness and scale length.

The surface brightness distribution falls off rather more steeply at the bright end.

If the assumption made about the independence of μ_0 and h fails, it has a predictable effect on the shape of $\Phi(\mu_0)$. If galaxies become systematically smaller as surface brightness decreases (as some unwritten lore presumes), then the surface brightness distribution will *rise* towards lower surface brightnesses. Since both surface brightness and size act against such dim, small galaxies, the required volume correction is larger than made here. On the other hand, if galaxies tend to be larger at lower surface brightnesses, the surface brightness distribution will fall more rapidly than shown. There is some hint of such a trend in Figure 1 (see also McGaugh & de Blok 1997). However, we base all this on incomplete data which is fraught with the very selection effects for which we would like to correct!

Different data in Figure 3 are consistent, given the large uncertainties. I do see some potential for controversy at the dim end. In an HI selected sample, Zwaan (these proceedings) reports a sharp turn down faintwards of $\mu_0 \approx 24$. Yet Dalcanton et al. (1997) not only detect galaxies dimmer than this, but also claim to measure a rather high density of them. One possibility is that the HI column densities of very low surface brightness galaxies are so low that the HI is ionized by the extragalactic UV radiation field. However, I do not think this can work. Only slightly brighter low surface brightness galaxies have abundant HI (McGaugh & de Blok 1997; Schombert et al. 1997), and the column densities are only a little lower than in high surface brightness spirals (de Blok et al. 1996). Moreover, by this reasoning, the $10^{11} \mathcal{M}_\odot$ of neutral hydrogen in Malin 1 should all be ionized.

That low surface brightness galaxies are *numerous* does not necessarily mean that they contribute a large amount to the integrated luminosity density (Figure 4). If the assumption I have made holds, low surface brightness galaxies are on average the same size as high surface brightness galaxies, and are therefore less luminous. The luminosity density thus declines with surface brightness.

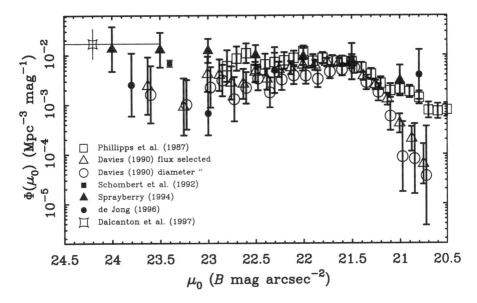

Figure 3. The surface brightness distribution, i.e., the number density of galaxies at each central surface brightness. There is a sharp feature at the bright end corresponding to the Freeman value, perhaps corresponding to a threshold for disk stability. At the dim end large numbers of low surface brightness galaxies are indicated. There is considerable uncertainty in $\Phi(\mu_0)$ at all surface brightnesses. Neither the shape nor the precise value of the Freeman limit are well determined.

This estimate of the luminosity density (Figure 4) as a function of surface brightness is more robust than the number density (Figure 3). If low surface brightness galaxies are larger on average than high surface brightness galaxies, they give away less in luminosity. However, they must then be less numerous in Figure 3, and the effects tend to offset. They do not offset perfectly (McGaugh 1996); because of the severity of the volume sampling function, it is easier to conceal luminosity in the small, dim population than in Malin-1-like giants.

3. Complications

While I have emphasized the severity of selection effects and the magnitude of the volume corrections required, I have so far actually painted a fairly rosy picture. I have made simple assumptions about the properties of $\Phi(\mu_0, h)$ and $V(\mu_0, h)$, and have further assumed that one can simply apply the volume sampling correction to galaxy samples which are complete as advertised. In reality, none of these things are likely to be realized, and there are a number of other important effects which have been completely neglected.

To mention a just few:

- Galaxies are not axially symmetric exponential disks.

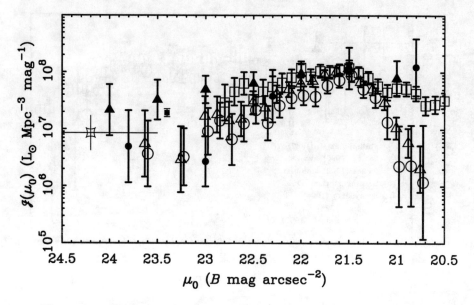

Figure 4. Similar to the previous figure, but now the luminosity density. The luminosity density declines with dropping surface brightness. This happens simply because lower surface brightness galaxies are fainter *on average* than high surface brightness galaxies. There is a significant fraction of the total luminosity density due to all disks in the low surface brightness tail ($< 30\%$), but it seems unlikely that a large or even divergent population remains hidden.

- The galaxy SED matters for both bandpass and redshift effects.
- There are systematic measurement, as well as selection, effects.
- Inclination and internal extinction matter.
- Large scale structure may render meaningless a 'fair' volume.

While it is convenient to assume the exponential profile as the next best thing to pretending galaxies are point sources, there is no guarantee that this is adequate for purposes of selection. Early type galaxies are generally bulge dominated. Though the disk may well dominate in the regions of the selection isophote, the steep inner profile of the bulge makes these galaxies more prominent than assumed, especially to flux limited surveys. For late type galaxies the exponential profile is a better description of the radial profile, but these galaxies tend to be irregular and have large deviations from axial symmetry. These effects tend to exacerbate further the already large difference in accessible volume between bright and dim galaxies. Moreover, features like spiral arms, star forming regions, and dust lanes can complicate matters in a way which is difficult to generalize.

Another issue is the actual spectral energy distribution (SED) of galaxies. While I have assumed a generic μ_0 in the survey band which defines μ_ℓ, one would really like to know the bolometric surface brightness. As $\mu_0^{bol} \to \mu_\ell$, the visibility of a galaxy depends quite sensitively on color (McGaugh 1996). This may be a partial explanation for why the multi-band CCD survey of O'Neil et al. (1997) turns up more red low surface brightness galaxies than had previous surveys conducted on blue sensitive photographic plates (Schombert et al. 1992; Impey et al. 1996). If typical galaxy redshifts for a survey are high enough for K-corrections to be important, things are even more complicated (Ferguson, these proceedings).

One crucial thing in trying to recover the galaxy luminosity function is the systematic underestimate of the luminosities of low surface brightness galaxies made by measuring at some limiting isophote (McGaugh 1994; Dalcanton 1998). This is a particularly pernicious effect, because it is severe only for a very small percentage of galaxies in a complete sample. Yet these are precisely the objects for which large volume corrections are necessary. This magnifies systematic errors which in turn have a disproportionate effect on the derived luminosity function. All fit parameters (L^*, ϕ^*, α) can be affected, but the faint end slope is particularly susceptible to underestimation. Worse, many analyses still assume galaxies are point sources, which is tantamount to drawing the flux selection boundary in Figure 1 as a straight line parallel to that for L^*.

So far, I have assumed all galaxies are face on. This is bad, but not too terrible *if* galaxies are optically thin. This is probably the case for low surface brightness galaxies (McGaugh 1994; Tully et al. 1998). It is probably not the case for high surface brightness spirals. Just how optically thick spirals may be remains controversial. For the purposes of selection, internal extinction can have a significant inclination dependent effect on the position of a galaxy in Figure 1 (see Lu 1998).

Finally, I'd like to mention that a fundamental assumption is that we can probe a large enough volume of the universe for it to be considered homogeneous and representative. Yet as redshift surveys press ever deeper, we discover ever larger features, with no clear convergence as yet (Huchra, these proceedings). While the selection effects embodied in the volume sampling function will always be important, it is not obvious that large scale structure will admit a straightforward correction. Yet again, this problem is particularly severe at the dim end. While large surveys are sensitive to high surface brightness L^* galaxies over impressive volumes, they still probe only modest volumes for small and low surface brightness galaxies.

4. Conclusions

In spite of some of the daunting problems, substantial progress has been made. There is great scope for further improvement. If any of the difficulties I have mentioned imposes some fundamental limitation, we are as yet far from reaching it.

Acknowledgments. I am grateful to many of the attendees of this meeting for discussing these topics over the years. I would particularly like to thank Jon Davies for organizing a great conference.

References

Bothun, G.D., Impey, C.D., Malin, D.F., & Mould, J.R. 1987, AJ, 94, 23
Dalcanton, J.J. 1998, ApJ, 495, 251
Dalcanton, J.J., Spergel, D.N., Gunn, J.E., Schmidt, M. & Schneider, D.P. 1997, AJ, 114, 635
Davies, J.I. 1990, MNRAS, 244, 8
Davies, J.I., Phillipps, S., Disney, M.J. 1988, MNRAS, 231, 69P
Disney, M.J. 1976, Nature, 263, 573
Disney, M.J., & Phillipps, S. 1983, MNRAS, 205, 1253
de Blok, W.J.G., van der Hulst, J.M., & Bothun, G.D. 1995, MNRAS, 274, 235
de Blok, W.J.G., McGaugh, S.S., & van der Hulst, J.M. 1997, MNRAS, 283, 18
de Jong, R. S. 1996, A&AS, 118, 557
de Jong, R. S. 1996, A&A, 313, 377
Freeman, K.C. 1970, ApJ, 160, 811
Impey, C.D., Sprayberry, D., Irwin, M.J., & Bothun, G.D. 1996, ApJS, 105, 209
Lu, N.Y. 1998, ApJ, in press
McGaugh, S.S. 1994, Nature, 367, 538
McGaugh, S.S. 1994, ApJ, 426, 135
McGaugh, S.S. 1996, MNRAS, 280, 337
McGaugh, S.S., & Bothun, G.D. 1994, AJ, 107, 530
McGaugh, S.S., Bothun, G.D., & Schombert, J.M. 1995, AJ, 110, 573
McGaugh, S.S., & de Blok, W.J.G. 1997, ApJ, 499, 66
McGaugh, S.S., & de Blok, W.J.G. 1998, ApJ, 499, 66
Milgrom, M. 1989, ApJ, 338, 121
Phillipps, S., & Disney, M.J. 1983, MNRAS, 203, 55
Phillipps, S., Disney, M.J., Kibblewhite, E.J., & Cawson, M.G.M. 1987, MNRAS, 229, 505
O'Neil, K., Bothun, G.D., Schombert, J., Cornell, M.E., & Impey, C.D. 1997, AJ, 114, 2448
Romanishin, W., Strom, K.M., & Strom, S.E. 1983, ApJS, 53, 105
Schombert, J.M., Bothun, G.D., Schneider, S.E., & McGaugh, S.S. 1992, AJ, 103, 1107
Schombert, J.M., Pildis, R.A., Eder J.A. 1997, ApJS, 111, 233
Sérsic, J.-L. 1969, Atlas de galaxias australes (Observatorio Astronomica, Córdoba)
Sprayberry, D. 1994, Ph.D. thesis, University of Arizona
Sprayberry, D., Impey, C.D., Bothun, G.D., & Irwin, M. 1995a, AJ, 109, 558
Tully, R.B., Pierce, M.J., Huang, J.-S., Saunders, W., Verheijen, M.A.W., Witchalls, P.L. 1998, AJ, 115, 2264

Selection Effects at 21cm

F.H. Briggs

Kapteyn Astronomical Institute, P.O. Box 800, 9700 AV Groningen, The Netherlands

Abstract. Surveys in the 21cm line of neutral hydrogen are testing the completeness of the catalogs of nearby galaxies. The remarkable observational fact is that the potential wells that confine gas to sufficient density that it can remain neutral in the face of ionizing radiation also provide sites for star formation, so that there are no known cases of neutral intergalactic clouds without associated star light.

1. Introduction

Invisible hydrogen clouds are common place in the vicinity of galaxies. This "low surface brightness" material lies in tidal debris from galaxy interactions (cf. Haynes et al 1979, van der Hulst 1979, Yun et al 1994, Putnam et al 1998), in the extended HI disks that facilitate kinematical studies (cf Broeils & van Woerden 1994), in outlying gas rings around galaxies and pairs of galaxies (Schneider 1989, van Driel et al 1988), in the high velocity clouds around the Milky Way (Wakker and van Woerden 1998), and in a number of other structures for which the nature of the extended LSB gas is less clear (cf. Fisher & Tully 1976, Simkin et al 1987, Giovanelli & Haynes 1989).

Despite vigorous searching, no examples of isolated neutral hydrogen clouds of a truly intergalactic nature have been discovered, and, in fact, for some years, the upper limit to the baryon content of intergalactic clouds with HI masses in the range comparable to normal galaxies has been known to be cosmologically insignificant (Fisher & Tully 1981). This finding is in sharp contrast to the recognition that the *ionized* intergalactic medium may form an important, perhaps dominant, reservoir of the Universe's baryons (Rauch et al 1997, Shull et al 1996).

What selection effects influence and limit the neutral hydrogen picture of the nearby universe? What properties could neutral gaseous systems have that would allow them to escape detection in the surveys that have been made so far?

2. Selection Effect '0': the need for neutrals

Clearly, a easy way to hide hydrogen from observers at 21cm wavelength is to ionize it. However, even ionized clouds will have some neutral fraction, and in principle, long integrations with sensitive receivers could detect the 21cm emission from predominantly ionized clouds. The optical depth in the 21cm line of a cloud with neutral H° column density N_{HI}, velocity width ΔV (FWHM) and excitation temperature T_{spin} is

$$\tau_{21} \approx \left(\frac{N_{HI}}{2\times 10^{21} \text{ cm}^{-2}}\right)\left(\frac{10 \text{ km s}^{-1}}{\Delta V}\right)\left(\frac{100 \text{ K}}{T_{spin}}\right) \quad (1)$$

Under the conditions for which emission is normally observed from HI clouds (i.e., the excitation temperature is significantly above the temperature provided by the background radiation flux ($T_{spin} \gg 2.7$ K) and $\tau_{21} \ll 1$), the brightness temperature of the emergent radiation from a cloud of neutral H° column density N_{HI} is

$$T_B \approx \tau_{21} T_{spin} \approx \left(\frac{N_{HI}}{2\times 10^{19} \text{ cm}^{-2}}\right)\left(\frac{10 \text{ km s}^{-1}}{\Delta V}\right) \text{ K} \quad (2)$$

Under these conditions, the integration time required for a radio telescope to detect a cloud (with $\Delta V \approx 20$ km s^{-1}) that fills the telescope beam is approximately one second for $N_{HI} = 2\times 10^{19}$, two minutes for $N_{HI} = 2\times 10^{18}$, and three hours for $N_{HI} = 2\times 10^{17}$ cm^{-2}. The latter column density is of interest because clouds with $N_{HI} \leq 3\times 10^{17}$ cm^{-2} are optically thin for wavelengths shortward of the Lyman limit ($\tau_{LL} \leq 1$) and are therefore vulnerable to ionizing radiation with the expectation that the hydrogen within them will be predominantly ionized. For example, it has long been expected that photoionization will cause the outer edges of gaseous galaxy disks to show abrupt declines in the column density of neutral atoms (Sunyaev 1969, Maloney 1993, Corbelli & Salpeter 1993); deep integrations in the outskirts of galaxies and in the intergalactic medium at large, may detect local analogs of the highest column density clouds in the Lyman-α forest and Lyman-limit population identified through quasar absorption-line studies (cf. Hoffman et al 1993, Charlton et al 1994).

Even neutral gas can be hidden from 21 cm emission-line surveys if the excitation temperature T_{spin} is low, as might occur if the gas density is so low that T_{spin} is not coupled to the gas kinetic temperature T_k or if T_k is intrinsically low. As indicated by Eq. 1, such low T_{spin} clouds would be good absorbers, which should appear in absorption against background continuum sources (cf. Corbelli & Schneider 1990). Of course, in instances when the 21 cm line is optically thick, application of Eq. 2 will underestimate the column density and the HI cloud masses (cf. Dickey et al 1994, Braun 1995, Braun 1997).

3. Selection effect 1: 'optical selection' of targets

An historically important selection effect is that the vast majority of extragalactic 21cm line observations have been made with telescopes pointed directly at optically selected objects. Only within the past decade has it become technically feasible to make "optically blind" surveys of sufficiently large volumes of space with adequate sensitivity to recover the known galaxy population, let alone identify new populations of intergalactic clouds or gas-rich LSB galaxies. However, if there were such populations containing HI masses comparable to those in ordinary galaxies and comparable in spatial number density, then the historically important attempts (Shostok 1977, Materne et al 1979, Lo & Sargent

1979, Haynes & Roberts) would have discovered them. The lack of detections in these surveys, combined with the absence of serendipitous detections in the studies targeted on optically selected galaxies, led Fisher and Tully (1981) to conclude that the mass in intergalactic HI clouds in the range 10^7 to $10^{10} M_\odot$ could only be a small fraction of the total mass in galaxies, and thus can only contribute a tiny fraction of the total cosmological mass density. Briggs (1990) performed an update of this sort of analysis to conclude that the amount of the HI mass contained in un-identified new populations is only a few percent of HI mass contained in catalogued galaxies over this same mass range.

There have been a few surprises during this period of extensive observation of optically selected targets, including discoveries of uncatalogued HI-rich clouds of undetectably low optical surface brightness (cf. Fisher & Tully 1976, Schneider 1989, Giovanelli & Haynes 1989). However, in every case, the neutral gas is clearly associated in position and redshift velocity with optically visible galaxies, implying that the gravitational potential belonging to the optical galaxy is responsible for the confinement of the hydrogen to sufficient density that it remains neutral.

Similar conclusions result from the recent blind surveys that have been successful at compiling samples of HI-selected galaxies and recovering the known late-type galaxy population (cf. Szomoru et al 1994, Henning 1995, Zwaan et al 1997, Spitzak & Schneider 1998, Kraan-Korteweg et al 1998). All the confirmed HI detections from these surveys are found to be associated with optical emission from star light, provided that the objects are located at sufficiently high galactic latitude that they are not heavily obscured and are not located close to a bright foreground star. This finding lends credence to the idea that a reasonably complete picture of the neutral gas content of the nearby universe can be obtained from a study of the HI gas in the large optically galaxy samples (cf. Rao & Briggs 1993, Briggs 1997).

4. Selection effects for spectroscopic features

Surveys that require the detection of spectroscopic emission features differ in several respects from flux- or magnitude-limited surveys. These differences are illustrated in Fig. 1 and discussed in the following paragraphs.

Until recently, the limitations to radio spectrometer bandwidth and the number of available spectral channels within the spectrometer have often caused surveys to be limited in redshift depth, since they were unable to cover large redshift ranges efficiently. Most of these surveys have sufficient sensitivity to detect large HI masses of $M_{HI} > 10^{10} M_\odot$ beyond this restricted survey depth. At the low mass extreme $M_{HI} < 10^7 M_\odot$ nearly all surveys are limited in depth by sensitivity, since integration times become very long to detect these tiny masses at distances of order 10 Mpc or more.

The velocity width of spectral features also influences the detection efficiencies. If two galaxies with the same HI mass M_{HI} are compared, the one with the narrower profile will be easier to detect; the narrower profile could be detected to greater distance. If all galaxies had the same velocity spread, then the limiting distance to which each galaxy could be detected d_c would be $\propto M_{HI}^{1/2}$, according to the inverse square law, and the volume within which each could be detected

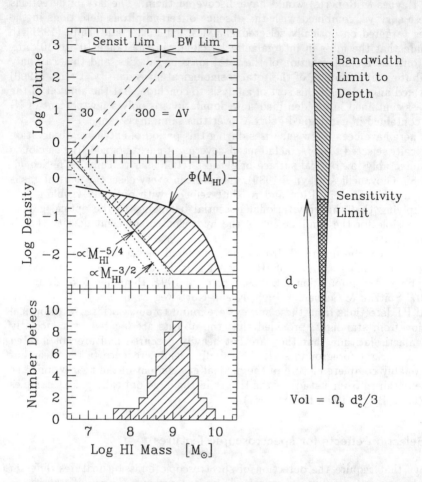

Figure 1. 21cm Line Detection Sensitivity. *Right:* The volume probed by a radio telescope beam depends on the beam solid angle Ω_b and the depth d_c to which a given HI mass can be detected. For large masses, the detection volume is often limited by the spectrometer bandwidth. *Left top:* In the "sensitivity limited regime" applicable for small HI masses, the survey volume also depends on velocity width ΔV shown here for $\Delta V = 30$, 100 and 300 km s^{-1}. *Left middle:* The quantity (survey volume)$^{-1}$ specifies a sensitivity function to objects whose space density is $\Phi(M_{HI})$, here given as number per volume per decade of mass. In the low-mass (sensitivity-limited) regime, including the trend of increasing rotation speed with mass implies a sensitivity function $\propto M_{HI}^{-5/4}$ rather than the $M_{HI}^{-3/2}$ dependence for a purely "flux limited" survey. *Left bottom:* Number of detections expected in bins of 0.2 Dex.

would be proportional to $M_{HI}^{3/2}$. The noise level in flux density, σ, in a spectrum that has been optimally smoothed to match the profile is $\sigma = \sigma_o \sqrt{\Delta v_o/\Delta V}$ for a spectrum originally recorded with channel spacing (resolution) of Δv_o for which the noise level is σ_o. The HI mass for a spectral feature of strength S_{Jy} is computed from the integral over the profile, $M_{HI} = 2 \times 10^5 d_{Mpc}^2 \int S_{Jy} dV_{km/s} M_\odot$. Thus, the minimum detectable HI mass is $M_{HI} \propto 5\sigma \Delta V d^2 \propto d^2 \sqrt{\Delta V}$, if the minimum detectable profile is modeled as a rectangle of height in flux density $\Delta S = 5\sigma$ and width ΔV. A sort of HI Tully-Fisher relation (cf. Briggs & Rao 1993) has $\Delta V \propto M_{HI}^{1/3} \sin i$ for galaxies with inclination i relative to the plane of the sky, leading to the result that $d_c \propto M_{HI}^{5/12} \sin^{-1/4} i$. Note that the $\sin^{-1/4} i$ factor is substantially different from unity for a only small fraction of a randomly oriented sample. The net result is that the survey volume in which a HI mass would be detected is more closely $volume \propto M_{HI}^{5/4}$ in the sensitivity-limited regime. Fig. 1 illustrates the difference in survey volumes for profiles of width 30, 100 and 300 km s^{-1}. These volumes are translated into survey "sensitivity functions" ($\propto 1/volume$) in units of Mpc^{-3} for comparison with HI mass functions; if a survey finds no masses of $10^7 M_\odot$ in a volume of 1 Mpc^3, then an estimate of the upper limit to the density of $10^7 M_\odot$ objects is ~ 1 Mpc^{-3}.

Figure 1 (left middle panel) illustrates a typical survey mass sensitivity in comparison to an HI mass function $\Phi(M_{HI})$ plotted as number of objects per Mpc^{-3} per decade of mass. For mass ranges where the survey sensitivity function lies below the mass function, it becomes probable that the survey will detect galaxies of that mass. The shaded area represents the mass range where detections should occur in the number as indicated in the lower plot, where detections are binned in 0.2 decade bins. In the regions below $\sim 10^{7.3}$ and above $\sim 10^{10.1}$, this survey would probably obtain no detections but could place upper limits on the space density of galaxies in these mass ranges.

The knowledge of the space density of tiny HI-rich extragalactic objects has remained highly uncertain because of the difficulty in surveying a sufficiently large volume with adequate sensitivity to detect them. A further complication is that the small masses have narrow velocity widths without the distinctive double horned profiles that characterize large spiral galaxies with flat rotation curves; this means that low mass systems can be more easily confused with radio interference.

The poor diffraction-limited angular resolution provided by most single-dish radio telescopes gives rise to confusion problems, and many detections in blind surveys may be multiple galaxy systems. Small galaxies in close proximity to large ones can easily be missed, if their redshift velocities fall within the range spanned by the emission profiles of bright, dominant galaxies.

4.1. Computation of the HI-mass function

A number of subtle effects enter when the space density of gas-rich galaxies, the HI-mass function, is constructed. These problems are especially acute for the small HI masses, which can only be detected in blind surveys if the objects are very close by. Some of these problems, including the effects of peculiar velocities, deviations from pure Hubble flow, and large-scale fluctuations in density, are illustrated in Fig. 2, which presents the recent Nançay optically-blind, 21cm

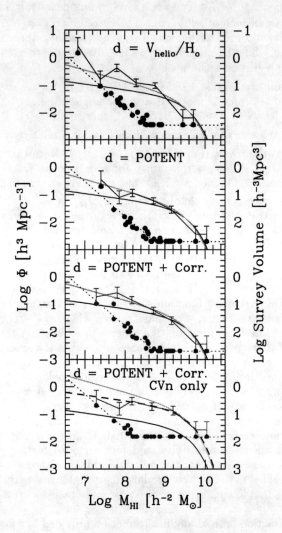

Figure 2. HI-mass function for the Canes Venatici survey volume (Kraan-Korteweg et al 1998), normalized to number of objects per decade of mass. The smooth solid curve is the analytic form derived by Zwaan et al (1997) with a slope of $\alpha = -1.2$, the grey line has a slope of $\alpha = -1.4$ (Banks et al 1998). The bottom panel shows the result restricted to the CVn-group regions (<1200 km s^{-1}), and the dashed curve represents the Zwaan et al HI-mass function multiplied by a factor of 4.5. The dotted line in each panel gives an indication of the volume probed as a function of mass (see right vertical axis); the points give $1/V_{max}$ for each of the galaxies in the sample, taking into account the different velocity widths.

line survey of the Canes Venatici region. First, the top panel shows the number density of galaxies as computed using distances, HI masses, and sensitivity volumes based on conversion of heliocentric velocities to distance V_{hel}/H_o. The mass functions are binned into half-decade bins, but are scaled to give number of objects per decade. The value for each decade is computed from the sum $\Sigma\ 1/V_{max}$, where V_{max} is the volume of the survey in which a galaxy with the properties M_{HI} and ΔV could have been detected. A steeply rising, low mass tail results from this calculation due to one galaxy, UGC 7131, which is treated in this naive calculation as a very nearby, but low mass object. Placed at the greater distance implied by independent distance measurement (Makarova et al 1998), it becomes more massive, and it finds its place among other galaxies of greater velocity width and higher HI mass in the higher mass bins, as shown in the lower panels.

An improved calculation based on POTENT distances, which include local deviations from uniform Hubble expansion (Bertschinger et al 1990), is displayed in the second panel. In the third panel, four galaxies from the Nançay sample with independent distance measurements have been plotted according to their revised distances. In the 4th panel we have restricted our sample to include only the overdense foreground region that contains the CVn and Coma groups, i.e. the volume within V_{hel} <1200 km s^{-1} and about 1/2 the RA coverage (about half the solid angle) of the full survey. Large HI-mass galaxies can be detected throughout the volume we surveyed, but small galaxies can be detected only in the front part of our volume. The volume normalization factors, which are used to compute the mass function, are sensitivity limited for the small masses to only the front part of our survey volume. For the large masses, the V_{max}'s include the whole volume, including the volume where the numbers of galaxies are much less. Hence, when restricting the "survey volume" we get a fairer comparison of the number of little galaxies to the number of big ones.

In all four panels the solid line represents the HI-mass function with a slope of $\alpha = -1.2$ as derived by Zwaan et al (1997), whereas the grey line represents an HI-mass function with a slope of $\alpha = -1.4$ as deduced by Banks et al (1998) for a survey in the CenA-group region. Restricting our volume to the dense foreground region including "only" the CVn and Coma groups, we find that the Zwaan et al HI-mass function with a scaling factor of 4.5 to account for the local overdensity (dashed line in the bottom panel) gives an excellent fit to the data.

Clearly the computation of mass functions for tiny dwarfs will remain vulnerable to subtleties, due to the small volumes in which they can be detected and the accompanying uncertainties in distance.

5. Conclusion: physical selection

The largest reservoirs of HI gas in the nearby universe are found in large galaxies. The bulk of the HI associated with low optical surface brightness falls in the outskirts of large galaxies. Apparently an additional consequence of the gravitational confinement that preserves gas neutrality is the production of stars, since no cases of free-floating HI clouds away from galaxies have been found. Deep integrations in the 21cm line over long sightlines should eventually detect the highest column density clouds of the ionized Lyman-α forest.

References

Banks, G.D., Disney, M.J., Knezek, P., et al 1998, in prep
Bertschinger, E. et al 1990, ApJ, 364, 370
Braun, R. 1997, ApJ, 484, 637
Braun, R. 1995, BAAS, 187, 6501
Briggs, F.H., & Rao, S. 1993, ApJ, 417, 494
Briggs, F.H. 1997, ApJ, 484, 618
Broeils, A.H., van Woerden, H. 1994, A&AS, 107, 129
Charlton, J.C., Salpeter, E.E., & Linder, S.M. 1994, ApJ, 430, L29
Corbelli, E., & Salpeter, E.E. 1993, ApJ, 419, 104
Corbelli, E., & Schneider, S.S. 1990, ApJ, 356, 14
Dickey, J.M., Mebold, U., Marx, M., et al 1994, A&A, 289, 357
Fisher, J.R., & Tully, R.B. 1976, A&A, 53, 397
Fisher, J.R., & Tully, R.B. 1981, ApJ, 243, L23
Giovanelli, R. & Haynes, M.P. 1989, ApJ, 346, L5
Haynes, M.P., & Roberts, M.S. 1979, ApJ, 227, 767
Haynes, M.P., Giovanelli, R., & Roberts, M.S. 1979, ApJ, 229, 83
Henning, P.A. 1995, ApJ, 450, 578
Hoffman, G.L., Lu, N.Y., Salpeter, E.E., et al 1993, AJ, 106,39
Kraan-Korteweg, R.C., van Driel, W., Briggs, F.H., Binggeli, B., & Mostefaoui, T.I. 1998, A&A, in press
Lo, K.Y., & Sargent, W.L.W. 1979, ApJ, 227, 756
Maloney, P. 1993, ApJ, 414, 41
Materne, J., Huchtmeier, W.K., & Hulsbosch, A.N.M. 1979, MNRAS, 18, 563
Putnam, M.E. Gibson, B.K., Staveley-Smith, L., et al 1998, Nature, 394, 742
Rao, S.M., & Briggs, F.H. 1993, ApJ, 419, 515
Rauch, M., et al 1997, ApJ, 489, 7
Schneider, S.E. 1989 , ApJ, 343, 94
Shostok, G.S. 1977, A&A, 54, 919
Shull, J.M., Stocke, J.T., & Penton, S. 1996, AJ, 111, 72
Simkin, S.M., Su, H-J., van Gorkom, J., Hibbard, J. 1987, Sci, 235, 1367
Spitzak, J., & Schneider, S.E. 1998, ApJS, in press
Sunyaev, R.A. 1969, Astrophys.Lett., 3, 33
Szomoru, A., Guhathakurta, P., van Gorkom, J.H., Knapen, J.H., Weinberg, D.H., & Fruchter, A.S. 1994, AJ, 108, 491
van der Hulst, J.M. 1979, A&A, 75, 97
van Driel, W., van Woerden, H., Schwarz, U.J., & Gallagher, J.S. 1988, A&A, 191, 201
Wakker, B.P., & van Woerden, H., 1997, ARA&A, 35, 217
Yun, M.S., Ho, P.T.P., & Lo, K.Y. 1994, Nature, 372, 530
Zwaan, M.A., Briggs, F.H., Sprayberry, D., & Sorar, E. 1997, ApJ, 490, 173

Constraints on the Unseen Galaxy Population from the Lyα Forest

Kenneth M. Lanzetta and Hsiao-Wen Chen

Department of Physics and Astronomy, State University of New York at Stony Brook, Stony Brook, NY 11794-3800, U.S.A.

John K. Webb

School of Physics, University of New South Wales, Sydney 2052, NSW, AUSTRALIA

Xavier Barcons

Instituto de Física de Cantabria (Consejo Superior de Investigaciones Científicas—Universidad de Cantabria), Facultad de Ciencias, 39005 Santander, SPAIN

Abstract. Here we describe results of our attempt to determine what types of galaxies are responsible for the Lyα forest absorption systems, based on our ongoing imaging and spectroscopic survey of faint galaxies in fields of HST spectroscopic target QSOs. Our primary conclusions are that the bulk of the Lyα forest arises in more or less normal galaxies (that span the normal range of luminosity and morphology) and that any "unseen" low surface brightness galaxies are unlikely to contribute significantly to the luminosity density of the universe.

1. Introduction

Our analysis of the relationship between galaxies and Lyα absorption systems at redshifts $z < 1$ has over the past few years led us to conclude that (1) most galaxies possess extended gaseous envelopes of $\approx 160\ h^{-1}$ kpc radius and (2) many or most Lyα absorption systems arise in extended gaseous envelopes of galaxies (e.g. Lanzetta et al. 1995; Barcons et al. 1995; Chen et al. 1998). These conclusions bear directly, of course, on questions concerning the nature and physical state of gaseous material at very large galactocentric distances. But what is of more immediate relevance to the topic of this meeting is that these conclusions also imply that the "forest" of Lyα absorption lines that are routinely observed in the spectra of background QSOs probe *galaxies* (rather than something else) to redshifts as large as $z \approx 5$.

The Lyα resonance transition is an extraordinarily sensitive tracer of very low column density material. Ordinary spectra of ordinary QSOs can easily detect neutral hydrogen column densities as low as $N \approx 5 \times 10^{13}$ cm^{-2}, which is many orders of magnitude below the column densities probed by star light or 21 cm emission. In this sense, the Lyα forest represents a more or less complete inventory of the baryonic constituents of the universe. Galaxies that are difficult to detect by "ordinary" means (i.e. by means of star light or 21 cm emission) are easy to detect by means of Lyα absorption lines, so it remains to establish what portion of the Lyα forest can be attributed to "normal" galaxies that are represented by normal galaxy luminosity functions in order to determine what portion of the Lyα forest is left over for "unseen" galaxies that are not represented by normal galaxy luminosity functions.

Here we describe the results of our attempt to determine what types of galaxies are responsible for the Lyα forest absorption systems, based on our ongoing imaging and spectroscopic survey of faint galaxies in fields of Hubble Space Telescope (HST) spectroscopic target QSOs. Using new observations of the galaxies of the survey, we have sought to establish just what factors play a role in determining the gaseous extent of galaxies. Our primary conclusions are that the bulk of the Lyα forest arises in more or less normal galaxies (that span the normal range of luminosity and morphology) and that any "unseen" low surface brightness galaxies are unlikely to contribute significantly to the luminosity density of the universe. More stringent conclusions along these lines are within reach, requiring only further observations and analysis. Throughout we adopt a standard Friedmann cosmological model of deceleration parameter $q_0 = 0.5$ and Hubble constant $H_0 = 100\ h$ km s^{-1} Mpc^{-1}.

2. Imaging and Spectroscopic Survey

Over the past several years, we have been conducting an ongoing imaging and spectroscopic survey of faint galaxies in fields of HST spectroscopic target QSOs. The goal of the survey is to determine the gaseous extent of galaxies and the origin of Lyα absorption systems by directly comparing the redshifts of galaxies and absorbers identified along common lines of sight. The observations have been and will be described elsewhere (e.g. Lanzetta et al. 1995; Chen et al. 1998), but in summary the observations consist of (1) optical images and spectroscopy of objects in the fields of the QSOs, obtained with various telescopes and from the literature, and (2) ultraviolet spectroscopy of the QSOs, obtained with the HST using the Faint Object Spectrograph (FOS) and accessed through the HST archive. The optical images and spectroscopy are used to identify and measure galaxy redshifts and impact parameters, and the ultraviolet spectroscopy is used to identify and measure absorber redshifts and equivalent widths.

A total of 352 galaxies and 230 absorbers toward 24 fields are included into the current analysis. The galaxies and absorbers are "matched" or "associated" using quantitative criteria that are set by the galaxy–absorber cross-correlation function $\xi_{\rm ga}(v, \rho)$, as it depends on the line-of-sight velocity separation v and the transverse impact parameter separation ρ. (Here we adopt the galaxy–absorber cross-correlation function measured previously by Lanzetta et al. 1997 on the basis of 3126 galaxy and absorber pairs.) In this way, "physical" galaxy and absorber pairs are quantitatively distinguished from "correlated" and "random" galaxy and absorber pairs. Galaxies and absorbers within 3000 km s^{-1} of the QSOs are excluded in order to focus the analysis on the "intervening" population.

Results of the survey are illustrated schematically in Figure 1. Figure shows in the top panel an image of the field surrounding 0454−2203 and in the bottom

Constriants from the Lyα forest   37

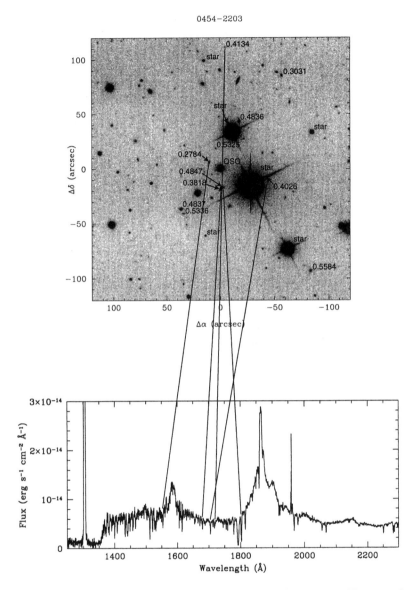

Figure 1. Schematic illustration of results of the survey. Top panel shows an image of the field surrounding 0454−2203. Bottom panel shows a spectrum of 0454−2203. Redshifts of the various faint galaxies so far identified by the survey are indicated in the top panel, and Lyα absorption lines matched with galaxies so far identified by the survey are indicated in the bottom panel. Redshift of 0454−2203 is $z = 0.54$, and the most prominent emission line in the spectrum is Lyα.

panel a spectrum of 0454−2203. Redshifts of the various faint galaxies so far identified by the survey are indicated in the top panel, and Lyα absorption lines matched with galaxies so far identified by the survey are indicated in the bottom panel. All matched galaxy and absorber pairs indicated in Figure 1 have cross-correlation amplitudes satisfying $\xi_{ga}(v,\rho) > 1$.

3. Gaseous Extent of Galaxies

One of the most striking results of the survey is that there exists a distinctive anti-correlation between Lyα absorption equivalent width W and galaxy impact parameter ρ. In particular, galaxies at impact parameters less than $\approx 160\ h^{-1}$ kpc are *almost always* associated with corresponding Lyα absorption systems while galaxies at impact parameters greater than $\approx 160\ h^{-1}$ kpc are *almost never* associated with corresponding Lyα absorption systems. The anti-correlation is statistically highly significant and persists even when various subsamples (e.g. absorption systems that exhibit metal absorption lines, or absorption systems that exhibit very strong Lyα absorption lines) are arbitrarily removed from the analysis. On the basis of this result, we conclude that galaxies are surrounded by extended gaseous envelopes of $\approx 160\ h^{-1}$ kpc radius. The anti-correlation between Lyα absorption equivalent width and galaxy impact parameter is shown in the top panel of Figure 2.

Yet it is clear from Figure 2 that the scatter about the mean relationship between Lyα absorption equivalent width and galaxy impact parameter is quite substantial. Evidently the amount of gas encountered along the line of sight depends on other factors besides galaxy impact parameter, including perhaps galaxy luminosity, size, or morphological type; the geometry of the impact (e.g. if tenuous gas is distributed around galaxies in flattened disks rather than in spherical halos); or disturbed morphologies or the presence of close companions (e.g. if tenuous gas is distributed around galaxies as a result of interactions). To determine these other factors, we initiated a program to obtain and analyze HST Wide Field Planetary Camera 2 (WFPC) images of galaxies identified in the survey. These observations were obtained (and are being obtained) in HST Cycles 5 and 6.

Using the WFPC2 images together with existing spectroscopic observations, we measured properties of galaxies identified in the survey, including rest-frame B-band luminosity L_B, effective radius r_e, average surface brightness $\langle\mu\rangle$, disk-to-bulge ratio D/B, redshift z, and inclination and orientation angles. We then applied multivariate analysis techniques to search for a "fundamental surface" in the multidimensional space that is spanned by various combinations of the measurements. Initial results of the analysis are described by Chen et al. (1998).

The primary result of the analysis is that the amount of gas encountered along the line of sight depends on the galaxy impact parameter and B-band luminosity but does not depend strongly on the galaxy average surface brightness, disk-to-bulge ratio, or redshift. Spherical halos cannot be distinguished from flattened disks on the basis of the current observations, and there is no evidence that galaxy interactions play an important role in distributing tenuous gas around galaxies in most cases. These results are presented in the bottom panel of Figure 2 and in Figure 3. The bottom panel of Figure 2 shows the

Constriants from the Lyα forest 39

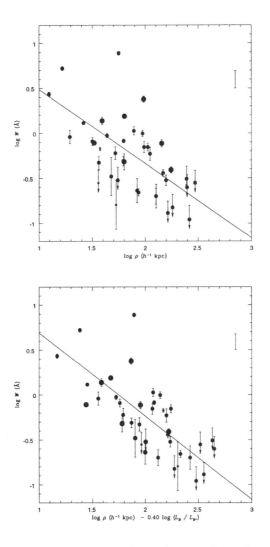

Figure 2. *Top panel:* Anti-correlation between Lyα absorption equivalent width W and galaxy impact parameter ρ. *Bottom panel:* Anti-correlation between Lyα absorption equivalent width W and galaxy impact parameter ρ as "corrected" by the best-fit scaling with galaxy B-band luminosity L_B. Symbol size indicates galaxy luminosity (larger symbols indicate larger luminosities) and symbol type indicates galaxy morphological type (circles for elliptical and S0 galaxies, triangles for early-type spiral galaxies, and squares for late-type spiral galaxies). The error bars in the upper right corners indicate "cosmic" scatter. Both panels are based on galaxies for which HST WFPC2 images have already been obtained.

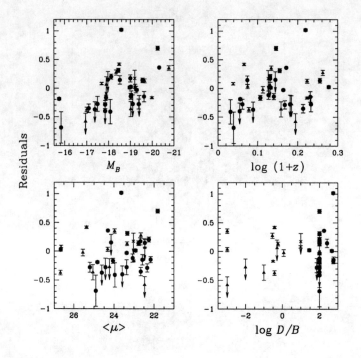

Figure 3. Residuals with respect to the best-fit Lyα absorption equivalent width versus galaxy impact parameter relation as functions of galaxy B-band absolute magnitude M_B, redshift $\log(1+z)$, average surface brightness $\langle\mu\rangle$, and disk-to-bulge ratio $\log D/B$. Symbols are as for Figure 2.

anti-correlation between Lyα absorption equivalent width and galaxy impact as "corrected" by the best-fit scaling with galaxy B-band luminosity. Figure 3 shows the residuals with respect to the best-fit Lyα absorption equivalent width versus galaxy impact parameter relation as functions of galaxy B-band absolute magnitude M_B, redshift $\log(1+z)$, average surface brightness $\langle\mu\rangle$, and disk-to-bulge ratio $\log D/B$.

We ascribe two especially important implications to the remarkably tight anti-correlation between Lyα absorption equivalent width and galaxy impact as corrected by the best-fit scaling with galaxy B-band luminosity: First, it indicates that we have generally matched the appropriate galaxies with the appropriate absorption systems. Second, it indicates that galaxies are surrounded by distinct gaseous "envelopes" and that these envelopes are tightly associated with the individual galaxies, rather than loosely associated with the "large-scale environments" of the individual galaxies, e.g. as inter-group or inter-cluster gas. The result that the amount of gas encountered along the line of sight does not depend strongly on the galaxy average surface brightness, disk-to-bulge ratio, or redshift apparently indicates that extended gaseous envelopes are a common and generic feature of galaxies spanning a wide range of luminosity and morphological type and therefore that the Lyα forest traces a representative portion of the galaxy population.

The scaling relationship between galaxy gaseous radius R and galaxy B-band luminosity L_B is well described by

$$\frac{R}{R_*} = \left(\frac{L_B}{L_{B_*}}\right) \quad (1)$$

This relationship is analogous to the Holmberg relationship between galaxy stellar radius and galaxy B-band luminosity. Based on a sample that is slightly larger than the sample analyzed by Chen et al. (1998), the best-fit parameter estimates are (Chen et al. 1999, in preparation)

$$t = 0.40 \pm 0.09 \quad (2)$$

and

$$R_* = 190 \pm 34 \; h^{-1} \; \text{kpc}, \quad (3)$$

which applies for Lyα absorption equivalent widths satisfying $W > 0.32$ Å. The most important implication of the scaling relationship of equation (1) is that it provides, for the first time, a means of quantitatively relating statistical properties of Lyα absorption systems to statistical properties of galaxies.

4. Constraints on the Unseen Galaxy Population

The predicted number density $n(z)$ of Lyα absorption systems that arise in extended gaseous envelopes of galaxies is

$$n(z) = \frac{c}{H_0}(1+z)(1+2q_0z)^{-1/2} \int_{L_{B_{\min}}}^{\infty} dL_B \Phi(L_B, z) \pi R^2(L_B), \quad (4)$$

where c is the speed of light, H_0 is the Hubble constant, z is the redshift $\Phi(L_B, z)$ is the galaxy luminosity function, $R(L_B)$ is the galaxy gaseous radius, and $L_{B_{\min}}$ is the minimum galaxy luminosity under consideration. By adopting the *known* (from equation 1) relationship between galaxy gaseous radius and galaxy B-band luminosity and a *known* galaxy luminosity function, comparison of the *predicted* and *observed* number densities of Lyα absorption systems constrains "unseen" galaxies that are not represented by the galaxy luminosity function.

Results of this comparison are shown in Figure 4. Figure 4 shows the observed number density of Lyα absorption systems (solid circles) and the predicted number density of Lyα absorption systems (open circles), based on a galaxy luminosity function determined from photometric redshifts of galaxies in the Hubble Deep Field (Fernández-Soto, Lanzetta, & Yahil 1998). Figure 4 also shows the predicted number density of Lyα absorption systems corrected for incompleteness at faint galaxy luminosities (crosses). The primary result of Figure 4 is that to within measurement error *known* galaxies of *known* gas cross sections can account for *all* Lyα absorption systems at redshifts $z < 2$ (and for most or all Lyα absorption systems at higher redshifts, after allowing for incompleteness at faint galaxy luminosities.) This suggests that the "unseen" galaxy

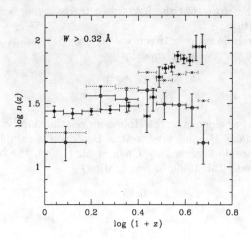

Figure 4. Observed (solid circles) and predicted (open circles) number densities of Lyα absorption sytems. Observed measurements are taken from Bechtold (1994) and Weymann et al. (1998). Crosses show predicted number density of Lyα absorption systems corrected for incompleteness at faint galaxy luminosities.

population produces at most a small fraction of the Lyα absorption systems. For the scaling relationship of equation (1) with $t = 0.4$, equation (4) indicates that the number density of Lyα absorption systems is roughly proportional to the B-band luminosity density of the universe (multiplied by weakly redshift-dependent factor). We thus conclude that any "unseen" low surface brightness galaxies are unlikely to contribute significantly the the luminosity density of the universe.

Acknowledgments. This research was supported by NASA grant NAGW–4422 and NSF grant AST–9624216.

References

Barcons, X., Lanzetta, K. M., & Webb, J. K. 1995, Nature, 376, 321
Bechtold, J. 1994, ApJS, 91, 1
Chen, H.-W., Lanzetta, K. M., Webb, J. K., & Barcons, X. 1998, ApJ, 498, 77
Fernández-Soto, A., Lanzetta, K. M., & Yahil, A. 1998, AJ, in press
Lanzetta, K. M., Webb, J. K., & Barcons, X. 1997, in Proc. of the 18th Texas Symp. on Relativistic Astrophysics, ed. A. Olinto, J. Frieman, & D. Schramm, in press
Lanzetta, K. M., Bowen, D. V., Tytler, D., & Webb, J. K. 1995, ApJ, 442, 538
Weymann, R. J., et al. 1998, ApJ, in press

Searching for LSB - II

It's here, can't you see, between my hands.

Low Luminosity Galaxies in Large Surveys

John P. Huchra

Harvard-Smithsonian center for Astrophysics, 60 Garden Street, Cambridge, MA 02138-1516 USA

Abstract. The debate about the slope and amplitude of the galaxy luminosity function at the faint end is discussed w.r.t. faint galaxies in large surveys, in particular the second CfA (CfA2) and the Las Campanas (LCRS) redshift surveys. Large surveys are necessary to determine the statistics of rare objects or objects that can only be seen out to limited volumes. Both surveys show excesses of faint galaxies over Schechter function fits, but the parent sample for the LCRS survey generally does not contain large or low surface brightness galaxies which do appear in the CfA2 survey. The objects that comprise the relatively large excess of faint galaxies in the CfA2 survey are shown to be primarily of low surface brightness and late morphological type and are generally emission line galaxies. Galaxy samples constructed like the LCRS will generally always be deficient in low luminosity galaxies and thus are not useful for constraining the faint end of the galaxy luminosity function.

1. Introduction

The study of the space density of galaxies or the galaxian luminosity function dates back to Hubble (1936). Based on his estimates of distances to nearby bright galaxies, Hubble described the luminosity function of galaxies as a Gaussian, translated into todays terms ($H_0 = 75$ km/s/Mpc and blue magnitudes), of FWHM about 2 magnitudes and mean about -20. Hubble's distribution is actually very close to what one would see in a very small sample of brcfa2.sb.psight galaxies drawn from a magnitude limited survey. Holmberg (1950) attempted to improve this by studying groups of galaxies, including the Local Group dwarves, and derived a somewhat broader distribution that still cut off at both the luminous and low-luminosity ends. The debate truly started with Zwicky (1957) who studied galaxy clusters and found, rather than evidence for a cutoff in the number of low-luminosity galaxies, almost a geometrical increase in their numbers to faint magnitudes.

Major advances in the study of luminosity functions (LF) came with Abell's (1962) characterization of the LF as two power laws, the development of a first-order theory of structure formation by Press & Schechter (1974) and Schechter's (1976) subsequent derivation of an integrable and analytic form for the LF, discussions of selection effects and biases in the determination of the LF by Kiang (1961), Felten (1977) and others, and the publication and analysis of the Revised Shapley Ames catalog (c.f. Sandage, Tammann & Yahil 1979) with excellent morphological types. The Schechter function is characterized by a faint end slope α, a normalization ϕ^*, and a characteristic luminosity or absolute magnitude, L^* or M^*. Good reviews of mathematical methods for the determination of the LF from galaxy samples can be found in EEP (1988) and Willmer (1997).

Although there were early hints about incompleteness of galaxy samples (Arp 1965; Disney 1976), the debate about the low luminosity end of the LF really came to the fore with the publication of two papers, the first a review by Binggeli, Sandage and Tammann (1988) where they argue for a general turndown, much like Hubble's, for normal galaxies but a large "excess" of faint dwarf ellipticals in clusters, most notably Virgo. While this presented a small conundrum — "why should such galaxies exist in large numbers in clusters?" — it was not viewed as cosmologically significant since clusters make up only a small percentage of all galaxies.

The second was the first determination of the LF from a truly large sample of 9,000+ galaxies (Marzke, Huchra and Geller 1994), which showed that a relatively flat Schechter function was a good fit to the LF of brighter galaxies, but that there exists a significant excess over the extrapolation of the standard LF at low luminosities. This excess had a slope as a function of magnitude which approached divergence of the luminosity density and which, if confirmed, could eliminate much of the need for large numbers of mergers at intermediate redshifts.

This result was almost immediately countered by the results of the LCRS survey (Lin et al. 1996) with a sample of over 20,000 galaxies which seemed to show a turndown at low luminosity, although they did detect an excess of galaxies above the simple Schecter function extrapolation. Low luminosity excesses had not been seen in some earlier but smaller surveys (Loveday et al. 1992; Efstathiou, Ellis and Peterson 1988), but several newer galaxy (Zucca et al. 1997) and cluster surveys (cf. Trentham 1998) do show evidence for an excess, although not all agree (Gaidos 1997). In addition, most of the samples containing low luminosity galaxies tended to show that these objects are morphologically late type (e.g. Marzke et al. 1994; Marzke & daCosta 1997), and, when split into spectroscopic classes, are almost all emission line galaxies (Bromley et al. 1998).

1.1. Problems

Despite essentially having nailed the shape of the LF for bright galaxies, nonetheless these problems still remain:

 I. What is the real space density of low luminosity galaxies?
   (a) Does either the luminosity or mass density diverge?
   (b) Why do surveys differ?
 II. What are the morphologies of the low-luminosity galaxies?
   (a) How Universal is the LF?
   (b) What does the answer tell us about galaxy formation?

Basically, is there really an analytical form for $\Phi(L)$ and how can we express it in terms of color, morphological type, density, etc.?

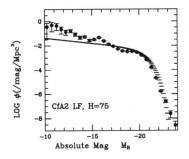 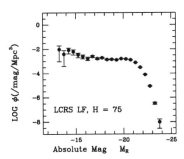

Figure 1. The LF for the 15,466 high galactic latitude galaxies in the CfA2 Survey, and the LF for the LCRS survey. In each case, the maximum likelihood Schechter function is also plotted.

2. The Program

One simple way to answer these questions is to examine more closely the low luminosity galaxies in the CfA2 survey and compare their properties to the low luminosity galaxies in the LCRS survey. We started the study of the low luminosity galaxies about 2 years ago as the 18,000+ galaxy CfA2 survey was drawing to a close in part to make sure we had the correct velocities and identifications for these interesting objects.

The luminosity function for the CfA2 survey estimated by the simple V/V_m technique is shown in Figure 1a along with the best fit Schechter function which has α=-1.22 and M_B^*=-20.3. The excess of faint galaxies above the Schecter function fit — even the fit including the faint end — is easily seen.

For comparison, we also plot the luminosity function we estimate for the 21,258 galaxies in the LCRS sample to m_R=17.5 in Figure 1b. We confirm the shallow faint end slope of Lin et al. (1996), the simple V/V_m technique and a maximum likelihood fit gives α=-0.66 and M_R^*=20.74. Again, there is an excess of low luminosity galaxies above the fit.

2.1. Why Large Surveys?

The first issue to confront is the need for such samples. One of the reasons earlier samples didn't detect any excesses is simply the numbers involved. It is dangerous to infer the faint end of the LF from the bright end and it is necessary to develop a "fair sample." For example, the 2 largest existing surveys still sample only a relatively small volume for low luminosity galaxies:

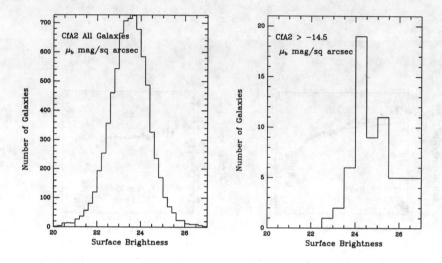

Figure 2. Surface brightness histograms for all the CfA2 survey galaxies with measured diameters and for only the galaxies fainter than $M_B = -14.5$.

$$\text{CFA2} \quad B > -15.0 \rightarrow V_{lim} 950 km/s$$

$$\text{Volume Surveyed} \sim 3000 Mpc^3$$

$$\text{LCRS} \quad R > -16.0 \rightarrow V_{lim} 3750 km/s$$

$$\text{Volume Surveyed} \sim 9000 Mpc^3$$

The CfA2 survey covers about 4.5 steradians to an apparent B limit of 15.5, the LCRS survey covers only .22 steradians but to R ~ 17.5. Even though these surveys contain tens of thousands of galaxies, 99+% of them are luminous. The faint end "excess" in the CfA2 survey is comprised of fewer than 100 galaxies; there are only 80 galaxies with $M_B \geq -14.5$ and 110 with $M_B \geq -15.0$. In the LCRS, the statistics are even poorer, despite the large effective volume surveyed (emphasizing again the basic difference between the surveys), with only 19 (!) galaxies with $M_R \geq -16.0$, and 31 with $M_R \geq -16.5$.

3. Properties of The Low Luminosity Galaxies

3.1. CfA2 Low Luminosity Galaxies

We have morphologically typed the 110 CfA2 galaxies fainter than -15.0 using either the DSS or CCD images obtained at FLWO. 9% are ellipticals or S0/s, usually dwarf ellipticals such as NGC 147 and NGC185. 7% are early to mid type spirals and 86% are late type ($T \geq 6$) spirals or irregulars.

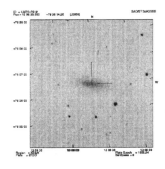 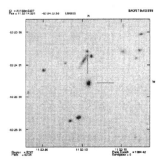

Figure 3. Typical low luminosity galaxies from the CfA2 (left) and LCRS surveys as seen on the DSS (from Space Telescope Science Institute and AURA). The LCRS galaxy is $M_R=-14.79$ and the CfA2 galaxy is $M_B=-13.3$, comparable absolute magnitudes for the typical galaxy B-R of ~ 1.4.

We are currently obtaining spectra for the faint CfA2 galaxies. We have high S/N spectra for 31 of the 80 galaxies fainter than -14.5. Using the criteria $W_A(6562$ or $5007) > 3$Å, 26 of those 31 show moderately strong emission, or 84%. This is high compared to the average for the CfA1 survey, approximately 70% (Burg 1987).

Because diameters exist from the Nilson or other catalogs for more than half of the CfA2 galaxies, we can examine the surface brightness distribution of the low luminosity galaxies relative to the population as a whole. For the $\sim$7,800 galaxies with diameters, the mean integrated surface brightness is $\sim$ 23.5 magnitudes/square-arcsec. For the 80 galaxies fainter than -14.5, the mean surface brightness is 24.8, and it is clear (Figure 2) that the surface brightness distributions are not the same.

3.2. LCRS Low Luminosity Galaxies

What are the low luminosity LCRS galaxies like? We have looked at all of the faint galaxies in both samples. *They are markedly different.* Figure 3 shows typical examples from each survey.

Generally the LCRS low luminosity galaxies are compact and high surface brightness. Less than 10% are the LSB galaxies typical of the CfA2 survey, and in fact, three of the nearest faint LCRS "galaxies" are actually HII regions in larger, normal galaxies — LCRS B101440.4-031359, LCRS B123733.9-053240, and LCRS B222917.3-412537 (1950 coordinates)!

The LCRS catalog contains few if any low surface brightness objects because it was constructed not to (Shectman et al 1996). The TDI CCD scans used to construct the parent galaxy catalog provided relatively short (60 second) effective exposures. Galaxies were chosen to have a minimum central surface brightness to facilitate the fiber spectroscopy. That essentially removed a major component of the low luminosity galaxies seen in samples that are just magnitude limited. The relative fraction of low luminosity galaxies in the LCRS survey is down a significant factor (~ 4) from the CfA2 survey for primarily this reason.

4. Conclusions

Its now clear that there is very strong evidence for an excess of low luminosity galaxies over the simple extrapolation of the Schechter function fit at bright magnitudes. Evidence points towards the Schechter form as a good fit to the LF of individual morphological types, so the lack of a single global fit probably argues for different processes dominating early and very late type galaxy formation and evolution. No surprise here!

Its also pretty clear that the deficiency of low luminosity galaxies in the LCRS, and other comparably selected samples, is due to the insensitivity of the parent catalogs to low surface brightness galaxies. This may also be exacerbated by selection in the red instead of the blue, although the presence of strong $H\alpha$ emission in most of the CfA2 faint galaxies argues that this is not the dominant effect. The LCRS catalog did not include large and/or low surface brightness galaxies, so tails off strongly at the low surface brightness end. What you see is what you get! Unfortunately, this may well be a characteristic of any multifiber survey — LSB and also low metallicity systems will not yield redshifts in short integration times unless one is lucky enough to hit an emission line region. Thus they will be doubly dammed, not likely to be in the sample and, if in, not likely to get a redshift. The LCRS and similarly constructed catalogs cannot be used to set limits on the faint end of the LF.

The CfA2 survey contains a reasonable selection of LSB galaxies, primarily because it includes the Nilson (1973) catalog and because Zwicky and his collaborators included both compact and low surface brightness objects in their catalog. Selection by-eye has some advantages over automatic algorithms which generally are tuned for some characteristic size. It still probably is incomplete at the faint end, but, for now, something is better than nothing.

The low luminosity galaxies in the CfA2 survey are generally 1. Low Surface Brightness, 2. Late morphological types, and 3. Emission line objects. We probably need more accurate distances (rather than just the redshift distances) to say more about their properties.

Better surveys are coming. HIPASS and other 21-cm surveys have been discussed here (c.f. Kraan-Korteweg 1999). SDSS is coming with multicolor data (although the main survey area does not go much deeper than the POSS-2 plates). 2MASS is coming and we have been running two galaxy finding algorithms, the standard processor which extracts high surface brightness objects, but also a low surface brightness algorithm specifically designed to find large, LSB objects (Schneider et al 1998). Stay tuned!

Acknowledgments. As always, I would like to thank my colleagues and coworkers without whom most of this work would not be possible, M Geller, J. Mader, L. Macri and S. Tokarz plus the tireless observers for the CfA Redshift Survey. This work is supported in part by NASA grant NAGW-201 and by the Smithsonian Institution and Harvard University. This research made use of the NASA/IPAC Extragalactic Database operated by JPL, Caltech under contract with NASA.

References

Abell, G. 1962, IAU Coll. #15, 213.
Arp, H. C. 1965, ApJL 142, 402.
Binggeli, B., Sandage, A. & Tammann, G. 1988, ARA&A 26, 509
Bromley, B., Press, W., Lin, H. & Kirshner, R. 1998, ApJ 595, 25.
Burg, R. 1987, PhD Thesis, Massachusetts Institute of Techology.
De Lapparent, V. Geller, M. and Huchra, J. 1989, ApJ 343, 1.
Disney, M. 1976, Nature 263, 573.
Eftstathiou, G., Ellis, R. and Peterson, B. 1988, MNRAS 232, 431.
Felten, J. 1977, AJ 82, 861.
Gaidos, E. 1997, AJ 113, 117.
Hubble, E. 1936, ApJ 84, 158; 84, 270.
Holmberg, E. 1950, Medd. Lund Series 2 #128.
Impey, C. & Bothun, G. 1997, ARAA 35, 267.
Kiang, T. 1961, MNRAS 122, 263.
Kraan-Korteweg, R. 1999, IAU Colloq 171, this volume.
Lin, H., Kirshner, R., Shectman, S., Landy, S., Oemler, A., Tucker, D & Schechter, P. 1996, ApJ 464, 60.
Loveday, J., Peterson, B., Efstathiou, G & Maddox, S. 1992, ApJ 390, 338.
Loveday, J. 1998, preprint. (Astroph 9805255)
Marzke, R., Huchra, J. & Geller, M. 1994, ApJ 428, 43
Marzke, R., Geller, M., Huchra, J. & Corwin, H. 1994, AJ 108, 437
Marzke, R. & daCosta, L. 1997, AJ 113, 185.
Nilson, *The Uppsala General Catalogue of Galaxies*, Ann. Uppsala Astron. Obs. Band 6, Ser. V:A. Vol. 1.
Press, W. and Schechter, P. 1974, ApJ 187, 425.
Sandage, A., Tammann, G. & Yahil, A. 1979, ApJ 232, 352.
Schechter, P. 1976, ApJ 203, 297.
Schneider, S., Rosenberg, J., Chester, T., Jarrett, T. & Huchra, J. 1998, in the DENIS Symposium.
Shectman, S., Landy, S., Oemler, A., Tucker, D. Lin, H., Kirhsner, R. & Schechter, P. 1996, ApJ 470, 172.
Trentham, N. 1998, MNRAS 294, 193
Willmer, C. 1997, AJ 114, 898.
Zucca, E. *et al.* 1997, A&A 326, 477.
Zwicky, 1957, *Morphological Astronomy*, circa p. 220.

The Low Surface Brightness Universe, IAU Col. 171
ASP Conference Series, Vol. 170, 1999
J. I. Davies, C. Impey and S. Phillipps, eds.

The Space Density of Spiral Galaxies as function of their Luminosity, Surface Brightness and Scalesize

Roelof S. de Jong[1]

Univ. of Durham, Dept. of Physics, South Road, Durham DH1 3LE, UK

Cedric Lacey

TAC, Juliane Maries Vej 30, DK-2100 Copenhagen O, Denmark

Abstract. The local space density of galaxies as a function of their basic structural parameters –like luminosity, surface brightness and scalesize– is still poorly known. Our poor knowledge is mainly the result of strong selection biases against low surface brightness *and* small scalesize galaxies in any optically selected sample. We show that in order to correct for selection biases one has to obtain accurate surface photometry *and* distance estimates for a large ($\gtrsim$1000) sample of galaxies. We derive bivariate space density distributions in the (scalesize, surface brightness)-plane and the (luminosity, scalesize)-plane for a sample of ~1000 local Sb-Sdm spiral galaxies. We present a parameterization of these bivariate distributions, based on a Schechter type luminosity function and a log-normal scalesize distribution at a given luminosity. We show how surface brightness limits and $(1+z)^4$ cosmological redshift dimming can influence interpretation of luminosity function determinations and deep galaxy counts.

1. Introduction

Knowing the space density of galaxies as function of their structural parameters (luminosity, surface brightness (SB) and scalesize) is important when:

1) making comparisons between different galaxy samples, because selection functions of extended resolved objects depend on at least two structural parameters. This becomes particularly relevant when comparing samples at different redshifts, where $(1+z)^4$ redshift dimming can give rise to strong SB biases.

2) testing galaxy formation and evolution models, as any successful galaxy formation theory will have to be able to explain the spread in structural parameters and their relative frequency in the local galaxy population.

Many papers have been devoted to the determination of the space density of galaxies as function of their luminosity, i.e. the galaxy luminosity function (for a recent review see Ellis 1997). In many of these papers one has conveniently

---

[1]Hubble Fellow, Steward Observatory, 933 N. Cherry Ave., AZ 85716, USA

ignored the possibility of strong SB biases. Determinations of scalesize distributions have been scarce (some notable exceptions van der Kruit 1987, Hudson & Lynden-Bell 1991, Sodré & Lahav 1993, de Jong 1995) and often diameter distributions are calculated. Diameter distributions are not very useful in sample comparisons, as diameters have to be measured at a certain SB level, which might differ from sample to sample. Realize for instance, that there may be many galaxies that do not have a D_{25}, because there SB is below $25\,B$-mag arcsec^{-2}.

Since the classical paper of Freeman (1970), many papers have been devoted to the distribution of SB of disks in spiral galaxies (for review see Impey & Bothun 1997). Freeman found that 28 galaxies in his incomplete sample of 36 had disk central SB values of $21.65\pm0.3\,B$-mag arcsec^{-2} (see his review in these proceedings). Disney (1976) showed that the limited range in disk central SB values might be the result of selection biases. Since then several authors have argued that there seems to be indeed an upper limit in the SB distribution near Freeman's value, but that the distribution stays nearly flat when going to lower SB (e.g. McGaugh et al. 1995; de Jong 1995). Recently this picture has been challenged by Tully & Verheijen (1997), who argued that the SB distribution is bimodal, based on K-band data of $\sim$60 galaxies in the Ursa Major cluster.

In this paper we show that one should not try to separate the distributions of luminosity, SB and scalesize, but combine two of these to make bivariate distributions, as any sample will have selection biases in at least two structural parameters.

2. Correcting for selection bias

Many methods have been devised to correct observed frequencies of object properties for selection bias in order to obtain true space density distributions. We will here concentrate on the $V_{\max}$ method, where each object gets a weight proportional to the inverse of its maximum sample inclusion volume (Felten 1976). This metod is only correct if the objects are distributed homogeneously in space, and therefore the smallest objects in the sample should be visible at distances greater than the largest large scale structures in the universe. Homogeneity and completeness can be checked with the $V/V_{\max}$ method (e.g. van der Kruit 1987 and references therein). Accurate $V_{\max}$ values can be derived for each object for the modern surveys with automatic detection algorithms on digitized data. Each object should be artificially blue- or redshifted and be Monte Carlo replaced at many positions in the original data set. The recovery fraction of the automatic detection routine supplies the volume searched at each redshift shell and provides information for confusion limits and Malmquist bias at the survey limits. For samples not selected by an automated routine from digitized data (e.g. eye-selected from photographic plates), we just have to assume that the selection criteria are well behaved when we imagine moving a galaxy in distance.

Moving more specifically to the distribution of structural parameters of spiral galaxies, we will use the case of perfect exponential disks in a diameter limited sample. More generalised descriptions can be found in Disney & Phillipps (1983) and McGaugh et al. (1995). For an exponential disk with physical scalelength h and central SB μ_0 we find for the maximum distance at which a galaxy can lie before dropping out of the sample

$$d_{\max} \propto (\mu_{\lim} - \mu_0)\, h/\theta_{\lim}, \tag{1}$$

with $\theta_{\lim}$ the sample angular diameter limit measured at SB limit $\mu_{\lim}$. As the volume where a galaxy is visible goes as $d_{\max}^3$, this shows the strong selection bias against small scalesize and low SB galaxies. The scalesizes of spiral galaxies vary easily by a factor of 10 (de Jong 1996). Therefore, if all scalesizes were equally abundant at a given SB, we would have a 1000 times more of the largest scalesize galaxies than the smallest scalesize galaxies in a diameter limited sample. Luckily nature has not been that cruel to us and there are many more small galaxies then large ones. In the case of the SB distribution we have not been so lucky, as the SB distribution stays rather constant –at a given scalelength– going to lower SB values. Equation (1) shows that, at fixed scalelength, the visible volume of a galaxy 1 mag above the SB limit is 125 times smaller than that of a galaxy 5 mag above the SB limit. In order to have some number statistics close to the selection limit, we had better observe hundreds of galaxies to determine a SB distribution. Because we do not *a priori* know whether the scalesize and SB distributions are uncorrelated, we had better make sure that we determine the SB distribution at different scalesizes, and so we need at least 1000 galaxies.

SB measurements are distance independent (at least on local scales); a property that sometimes has been used to argue that one can determine SB distributions without knowing distances. If the distribution of h is the same at each SB level, the $h/\theta_{\lim}$ factor in Eq. (1) cancels out on average and one can make relative volume corrections without having to know physical scalesizes/distances. Likewise, using total magnitude of an exponential disk $M \propto \mu_0 - 5\log(h)$, Eq. (1) can be rewritten as

$$d_{\max} \propto (\mu_{\lim} - \mu_0)\, 10^{-0.2(M-\mu_0)}. \tag{2}$$

Again, assuming the SB distribution is the same for each luminosity, one can make relative volume corrections (very different from Eq. (1)!) to calculate a SB distribution without knowing distances. There is no reason for the SB distribution to be independent of either scalesize or luminosity (and we will show this is indeed not the case) and therefore Eq. (1) & (2) show that we need to know the distribution of at least one other *distance dependent* structural parameter to determine the SB distribution of galaxies. The reverse is also true: to measure the distribution of scalesizes or luminosities we also need to determine the distribution of one of the other structural parameters. In order to do so we will need surface photometry and distances for a sample of at least $\sim$1000 galaxies.

In this paper we will use the effective radius (r_e, the radius enclosing half of the total light of the galaxy) and the average effective SB within this radius ($<\mu>_e$) instead of the more conventional parameters for disks, scalelength and central SB. Using the effective parameters has the virtue that one does not have to make assumptions about the light distribution in the galaxy (all galaxies have an r_e, even irregular ones) and avoids complicated bulge/disk decomposition issues. The distributions presented here have also been calculated for disk parameters alone with very similar results, because most of the objects are of late spiral type with insignificant bulge contributions.

3. Local space density distributions

As described in the previous section, one needs accurate surface photometry and distance estimates for a sample of at least 1000 galaxies to create bivariate distributions. The galaxies should be selected in a well defined, reproducible and complete way. Data sets obeying all of these criteria are not available at the moment, but fortunately peculiar motion studies have produced large data sets with accurate photometry and redshifts. We have used the Mathewson, Ford & Buchorn (1992, 1996, MFB hereafter) data set, which was selected from the ESO-Uppsala catalog, a catalog with galaxies selected and classified by eye from photographic plates. We reselected a sample close to the MFB criteria from the ESO-Uppsala catalog, allowing us to evaluate incompleteness in the MFB sample (some galaxies were not observed due to foreground stars or inability to obtain a velocity width). We selected all galaxies from the ESO-Uppsala catalog with type $3\leq T\leq 8$ (Sb-Sdm), angular diameter $1.7'\leq \theta_{maj}\leq 5'$, axis ratio $0.174\leq b/a\leq 0.776$ and galactic latitude $|b|>11°$. This resulted in a sample of 1007 galaxies, of which about 850 have I-band surface photometry and redshifts.

The luminosity, r_e and $<\mu>_e$ values of the galaxies were derived from the luminosity profiles and corrected for Galactic foreground extinction using the prescription of Schlegel et al. (1998). Corrections for inclination and internal extinction were performed following a method similar to Byun (1992). Distance estimates were obtained from the Mark III catalog (Willick et al. 1997) if available, otherwise computed from the Hubble distance, with $H_0 = 65\,\mathrm{km\,s^{-1}\,Mpc^{-1}}$.

Using the $V_{\max}$ method described in the previous section, we have calculated the bivariate density distribution in the $(r_e,<\mu>_e)$-plane, which is presented on a logarithmic scale in Fig. 1. The paucity of galaxies in the top-right corner of the diagram is real, large, high SB galaxies are readily visible. To the bottom-left of the indicated 20 Mpc visibility line we are hit by low number statistics; for such small, low SB galaxies we are sampling too small a volume to have reliable statistics. The distribution shows a dramatic increase in galaxy space density going to smaller scalesizes. At a given scalesize, the SB shows a broad distribution, peaking at about $<\mu>_e=21.5\,I$-mag arcsec^{-2}. There is some indication that the peak in the distribution shifts to lower SB at smaller scalesizes.

4. Parametrization of the distributions

In this section we will define a parametrization of the bivariate distributions, as an aid to compare distributions derived from differently selected samples or to study redshift evolution. We will follow the most simple form of the Fall & Efstathiou (1980) disk galaxy formation theory to derive such a parametrization (for extended versions of the theory see e.g. van der Kruit 1987; Dalcanton et al. 1997; Mo et al. 1998; van den Bosch 1998). Galaxies form in this theory in hierarchically merging Dark Matter (DM) halos, giving rise to a distribution of DM halo masses described by the Press & Schechter (1974) theory, which formed the inspiration for the Schechter (1976) luminosity function (LF). We will use a Schechter LF to describe the luminosity dimension of our distribution function.

In the Fall & Efstathiou (1980) model, the scalesize of a galaxy is determined by its angular momentum, which is acquired by tidal toques from neighbouring

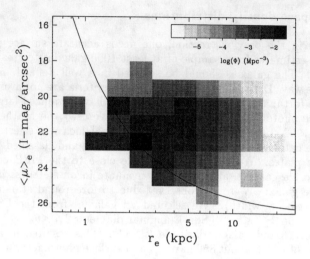

Figure 1. The space density distribution of Sb-Sdm galaxies as function of effective radius and average SB within that radius. Galaxies with exponential disk, having structural properties indicated by the line, can be seen out to 20 Mpc before dropping out of the sample.

DM halos in the expanding universe. The total angular momentum of the system is usually expressed in terms of the dimensionaless spin parameter (Peebles 1969)

$$\lambda = J|E|^{1/2}M_{\rm tot}^{-5/2}G^{-1}, \qquad (3)$$

with J the total angular momentum, E the total energy and $M_{\rm tot}$ the total mass of the system, all of which are dominated by the DM halo. N-body simulations (e.g. Warren et al. 1992) show that the distribution of λ values acquired from tidal torques in an expanding universe can be well be approximated by a log-normal distribution with a dispersion $\sigma_\lambda \sim 0.5$ in $\ln(\lambda)$.

A few simplifying approximations allow us to relate each of the factors in Eq. (3) to our observed bivariate distribution parameters. A perfect exponential disk of effective size r_e, mass M_d, rotating with a flat rotation curve of velocity V_c has $J_d \propto M_d r_e V_c$. We assume that the specific angular momentum of the disk is equal to the specific angular momentum of the dark halo. From the virial theorem we get $E \propto V_c^2 M_{\rm tot}$. If we assume that light traces disk mass ($M_d \propto L$) and that disk mass is proportional to total mass ($M_{\rm tot} \propto M_d$), we only need the Tully & Fisher (1977) relation ($L \propto V_c^\beta$, with $\beta \sim 3$ in the I-passband) to link the spin parameter λ to our observed bivariate distribution parameters.

These approximations give $\lambda \propto r_e L^{(2/\beta-1)} \simeq r_e L^{-1/3}$. As λ is expected to have a log-normal behavior, this means that, *at a given luminosity, we expect the distribution of scalesizes to be log-normal, and that the peak in the r_e distribution shifts with $\sim L^{-1/3}$*. This is exactly the behavior that is shown in Fig. 2, where the function over-plotted on the data shows the log-normal behavior at each luminosity bin, shifting by $L^{-1/3}$ between the luminosity bins and where the height is determined by the Schechter LF.

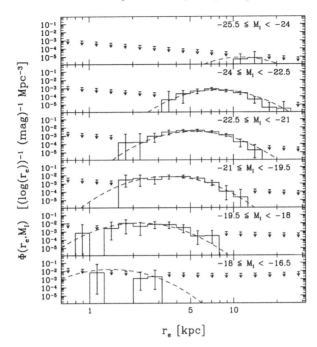

Figure 2. The space density distribution of effective scalesizes in different bins of total I-band luminosity as marked in the top-right of each panel. The histograms represent the data with errorbars showing the 95% confidence limits due to distance and Poisson errors. The 95% confidence upper limits were calculated using exponential disks and the survey limits determined from the photometry. The dashed line shows the bivariate distribution function described in the text.

The function plotted is the result of the well known χ-by-eye fitting method, and the detailed parameters will definitely change when a full fitting technique has been developed that takes the Poisson errors on the data points into account. For reference we list here the full bivariate function in magnitudes, and the parameter values giving a good approximation to the data:

$$\Phi(r_e, M)\, d\log r_e\, dM = \frac{\Phi_0}{\sigma_\lambda \sqrt{2\pi}} \exp\left(-\frac{1}{2}\left[\frac{\log r_e/r_{e*} - 0.4(M - M_*)(2/\beta - 1)}{\sigma_\lambda/\ln(10)}\right]^2\right)$$
$$10^{-0.4(M-M_*)(\alpha+1)} \exp(-10^{-0.4(M-M_*)})\, d\log r_e\, dM,$$

with the first line representing the log-normal scalesize distribution and the second line the Schechter LF in magnitudes (M). The χ-by-eye parameters are:

$\Phi_0 = 0.002\,\text{Mpc}^{-3}$ $\alpha = -1.25$ $\beta = 3.0$ (slope Tully-Fisher relation)
$M_* = -22.3\,I$-mag $r_{e*} = 6.7\,\text{kpc}$ $\sigma_\lambda = 0.3$

The width of the spin parameter distribution (σ_λ) we need is less than what is typically found in N-body simulations.

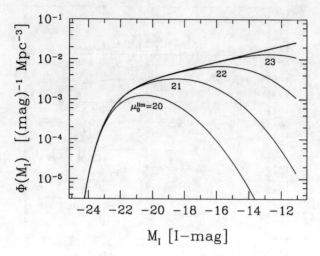

Figure 3. Luminosity functions computed by integrating the bivariate distribution function down to the indicated I-band central surface brightness limits. The uppermost line shows the integrated total LF.

5. Discussion & conclusions

Using our parameterization we can now estimate the effects of SB limits on determinations of LFs and on deep galaxy counts, especially in the context of $(1+z)^4$ redshift dimming. If we were to determine the LF from a galaxy sample with low SB galaxies cut out, we would underestimate the faint end of the LF, as most low SB galaxies are also low luminosity systems. SB cuts are relevant for redshift surveys selected from shallow (photographic plate) material (resulting in implicit cuts) and for most fiber based redshift surveys which often have explicit SB cuts (e.g. the Las Campanas and Sloan Surveys). How SB cuts can effect LF determinations is shown in Fig. 3, where we have integrated our parameterization down to the indicated I-band *central* SB.

Figure 3 suggests that the SB cuts typically present in local surveys do not dramatically effect LF determinations (using for a typical spiral $B-I \simeq 1.7$), especially taking into account that most spirals have some central light enhancement due to the bulge, making detections easier. The situation changes however when we move to higher redshifts and have to take $(1+z)^4$ cosmological redshift dimming into account. At $z=1$ our SB limit has already shifted 3 magnitudes up, and 6 magnitudes by the time we reach $z=3$. This means that even for the deepest image available at the moment –the Hubble Deep Field– the SB cut at $z=3$ (the U-band dropouts) runs at about 21 I-mag arcsec^{-2} (using a K-correction of an unevolved Sb galaxy). This limit makes a considerable fraction of galaxies in Fig. 1 undetectable, if we put this local galaxy population unevolved at $z=3$.

Tully & Verheijen (1997, these proceedings) have argued that the central SB of galaxies shows a bimodal distribution, in particular when looking at K-band data. We do not see such bimodality, independent whether we use their proposed bimodal dust extinction correction, we use only the 200 most face-on galaxies with the smallest extinction correction, we use bulge/disk decomposed parameters or effective parameters. In the many ways we have looked at the MFB data set, we have never seen any bimodality in the SB distributions. Whether the bimodal effect is the result of the special Ursa Major cluster environent that was studied or an unlucky case of low number statistics remains to be seen.

The simple parametrization presented in this paper gives an accurate representation of the observed bivariate distributions, independently of whether one believes in hierarchical galaxy formation models or in CDM-like universes. A detailed analysis of galaxy formation in CDM-like universes paying attention to bivariate space density distributions will appear in Lacey et al. (1999).

Acknowledgments. Support for R.S. de Jong was provided by NASA through Hubble Fellowship grant #HF-01106.01-98A from the Space Telescope Science Institute, which is operated by the Association of Universities for Research in Astronomy, Inc., under NASA contract NAS5-26555.

References

Byun, Y.-I. 1992, PhD. Thesis, The Australia National University
Dalcanton J. J., Spergel, D. N. & Summers, F. J. 1997, ApJ, 482, 676
de Jong, R. S. 1996, A&A, 313, 45
Disney, M. J. 1976, Nature, 263, 573
Disney, M. J. & Phillipps, S. 1983, MNRAS, 205, 1253
Ellis, R. S. 1997, ARA&A, 35, 389
Fall, S. M. & Efstathiou, G. 1980, MNRAS, 193, 189
Felten, J. E. 1976, ApJ, 207, 700
Freeman, K. C. 1970, ApJ, 160, 811
Hudson, M. J. & Lynden-Bell, D. 1991, MNRAS, 252, 219
Impey, C. & Bothun, G. 1997, ARA&A, 35, 267
Lacey, C., Cole, S., Baugh, C. & Frenk, C. S. 1999, in preparation
Mathewson, D. S. & Ford, V. L. 1996, ApJS, 107, 97
Mathewson, D. S., Ford, V. L. & Buchorn M. 1992, ApJS, 81, 413
McGaugh, S.S., Bothun, G. D. & Schombert, J. M. 1995, AJ, 110, 573
Mo, H. J., Mao, S. & White, S. D. M. 1998, MNRAS, 295, 319
Peebles, P. J. E. 1969, ApJ, 155, 393
Press, W. H., Schechter, P. 1974, ApJ, 187, 425
Schechter, P. 1976, ApJ, 203, 297
Schlegel, D. J., Finkbeiner, D. P. & Davis, M. 1998, ApJ, 500, 525
Sodré, L. & Lahav, O. 1993, MNRAS, 260, 285
Tully, R. B. & Verheijen, M. A. W. 1997, ApJ, 484, 145
van den Bosch, F. C. 1998, submitted to ApJ, astro-ph/9805113
van der Kruit, P. C. 1987, A&A, 173, 59
Warren, M.S., Quinn, P. J., Salmon, J. K. & Zurek, W. H. 1992, ApJ, 399, 405
Willick, J. A. et al. 1997, ApJS, 109, 333

The Faint End of the Galaxy Luminosity Function in Rich Clusters

R.M. Smith

Department of Physics and Astronomy, University of Wales Cardiff, UK

S. Phillipps

Department of Physics, University of Bristol, Bristol, UK

S.P. Driver

Dept. of Physics and Astronomy, University of St. Andrews, St. Andrews, UK

W.J. Couch

School of Physics, University of New South Wales, Sydney, Australia

Abstract. Recent results on the determination of the shape of the faint end of the galaxy luminosity function in rich clusters are discussed. There is increasing evidence that in many cases the faint end of the function is steep, indicating a large population of dwarf, possibly low surface-brightness, galaxies. In addition, the magnitude at which the turn-up appears is approximately constant with richness and distance. However, it is clear that not all clusters show such a feature.

1. Introduction

It is now almost 25 years since Schechter (1976) empirically fitted an analytic function of the form $\phi_L = \phi^*(L/L^*)^\alpha exp-(L/L^*)$ to the galaxy luminosity function (LF). In this equation, ϕ^* is a normalisation factor related to the number density of galaxies, L^* is the characteristic luminosity and α is the slope of the power-law faint end of the LF. Observations since then have mostly agreed with this general form, with a sharp drop-off at brighter magnitudes leading to very few bright ($M_B < -21$) galaxies whilst the faint end is a power-law of slope α. There is a limitation on the value of α in that if the faint end of the LF is steep then this conflicts with measurements of the intracluster light. In addition, if $\alpha \leq -2$ the Schechter LF diverges and the luminosity density of the Universe becomes infinite. It is expected, however, that the LF must turn-over at very faint magnitudes due to the difficulty of star-formation in very low mass systems, possibly due to photoionization by the ultraviolet background (e.g. Thoul & Weinberg 1996).

The LF is an observable property of the inherent galaxy distribution that can be directly compared to theoretical predictions. Hierarchical clustering models (e.g. White & Frenk 1991, Frenk et al. 1996), for example, appear to naturally produce a steep faint end slope of the LF whereas dynamical effects in dense regions may flatten it (e.g. Moore et al. 1996). Due to this close interface between observation and theory, much work has gone into the measurement of the LF, both in the field and in clusters. Original estimates, using magnitude-limited redshift surveys, of α both in the field and clusters were typically around -1.1 (e.g. Efstathiou 1988, Loveday 1992), in direct conflict with the theoretical models proposed at the time (e.g. White and Frenk 1991). However, it is very difficult observationally to measure the dwarf population and hence determine the shape of the faint end of the LF. Selection effects and the form of the LF lead to a strong bias in magnitude-limited samples towards the predominance of galaxies with luminosities close to L^*. Hence to obtain a significant result at magnitudes well away from L^* requires the observation of a very large sample of galaxies. Redshift surveys also have their own inherent problems. Technically it is very difficult to measure the redshift of a faint or low surface-brightness galaxy, especially if its spectrum does not contain emission lines. This will lead to an incompleteness particularly at the faint end of the LF. Due to such difficulties, redshift studies have, until recently, not reached far in to the dwarf regime ($M_B > -17$). Yet it is precisely here that the theoretical models make definite and varying predictions.

2. Photometric studies

The inherent problems with redshift surveys led several groups to search for other techniques to determine the shape of the LF without resort to spectroscopic observations. One possibility is by observing galaxies in clusters. There are several techniques by which cluster membership can be derived, even for the fainter galaxies, without the need for spectroscopy. As the redshift of the cluster can be measured from the brighter galaxies, the faint end of the LF can then be determined.

It must be remembered in the interpretation of results on the LF of clusters that they are dense environments, where the crossing time is less than, or of the order of, the Hubble time. It is well known that environmental effects play a crucial role in the evolution of giant galaxies. At bright magnitudes the dominant population in clusters consists of giant ellipticals whereas in the field late-type galaxies predominate. It is therefore very likely that dynamical effects also have a pronounced effect on the evolution of dwarf galaxies in dense regions, such as through the proposed process galaxy 'harassment' (Moore et al. 1996, these proceedings),. Hence measurement of the LF in clusters might therefore tell us more about cluster-related processes than galaxy formation and evolution in general. This is especially true in the richest clusters, where such effects are expected to be most dominant. Yet it is in the richest clusters where the excess over the background is greatest and therefore the greatest hope of deriving a reliable LF.

The earliest studies of the galaxy population within clusters were necessarily photographic and of nearby clusters - notably Virgo, Fornax and Coma. Photographic plates covered a large enough area of these clusters to obtain a sufficiently large sample of galaxies and also deep enough to reach the dwarf population. Also, by observing nearby clusters the larger angular size of the dwarfs compared to the background population enabled cluster membership to be determined visually, without the need for any subtraction of the contaminating populations. The initial studies of the three clusters detected large numbers of dwarfs, outnumbering the giants within the clusters (e.g. Binggeli et al. 1985, Ferguson 1989 and Godwin et al. 1983). This implied a steeper faint end of the LF than originally estimated. For example, Sandage et al. (1985) found a value of α of -1.4 for the Virgo cluster. As surveys went deeper and to lower surface brightnesses the number of faint galaxies detected in these clusters increased. Typical values found for the Virgo cluster, for example, increased to $\alpha \sim -1.7$ (Impey et al. 1988). With the advent of more sensitive emulsions and the development of digital stacking techniques it is now possible to push photographic investigations of the dwarf population of nearby clusters down to very faint absolute magnitudes. An example of this technique is presented by Jones et al. in these proceedings. They probe the dwarf population of the Virgo cluster down to a limiting magnitude of $M_R \sim -11$. Unfortunately, it is only in the nearest clusters where the attainable resolution is sufficient to distinguish between cluster and background galaxies using their morphology. To investigate more distant clusters requires a different, statistical, approach.

Using the knowledge that the excess of galaxies seen in the direction of a cluster over that of neighbouring fields is due to cluster members it is possible to derive the LF statistically. Although losing the information as to which galaxies are members such a technqiue allows the LF to be measured for more distant clusters than previous methods. Coupled with the advent of large-format and sensitive, CCD, detectors it is now possible to observe the dwarf population down to faint, $M_R \sim -13$, moderately distant $z \sim 0.2$ clusters. It is then possible to investigate whether the faint end of the LF is generally steep in all clusters and derive any correlation with cluster properties and/or distance. Such studies, however, have their own problems. Even with photographic studies of nearby clusters the removal of the background galaxy population is crucial and this becomes more difficult when there is no morphological information. With $N(M) \propto 10^{-0.4(\alpha+1)}$ from the Schechter LF and, for the number counts, $N(m) \propto 10^{0.4m}$ it is easily possible to get a slope of $\alpha \sim -2$ if the background is not subtracted correctly. There are generally two different techniques for background subtraction - either by observing, with the same instrumental set-up, a background field close to the cluster or using published number counts (e.g. Metcalfe et al. 1995). Each of these have their problems. With the number counts a strongly-varying function of magnitude, an error in the zero-point of the magnitude scale can lead to a significant error. In addition, seeing variations, variable galactic extinction, image-detection algorithm errors can all lead to errors in the background subtraction. Driver et al. (1998a) have used extensive simulations in an effort to quantify the observational limitations to deriving a reliable LF. They find that the reliability is a strong function of cluster richness, seeing and redshift but is relatively independent of the shape of the LF. Driver et al. (1998b) have also shown that both methods of background subtraction (mean number counts or nearby field subtraction) lead to similar shapes of the LF.

3. Recent Results

In the past couple of years there has been an explosion in the number of LFs measured down to faint magnitudes (e.g Driver et al. 1994, Smith et al. 1997, Wilson et al. 1997, Trentham 1997a, 1997b,1998, Driver et al. 1998b). As an example, the cluster Abell 2554 was observed by Smith et al. (1997) using the Thomson CCD on the AAT. This cluster, at a redshift of 0.106, is of Abell richness class 3 and Bautz-Morgan type II. The LF was derived by both methods of background removal - subtraction using the mean number counts of Metcalfe et al. (1995) and also using a nearby offset field observed with the same instrumental set-up. Both techniques produce a LF of similar shape, which is shown in Figure 1. The LF of Abell 2554 is best fit by 2 functions, firstly the giants by a Schechter LF with parameters $M^* = -22.5$ and $\alpha = -1.0$ whilst the dwarfs have $\phi(dwarfs) = 2 \text{x} \phi(giants)$, $M_R^* = -19.5$ and $\alpha = -1.8$.

Comparison with other published results is difficult. Different passbands, different telescopes and instruments, different parameters used in different object detection algorithms, and many other variations in the observational and analysis techniques all lead to uncertainties in any comparison. However, also plotted in Figure 1 are the LFs of two other clusters - Coma and Abell 963. The data for Abell 963 comes from the observations of Driver et al. (1994) using the HitchHiker camera on the WHT whilst that for Coma is from Thompson and Gregory (1993) and Godwin and Peach (1977).

It is interesting to note that the LFs of the three clusters presented here are remarkably similar. The magnitude at which the LF steepens is approximately constant, even though the redshifts of the clusters covers a range of 0.2. Comparison with several other published R-band cluster LFs (e.g. Trentham 1997a,1997b,1998) supports this result. Thus there is tentative evidence that there is little evolution in the dwarf population since z=0.2. How much evolution would be measurable? If the dwarfs in Abell 2554 were all fainter by 0.4 magnitudes then the difference between the measured LFs would be detectable. Is such a variation expected from the various evolutionary scenarios? Two models can be considered. Firstly, if the dwarfs are predominantly dwarf ellipticals, with an old stellar population, then the models for elliptical galaxy evolution (e.g. Gunn & Tinsley 1972, Bruzual & Charlot 1993) can be applied. In elliptical galaxies most of the light originates from red giants and thus their luminosity depends on how many stars have turned off the main sequence on the giant branch. Assuming a Salpeter initial mass function, the luminosity as a function of time is given by $L \propto (t - t_{form})^{2/3}$ or, assuming an early epoch of formation, $\Delta M \simeq -2.5 log(1+z)$. Thus by a redshift of 0.2 a brightening of 0.2 magnitudes would be expected. Such a shift would be undetectable in the present data. However, if the dwarf galaxy population is dominated by dwarf irregulars more evolution would be expected. If the last burst of star formation occurred at $z \simeq 0.5$ then $\simeq 0.6$mag of evolution would be expected. The effect would be more noticeable in the blue but the benefit of more sensitive detectors and a small and better known k-correction in the red would be lost. Although this suggests that the dwarfs within rich clusters are primarily ellipticals it is possible that the form of the LF is conspiring to hide any evolution of the dwarfs. For example, if the dwarf component was brighter but less numerous in the past then the two effects would cancel each other out in the observed LF.

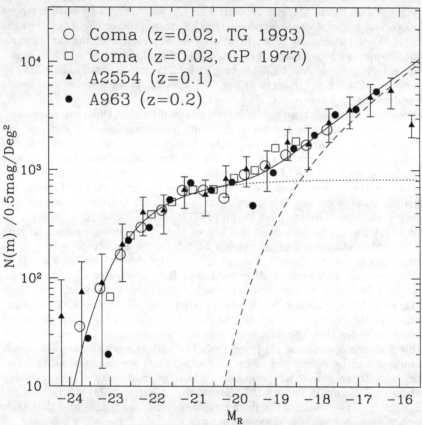

Figure 1. The LF of Abell 2554 compared to that of Coma (Abell 1654) and Abell 963. The dotted line shows a Schechter LF with parameters $M^* = -22.5$ and $\alpha = -1.0$, whilst the dashed line is a Schechter LF with $M^* = -19.5$ and $\alpha = -1.8$. The solid line is the combination of the two LFs.

The fading irregular model has been successfully applied to explain the faint blue galaxy problem (e.g Phillipps & Driver 1995) and also evolution of the field galaxy LF has been observed in deep redshift surveys (e.g Lilly et al. 1995, Ellis et al. 1996). The similarity between the shape of the cluster LF at varying redshifts hence suggests that there may be a different evolution path for dwarf galaxies in clusters and the field. This has been proposed by Moore et al. (1996, and these proceedings) where the lower luminosity systems are most affected by the strong tidal effects that occur within dense regions leading to galaxy 'harassment'.

The Coma LF plotted in Figure 1 is inconsistent with the results of Bernstein et al. (1995) and Biviano et al. (1995). These studies only sampled the core of the cluster whilst the Thompson and Gregory (1993) and Godwin and Peach (1977) surveys studied a similar area of the cluster to that of Abell 2554. It is therefore likely that the shape of the LF may vary within the cluster. Driver et al. (1998b) have also derived the LFs of a sample of seven Abell/ACO clusters using a very similar technique to that used for Abell 2554. Their results are presented in Figure 2. Although several of their clusters do have a steepening of the LF slope at $M_R \sim -19$ it is clear that it is not a universal feature of all clusters. It is thus apparent that there is not a ubiquitous LF as had been proposed by Smith et al. (1997) and Trentham (1997a, 1997b, 1998) and another parameter is crucial in determining the number of dwarfs within a cluster. Phillipps et al. (1998 and these proceedings) consider this possibility further.

4. Summary

In conclusion, both photographic and CCD studies of rich clusters of galaxies suggest that they contain a very large number of dwarf galaxies. In many clusters, the luminosity function of is not well fitted by a Schechter function as there is a steepening to a slope of $\alpha \sim -1.7$ at about $M_R \sim -19$. From a small sample of clusters, the position of this turn-up is independent of redshift. Comparison with evolutionary models leads to the tentative conclusion that the dwarfs within these clusters are primarily dwarf ellipticals that have undergone little evolution since $z \sim 0.2$. As more observations of rich clusters have been obtained, it has become clear that there is not a ubiquitous form of the LF.

References

Bernstein, G.M., Nichol, R.C., Tyson, J.A., Ulmer, M.P. & Wittman, D., 1995, AJ, 110, 1507

Binggeli, B., Sandage, A. and Tammann, G.A., 1985, AJ, 90, 1681

Biviano, A., Durret, F., Gerbal, D., Le Fevre, O., Lobo, C., Mazure, A. & Slezak, E., 1995, A&A, 297, 610

Bruzual, G.A. & Charlot, S., 1993, ApJ, 405, 538

Driver S.P., Phillipps S., Davies J.I., Morgan I. & Disney M.J., 1994, MNRAS, 268, 393

Driver, S.P., Couch, W.J., Phillipps, S. & Smith, R.M., 1998a, MNRAS, in press

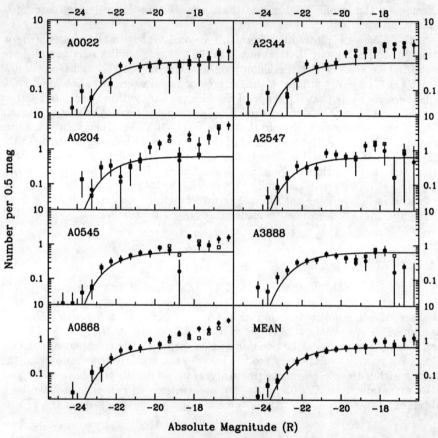

Figure 2. The luminosity functions of 7 Abell/ACO clusters as measured by Driver et al. (1998b). The solid dots and open dots represent differing methods of background/foreground galaxy subtraction.

Driver S.P., Couch W.J., Phillipps S., 1998b, MNRAS, in press
Efstathiou G., Ellis R.S., Peterson B.A., 1988, MNRAS, 232, 431
Ellis, R.S., Colless, M., Broadhurst, T.J., Heyl, J. & Glazebrook, K., 1996, MNRAS, 280, 235
Ferguson H.C., 1989, AJ, 98, 367
Frenk C.S., Evrard A.E., White S.D.M., Summers F.J., 1996, ApJ, 472, 460
Godwin, J.G., Metcalfe, N. & Peach, J.V., 1983, MNRAS, 202, 113
Godwin, J.G. & Peach, J.V., 1977, MNRAS, 181, 323
Gunn, J.E. & Tinsley, B.M., 1972, ApJ, 203, 52
Impey, C.D., Bothun, G. & Malin, D., 1988, ApJ, 330, 634
Jones J.B., Phillipps S., Schwartzenberg J.M., Parker Q.A., 1998, The Low Surface Brightness Universe, p.xxx
Lilly, S.J., Tresse, L., Hammer, F., Crampton, D. & Le Fevre, O., 1995, ApJ, 455, 108
Loveday J., Peterson B.A., Efstathiou G., Maddox S.J., 1992, ApJ, 390, 338
Metcalfe, N., Shanks, T., Fong, R. & Roche, N., 1995, MNRAS, 273, 257
Moore, B., Katz, N. & Lake, G., 1996, Nature, 379, 613
Phillipps, S., Jones, J.B., Smith, R.M., Couch, W.J., & Driver, S.P., 1998, The Low Surface Brightness Universe, p.xxx
Phillipps S., Driver S.P., 1995, MNRAS, 274, 832
Phillipps S., Driver S.P., Couch W.J., Smith R.M., 1998b, ApJ, 498, L119
Sandage, A., Binggeli, B. and Tammann, G.A., 1985, AJ, 90, 1759
Schechter P., 1976, ApJ, 203, 297
Smith R.M., Driver S.P., Phillipps S., 1997, MNRAS, 287, 415
Thompson L.A., Gregory S.A., 1993, AJ, 106, 2197
Thoul, A.A. & Weinberg, D.H., 1996, ApJ, 465, 608
Trentham N., 1997a, MNRAS, 286, 133
Trentham N., 1997b, MNRAS, 290, 334
Trentham N., 1998, MNRAS, 293, 71
White S.D.M., Frenk C.S., 1991, ApJ, 379, 52
Wilson G, Smail I., Ellis R.S., Couch W.J., 1997, MNRAS, 284, 915

Optical and Near-IR Field Luminosity Functions

Jon Loveday

Astronomy & Astrophysics Department, University of Chicago, 5640 S Ellis Ave, Chicago, IL 60637, USA

Abstract. We present preliminary measurements of the b_J and K-band luminosity functions (LFs) of field galaxies obtained from optical and K-band imaging of a sample of galaxies selected from the Stromlo-APM Redshift Survey. The b_J LF is consistent with that previously published from photographic data. The K-band LF has been estimated over a range of 12 magnitudes and is reasonably well fit by a Schechter function with faint-end slope $\alpha = -1.2$.

1. Introduction

Deep, near-infrared K-band (2.2μm) galaxy surveys are a powerful tool for studying galaxy evolution (eg. Gardner et al. 1993, Cowie et al. 1994, Glazebrook et al. 1995). Compared to blue-optical light, near-infrared light is a better tracer of mass in evolved stars and the correction for redshift dimming (the "k-correction") is approximately independent of morphological type. The rapid evolution in galaxy luminosity apparent in the b_J band is not seen in the K band. However, it is vital to have a reliable determination of the K-band luminosity function for *nearby* galaxies in order to interpret faint galaxy counts and to calculate the clustering of K-selected galaxy samples. The largest local K-band sample of galaxies with redshifts is that of Gardner et al. (1997). They measured redshifts for 510 galaxies selected from a K-band limited survey covering 4.4 square degrees. Since their survey is flux-limited, the majority of galaxies have K-band luminosities close to L_K^*. Thus they are able to measure the K-band luminosity function over a range of only 5 magnitudes, and the faint-end slope of their best-fit Schechter function, so important for predicting galaxy number counts, is poorly constrained (their Figure 1).

One can improve on current estimates of the K-band LF without a huge investment of telescope time by observing galaxies selected by their intrinsic luminosity rather than their apparent flux. The Stromlo-APM galaxy survey (Loveday et al. 1996) is an ideal source for a new determination of the joint optical/near-infrared luminosity function $\phi(L_{b_J}, L_K)$ since redshifts have already been measured for 1797 galaxies with $b_J < 17.15$ over a very large volume of space. The solid angle of the survey is 1.3 sr and the median redshift is about 15,300 km/s. One can make use of the fact that K and b_J luminosities are correlated, so that we can preferentially select galaxies of high and low luminosity and thus sample the luminosity range more evenly than a flux-limited sample.

We are thus able to measure the luminosity function to fainter luminosities than from a flux-limited sample of similar size.

2. Sample Selection

Our aim in selecting a subset of Stromlo-APM galaxies for which to obtain K-band photometry was to sample the magnitude range $-22 \leq M_{b_J}{}^1 \leq -13$ as uniformly as possible. An added complication in defining the sample arose because we wished to obtain optical CCD images for the same sample of galaxies. One planned use of this optical imaging is to measure morphological parameters for a representative sample of galaxies at low redshift in order to compare with HST observations of galaxies at high redshift ($z > 0.4$). To obtain comparable linear resolution to the HST data required observing galaxies at $z < 0.04$, assuming ground-based seeing of 1.3 arcsecond. Our "primary" sample thus consists of galaxies at redshifts $z < 0.04$. We divided the magnitude interval $-22 \leq M_{b_J} \leq -13$ into 90 bins each of width 0.1 mag. We then randomly selected up to six galaxies from the Stromlo-APM survey with $z < 0.04$ in each bin. Due to its redshift limit of $z < 0.04$, this primary sample contains rather few galaxies brighter than $M_{b_J} = -20$. We therefore formed a supplementary sample, consisting of galaxies at $z > 0.04$ to "top up" each magnitude bin, where possible, to six galaxies. This supplementary sample consists entirely of galaxies with $M_{b_J} < -20$. The primary sample contains 283 galaxies, and the supplementary sample contains 80 galaxies, giving a total sample size of 363 galaxies.

3. Observations

K-band imaging of the above sample of galaxies was carried out at the Cerro Tololo Interamerican Observatory (CTIO) 1.5m telescope using the CIRIM infrared array over the nine nights 1996 August 31 – September 4 and 1997 October 19–22. The pixel size at $f/7.5$ is $1.16''$, allowing most galaxies to be observed at 9 non-overlapping positions on the 256×256 array. Total integration time for each galaxy was 300s. The infrafred frames were reduced using IRAF, and photometry was performed using SExtractor 2.0.8 (Bertin & Arnouts 1996) with the "mag_best" option. Magnitude errors were estimated by combining in quadrature SExtractor's estimate of the error from photon statistics and the difference between magnitudes measured using local and global estimates of the sky background. 343 galaxies were observed with an estimated K-band magnitude error of less than 0.3 mag (rms mag error = 0.06 mag) and were calibrated using standard stars observed from the list of Elias et al. (1982).

Figure 1 shows the rest-frame $(b_J - k)$ versus M_K colour-magnitude relation for our data. The fit to all galaxies is given by

$$(b_J - k) = -0.260 \times M_K - 2.11, \quad \sigma = 0.86. \qquad (1)$$

We assume a K-band k-correction of $-2.5z$ for all galaxy types.

---

[1]Throughout, we assume a Hubble constant of $H_0 = 100$ km/s/Mpc.

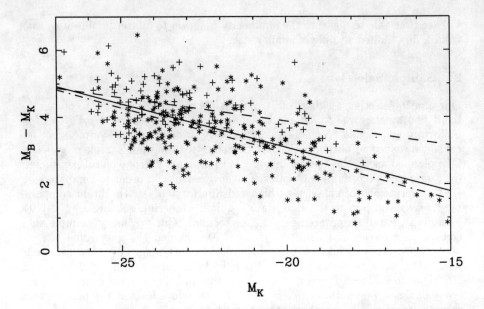

Figure 1. $b_J - k$ versus M_k colour-magnitude plot using APM b_J magnitudes. Plus signs represent early-type galaxies, asterisks late-type galaxies and dots represent unclassified galaxies. The solid line shows a least-squares fit to all galaxies, the dahsed line a fit to early types and the dot-dashed line a fit to late type galaxies.

Optical imaging was performed in the U, B and R bands using the CTIO 1.5m telescope with a Tex 2048^2 CCD over the ten nights 1996 September 7–16. Integration times were 120s in R, 240s in B and 120s in U. The U-band exposures were too short to provide accurate galaxy photometry but were taken under photometric conditions, allowing the possibility of later calibration of deeper, non-photometric U-band observations. Galaxy photometry was done using an earlier beta-release (1.2b9b) of SExtractor and so the photometry presented here is only preliminary. 300 galaxies have reliable B and R magnitudes, and are calibrated with Landolt (1992) standards. The colour equations of Couch & Newell (1980) were used to obtain a b_J magnitude from B and R. Figure 2 plots these CCD b_J magnitudes against APM b_J magnitudes. The mean and rms APM − CCD magnitude is $\Delta m = 0.09 \pm 1.05$. This scatter is larger than the 0.3 mag scatter for the full, $m > 15$ Stromlo-APM sample (Loveday et al. 1992) since 1) galaxies with APM magnitude brighter than 15 are included in the current sample, these are badly saturated on the photographic plates, and 2) the preferential sampling of galaxies of very high and low luminosity means that galaxies with poor APM magnitudes are more likely to be included. For example, the outlying galaxies in the lower-right of Figure 2 are due to the APM machine measuring just a small part of a large spiral galaxy, and thus grossly underestimating the galaxy's luminosity.

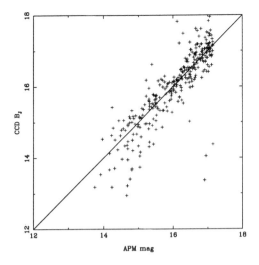

Figure 2. CCD b_J magnitudes plotted against APM magnitudes.

4. Luminosity Functions

When one has a sample selected on one quantity m_1 (in our case APM b_J magnitude) and wishes to estimate the luminosity function for another quantity m_2 (eg. CCD b_J or K magnitude), the best way to proceed is to calculate a bivariate luminosity function (BLF) $\phi(L_1, L_2)$ allowing for any selection effects in L_1 and then to integrate over L_1 to obtain $\phi(L_2)$. One can estimate the shape of $\phi(L_1, L_2)$, independently of inhomogeneities in the galaxy distribution using the maximum likelihood method of Sandage, Tamman and Yahil (1979). The probability of seeing a galaxy with luminosities L_1^i and L_2^i at redshift z_i is given by

$$p_i = \phi(L_1^i, L_2^i) S(L_1^i) \Big/ \int_{L_{2\min}(z_i)}^{L_{2\max}(z_i)} \int_{L_{1\min}(z_i)}^{L_{1\max}(z_i)} \phi(L_1, L_2) S(L_1) dL_1 dL_2. \quad (2)$$

The function $S(L_1)$ accounts for the known selection in L_1 and the luminosity limits $L_{1\min}(z_i)$ and $L_{1\max}(z_i)$ are the minimum and maximum luminosities observable at redshift z_i in a sample limited by apparent m_1 magnitude. If there are no flux limits in the m_2-band, then the integral over L_2 runs from 0 to $+\infty$. The maximum-likelihood shape of the BLF $\phi(L_1, L_2)$ is estimated by maximizing the likelihood $\mathcal{L} = \prod_{i=1}^{N_g} p_i$ (the product of the individual probabilities p_i for the N_g galaxies in the sample) with respect to the parameters describing the BLF.

In practice, we do not have a good *a priori* parametric model for $\phi(L_1, L_2)$, and so instead we measure $\phi(L_1, L_2)$ in a non-parametric way using an extension of the Efstathiou, Ellis and Peterson (1988) stepwise maximum likelihood (SWML) method. Sodré and Lahav (1993) have extended the SWML method to estimate the bivariate diameter-luminosity function and to allow for sample incompleteness. We adopt their extension of the SWML estimator here, including the sampling function $S(L_1)$ separately for the primary and supplementary galaxy samples. We normalise our LFs to the mean density of galaxies

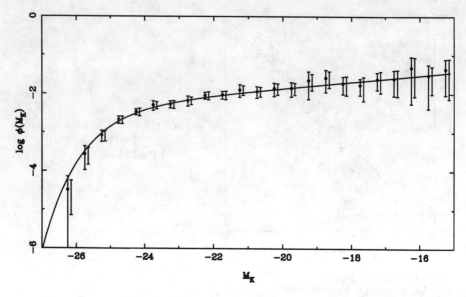

Figure 3. The K-band luminosity function estimated from nine Soneira-Peebles simulations (points) along with the input LF (curve).

with $-22 \leq M_{b_J} \leq -13$ in the full Stromlo-APM sample, $\bar{n} = 0.071 h^3 \mathrm{Mpc}^{-3}$, calculated as described by Loveday et al. (1992).

Once one has obtained the SWML estimate of $\phi(L_1, L_2)$, one can integrate over L_1 to obtain $\phi(L_2)$ and then fit a given functional form, eg. a Schechter (1976) function, by least-squares.

4.1. Test of the Method

We have tested the above procedure by using it to estimate the K-band luminosity function from a set of Monte Carlo simulations. We generated nine mock Stromlo surveys by a Soneira and Peebles (1978) hierarchical clustering simulation. Each galaxy in the simulation was assigned a K-band luminosity drawn at random from a Schechter function with $\alpha = -1.21$ and $M_K^* = -24.7$. Each galaxy was then assigned a b_J magnitude according to our observed colour-luminosity relation (1). Galaxies were selected on their apparent b_J magnitude, $b_J < 17.15$. This process was repeated until each simulation contained 2000 galaxies. We then sampled each simulation by absolute M_{b_J} magnitude as described in §2., finally yielding an average of 359 galaxies per simulation. We calculated the K-band luminosity function $\phi(L_K)$ for each simulation as described in §4. and fit a Schechter function to each by least squares. Averaging over the nine simulations, and estimating the BLF in bins of width 0.5 mag, we measure mean and rms Schechter function parameters $\alpha = -1.16 \pm 0.06$, $M^* = -24.4 \pm 0.6$. The errors on the mean values are $\sqrt{9}$ times smaller than the quoted rms scatter between the simulations. Thus our estimate of M^* is biased 1.5σ too faint and α is overestimated (too shallow) by about 2.5σ. However, our estimates are within the 1σ error from a single realisation.

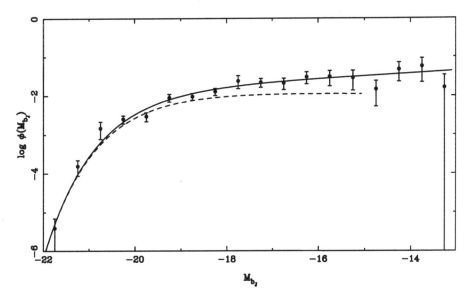

Figure 4. The b_J luminosity function estimated from our CCD magnitudes (filled symbols). Also shown (dashed line) is the Schechter function fit by Loveday et al. (1992) from photographic magnitudes.

The SWML estimates of the K-band LF for the simulations are shown in Figure 3. The points show the mean SWML estimate of $\phi(L_K)$ from the nine simulations and the curve shows the input Schechter function with shape $\alpha = -1.21$, $M_K^* = -24.7$. The error bars going through the data points show the rms scatter between realisations. The error bars offset slightly to the right show the mean error predicted by the covariance matrix (see Efstathiou et al) and are in reasonable agreement with the rms scatter between realisations.

Overall, we find that our procedure for estimating $\phi(L_K)$ from a sample limited by apparent b magnitude and futher selected by absolute B magnitude provides a robust and only weakly biased estimate of the K-band LF over a wide range of absolute magnitudes.

4.2. $\phi(L_{b_J})$

Our estimated b_J-band luminosity function is shown in Figure 4. The curve shows a Schechter function fit to the SWML estimate using least squares and allowing for finite bin width. The best-fit Schechter parameters are $\alpha = -1.16$, $M_{b_J}^* = -19.52$ and $\phi^* = 0.018 h^3 \mathrm{Mpc}^{-3}$. Also shown in this figure is the Schechter function fit by Loveday et al. (1992) from photographic magnitudes. Despite the large scatter in our CCD versus APM b_J magnitudes (Fig. 2), we find that the two estimates of the shape of $\phi(M_{b_J})$ are in reasonable agreement. Although the new estimate has slightly steeper faint-end slope, the slopes are in fact consistent within the 1σ uncertainties.

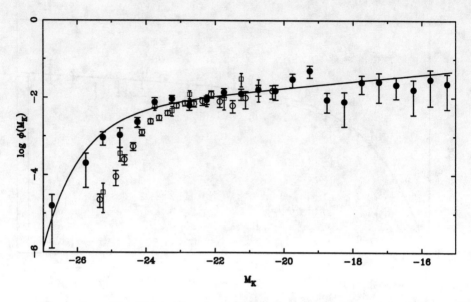

Figure 5. The K-band luminosity function estimated from our sample (filled symbols). Also shown are the results of Gardner et al. (open circles) and Szokoly et al. (open squares).

4.3. $\phi(L_K)$

Our estimated K-band luminosity function is shown in Figure 5. The curve shows a Schechter function fit to the SWML estimate using least squares and allowing for finite bin width. The best-fit Schechter parameters are $\alpha = -1.22$, $M_K^* = -24.73$ and $\phi^* = 0.0073 h^3 \text{Mpc}^{-3}$. Also shown in this figure are recent estimates of the K-band LF from the K-selected samples of Gardner et al. (1997) and of Szokoly et al. (1998). These estimates are in good agreement with ours at $M_K \approx -22$ but fall below our new estimate at the bright end. The reason for this is almost certainly the fact that our sample is selected in b_J, whereas the Gardner et al. and Szokoly et al. samples are K-selected. In a blue-selected sample, the bright end of the K-band LF will be dominated by red galaxies and the faint end dominated by blue galaxies. Since red galaxies tend to be luminous ellipticals, this would explain why we see a larger $\phi(M_K)$ at the bright end than seen in K-selected samples.

5. Conclusions

We have presented preliminary estimates of the b_J and K-band luminosity functions obtained from CCD and infrared array imaging. Our b_J LF is consistent with an earlier estimate from photographic data. We measure $\phi(L_K)$ over a range of 12 magnitudes, a significantly greater range of luminosities than has been measured until now. We are thus able to place much tighter constraints on the faint-end slope of the K-band LF, albeit for a b_J-selected sample. Planned future work includes measurement of the bivariate LF $\phi(L_{b_J}, L_K)$ using CCD b_J

magnitudes, the R-band LF and luminosity functions of galaxies subdivided by colour and morphological type. Given the topic of this meeting, it will also be of great interest to calculate bivariate luminosity-surface brightness distributions.

Acknowledgments. I thank my collaborators Simon Lilly and George Efstathiou for allowing me to show results in advance of publication, the CTIO staff for their excellent support and Jon Gardner and Gyula Szokoly for sending me their K-band LF data points.

References

Bertin, E. & Arnouts, S., 1996, A&AS, 117, 393

Cowie, L.L., Gardner, J.P., Hu, E.M., Songaila, A., Hodapp, K.-W., & Wainscoat, R.J, 1994, ApJ, 434, 114

Efstathiou, G., Ellis, R.S. & Peterson, B.A., 1988, MNRAS, 232, 431

Elias, J. H., Frogel, J. A., Matthews, K., & Neugebauer, G., 1982, AJ, 87, 1029

Gardner, J.P., Cowie, L.L. & Wainscoat, R.J., 1993, ApJ, 415. L9

Gardner, J.P., Sharples, R.M., Carrasco, B.E. & Frenk, C.S., 1997, ApJL, 480, 99

Glazebrook, K., Peacock, J.A., Miller, L. & Collins, C.A., 1995, MNRAS, 275, 169

Landolt, A.U., 1992, AJ, 104, 340

Loveday, J., Peterson, B.A., Efstathiou, G. & Maddox, S.J., 1992, ApJ, 390, 338

Loveday, J., Peterson, B.A., Maddox, S.J., & Efstathiou, G., 1996, ApJS, 107, 201

Sodré, L. & Lahav, O., 1993, MNRAS, 260, 285

Soneira, R.M. & Peebles, P.J.E., 1978, AJ, 83, 845

Szokoly, G.P., Subbarao, M.U., Connolly, A.J. & Mobasher, B., 1998, ApJ, 492, 452

The Low Surface Brightness Universe, IAU Col. 171
ASP Conference Series, Vol. 170, 1999
J. I. Davies, C. Impey and S. Phillipps, eds.

Low Surface Brightness Galaxies in Deep Surveys

Henry. C. Ferguson

Space Telescope Science Institute, 3700 San Martin Drive, Baltimore MD 21218, USA

Abstract.
We examine the constraints that can be placed on the space density of low-surface-brightness galaxies from deep HST images. Such images, while covering only a small solid angle, provide enough depth and spatial resolution to detect LSB galaxies at moderate redshift and distinguish them from galaxies of higher surface brightness.

We consider five simple models of the non-evolving or slowly-evolving population of LSB galaxies, motivated by various discussions in the recent literature. The basic results are (1) models with a large space-density of giant LSB galaxies at moderate redshift do not look like the real world and, (2) models with a large space-density of dwarf LSB galaxies are consistent with HST data (that is, they do not produce more faint LSB galaxies per unit solid angle than are detected at magnitudes $I \gtrsim 23$), but these LSB dwarf galaxies do not contribute much to faint galaxy counts unless they formed their stars in a rapid burst.

1. Background

The "faint blue galaxy problem" has garnered much attention over the last 10 years. The problem is basically that counts of galaxies rise more steeply toward faint magnitudes than expected from models with $\Omega_{\text{Matter}} \approx 1$. Reconciling the counts to $\Omega_M = 1$ within the context of traditional "pure luminosity evolution" models for galaxies would require a fairly late ($z \lesssim 3$) and bright formation epoch for galaxy spheroids (Yoshii and Takahara 1988). Such evolution should in principle show up in the redshift distribution, and was not observed. Only 5 years ago, when $\Omega_M = 1$ was the preferred model, this discrepancy was considered something of a crisis. While solutions involving a cosmological constant were proposed (Yoshii and Peterson 1994; Fukugita et al. 1990), and worked reasonably well, solutions involving additional populations of galaxies were also considered. Among these hypothesised galaxy populations were LSB galaxies and dwarf galaxies. In the last few years the landscape has changed and models with low Ω and/or a cosmological constant no longer seem far-fetched. Large populations of LSB or dwarf galaxies may no longer be necessary to "solve" the faint blue galaxy problem. Nevertheless the physical motivations for postulating their existence are still valid. It is thus worth revisiting the issue to see if faint galaxy surveys themselves can provide limits on the allowable distribution of galaxy surface brightnesses and luminosity at low redshift.

McGaugh (1994) was among the first to suggest that LSB galaxies could dominate faint galaxy counts. He pointed out that the isophotal detection limits of deep imaging surveys are far below those of the photographic surveys on which estimates of the local luminosity function were based. With isophotal detection thresholds of roughly 24.5 mag arcsec^{-2}, the photographic surveys would miss even relatively bright LSB galaxies. Furthermore, the isophotal thresholds affect the photometry of galaxies near the detection limit, leading to systematic biases in estimates of the luminosity function. The detection thresholds of deep CCD surveys (e.g. Tyson 1988) are about three magnitudes fainter and could thus pick up LSB galaxies at low to moderate redshift. The blue colors and weak clustering of local LSB galaxies also seemed to make them attractive candidates for faint-blue galaxies. Ferguson and McGaugh (1995) and McLeod and Rieke (1995) considered more detailed models of LSB galaxies and concluded that while they could not explain the galaxy counts for $\Omega_M = 1$, they could reduce the size of the discrepancy.

2. Recent Developments

The possibility that LSB galaxies could *dominate* the counts at $B \sim 25$ has been ruled out by improved angular-diameter measurements from the HST Medium Deep Survey and Hubble Deep Field projects (e.g. Roche et al. 1996; Williams et al. 1996). Typical half-light radii of galaxies at $B = 25$ are about 0.3 arcsec, significantly smaller than predicted by the Ferguson and McGaugh (1995) model. The HDF size distribution and selection boundaries are described in more detail by Ferguson (1998).

Nevertheless, at roughly the same time as these new measurements have been ruling out large populations of LSB galaxies at faint magnitudes, deeper wide-area surveys have been improving estimates of the surface-brightness distribution of nearby galaxies, and confirming that it is broad (Bothun, Impey, and McGaugh 1997). Indeed, the number of galaxies per unit surface brightness seems to be nearly constant, or only slowly declining, to central surface brightnesses as faint as $\mu_B = 24$. Of course, the surface brightness distribution is just one projection of the bivariate luminosity–surface-brightness distribution. It hard to tell from the exisiting data whether the LSB galaxies are mostly dwarf galaxies, or whether the SB distribution for giant galaxies is similarly broad. On the one hand the bivariate (μ_0, M) distribution derived by de Jong (1996) suggests that the distribution of μ_0 at fixed absolute magnitude widens toward low luminosities, while on the other hand many of the LSB galaxies detected in the POSS-II survey (Schombert et al. 1992) turn out to have HI line widths larger than 100 km s^{-1}.

The other projection of the bivariate brightness distribution, the luminosity function (LF), has also seen some recent developments (Marzke et al. 1998; Loveday 1997; Smith, Driver, and Phillipps 1997; Sawicki, Lin, and Yee 1997; Lin et al. 1996). In particular there appears to be an emerging consensus that the faint end slope of the LF is steep: $N(L) \propto L^{-\alpha}$ with $\alpha \lesssim -1.2$. Values as steep as $\alpha = -2.8$ have been reported for dwarf galaxies with $-12 \lesssim M_B \lesssim -16$ (Loveday 1997). The Hubble Deep Field (HDF), with a detection limit $V \approx 29$, can detect a flat-spectrum galaxy with $M_B = -14$ out to $z = 0.7$ if it has sufficiently high surface brightness. Thus the HDF number counts can be used to test whether such estimates of the faint end slope make sense when coupled with a broad SB distribution.

Another development is the attempt to link the SB distribution to the angular momentum distribution of galaxies predicted from tidal torquing in the early universe (Dalcanton, Spergel, and Summers 1997; Mo, Mao, and White 1998). The Dalcanton et al. (1997) model starts with a log-normal distribution of spin angular momenta. The dark matter is assumed to collapse initially into a halo with a Hernquist density profile, which is later modified in response to the cooling and condensation of the baryons into a disk near the center of the potential. This *ab initio* prediction of the bivariate brightness distribution provides a physically motivated hypothesis that can now be tested against the observations.

3. Five Easy Models

In the remainder of this presentation, I will explore the implications of these recent SB and LF distributions for the predicted properties of galaxies in the HDF. The purpose here is not to try to reproduce the galaxy counts, but instead to see if any of the recently inferred distribution functions Bothun, Impey, and McGaugh (1997, Dalcanton, Spergel, and Summers (1997) *overpredict* the counts of LSB galaxies in the HDF.

The modeling procedure, which involves Monte-Carlo sampling of the assumed bivariate brightness distribution, is described in detail by Ferguson and McGaugh (1995) and Ferguson and Babul (1998). For all models, the assumed cosmology has $H_0 = 65 \,\mathrm{km\,s^{-1}\,Mpc^{-1}}$, $\Omega_M = 0.1$ and $\Omega_\Lambda = 0$. The luminosity function has the Schechter (1976) form

$$\phi(L)dL = \phi^*(L/L^*)^\alpha e^{-L/L^*} d(L/L^*). \qquad (1)$$

The LF normalization, $\phi^* = 2.8 \times 10^{-2} h_{65}^3 \,\mathrm{Mpc}^{-3}$, is set to correspond to half the total galaxy population for the luminosity function normalization recommended by Ellis (1997). The luminosity function runs from $0.003L^*$ to $10L^*$ for all models. For the non-evolving models, the galaxies are given spectral energy distributions that correspond to a model with a star-formation e-folding time $\tau = 10^{10}$ yr and a metallicity 0.05 solar. This model matches the typical colors for LSB galaxies locally. Galaxies are distributed with constant space density, and, for models 1-3 and 5, are assumed not to evolve.

4. Model 1

The first simulation is a *non-evolving* model using the McGaugh (1996) SB distribution function:

$$\log \phi(\mu_0) = m(\mu_0 - \mu_0^*); \qquad (2)$$

with

$$m = -0.3 \text{ for } \mu_0 > \mu_0*, \qquad (3)$$

$$m = 2.6 \text{ for } \mu_0 < \mu_0*, \qquad (4)$$

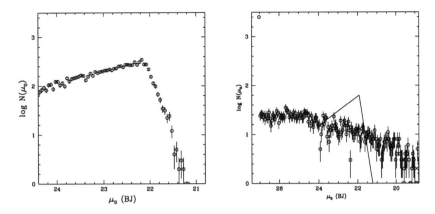

Figure 1. SB distributions for models 1-4 (Left), and model 5 (right). The SB distribution in the left panel is reproduced as thin lines in the right panel. The "data" points are galaxies drawn from Monte-Carlo realizations of the analytic distribution functions.

and
$$\mu_0* = 21.9 B_J \text{ at } z = 0. \tag{5}$$

The SB distribution is shown in Fig. 1. The luminosity function (Fig. 2) is given a steep slope $\alpha = -1.8$, and a characteristic absolute magnitude $M_{BJ}^* = -20.5 + 5 \log h_{65}^2$. The model dramatically overpredicts the number of LSB in the HDF (Fig. 3). This is not surprising given that this model is also strongly inconsistent with the Loveday et al. (1992) luminosity function.

5. Model 2

The second simulation is the same as the first, but the LF is assumed to have a faint-end slope $\alpha = -1$, consistent with Loveday et al. (1992). While such a model falls far short of matching galxy counts at HDF depths, nevertheless, it predicts more large LSB galaxies than are seen. Specifically, the model predicts that there should be about 12 galaxies in the HDF with isophotal areas (above the detection threshold) of more than 1.6 square arcsec, with a mean I-band surface brightness within the central 0.2" fainter than $\mu_{0,I} = 24.5$. The actual HDF has only two galaxies meeting these criteria. The likely explanation for the discrepancy is that surface brightness is correlated with luminosity, and that the bulk of the LSB galaxies are dwarfs.

6. Model 3

The third simulation adopts the same $\alpha = -1.8$ slope as model 1, but uses a characteristic absolute magnitude $M_{BJ}^* = -16$. Thus in this model, the LSB galaxies are almost all dwarfs. The inputs and results are not shown in the figures as they are nearly identical to those of model 4.

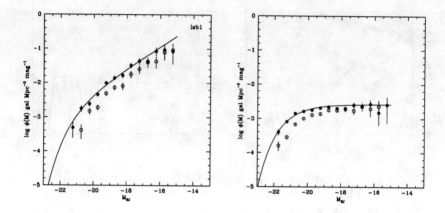

Figure 2. Luminosity functions at $z = 0$ for 1 and 2. The solid curve is the adopted model. The black dots are the luminosity function in a Monte-Carlo realization of the model not accounting for selection biases. The open circles are the luminosity function that would be recovered in a survey with the selection function of the Loveday et al. (1992) survey.

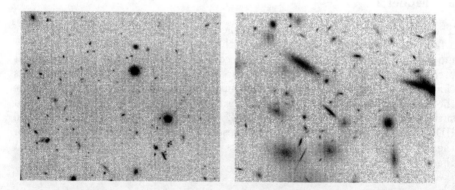

Figure 3. Left: A portion of the HDF image (WF chip 4). Right: simulated portion of the HDF image from model 1.

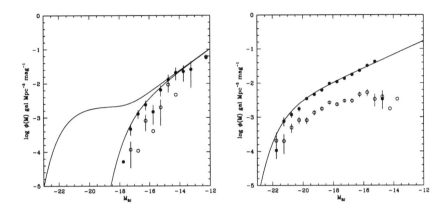

Figure 4. Luminosity functions at $z = 0$ for models 4 and 5. Symbols have the same meaning as Fig. 2.

7. Model 4

This model is the same as model 3, but the stellar populations are allowed to evolve. Galaxies are assumed to form at redshifts between $z = 1$ and $z = 2.5$, with a constant rate of formation in this interval. The spectral energy distributions evolve according to the $\tau = 10$ Gyr low-metallicity model used for the non-evolving models. The luminosity function for this model is shown in Fig. 4 and the simulated HDF image in Fig. 5. Models 3 and 4 do not overpredict the number of LSB galaxies in the HDF. That is, a large population of LSB dwarfs could exist out to arbitrary redshift and not violate the constraints of the HDF.

On the other hand, the models show that dwarf galaxies drawn from the McGaugh (1996) SB distribution contribute almost negligibly to the deep counts, *even with a steep LF slope* $\alpha = -1.8$. The galaxies in this model make up less than 10% of the galaxy population fainter than $V_{606} = 23$.

8. Model 5

Model 5 uses the Dalcanton, Spergel, and Summers (1997) SB distribution, which depends on the baryon fraction F, the specific angular momentum λ, the logarithmic spread in specific angular momentum σ_λ, and several other parameters. To first order, the model gives a central surface brightness $\Sigma_0 \propto FM_{tot}^{1/3}\lambda^{-(2+6F)}$. We have taken $F = 0.05$, $<\lambda> = 0.06$ and $\sigma_\lambda = 0.7$, and adjusted the other parameters so that the peak of the SB distribution at fixed scale length $r_0 = 2$ kpc is $\mu_0 = 24$. The SB distribution predicted by this model (Fig. 1) is broader than the empirical one of McGaugh (1996), but in this case there is some dependence on luminosity. The luminosity function slope is taken to be $\alpha = -1.5$ with $M_{BJ}^* = -20.5$. The luminosity function for this model is shown in Fig. 4.

Figure 5. Left: Model 4. Right: Model 5.

As was the case for models 1 and 2, this model overpredicts the number of large-angular-size LSB galaxies in the HDF (Fig. 5).

9. Summary

Galaxy counts for the different models (determined by running FOCAS on the simulated images) are shown in Table 1. Model 1 is the only one that predicts LSB galaxies are significant contributors to the counts, but it shows the clearest disagreement with the observed HDF size distribution. Models 2 and 5 are less strongly ruled out, and models 3 and 4 are consistent with the data. But in these cases the LSB galaxy population makes up at most about 20% of the counts fainter than $V_{606} = 23$.

Table 1. Galaxy Counts for the different Models.

V_{606}	HDF	Model 1	Model 2	Model 4	Model 5
23-25	128	126	10	4	24
25-27	466	404	40	25	108
27-29	1078	915	106	118	209

The surface brightness distributions considered here are relatively "flat," with nearly constant numbers of galaxies per logarithmic interval of surface brightness. The simulations indicate that flat distributions with many of *giant* LSB galaxies produce too many faint LSB galaxies to be consistent with the HDF. The discrepancy could be avoided if giant LSB galaxies formed only recently (e.g., at $z \lesssim 0.3$).

Flat SB distributions with only dwarfs are consistent with the HDF, but a *slowly evolving* popuation of LSB dwarfs probably contributes less than 20% of the faint-galaxy counts for any plausible luminosity function.

References

Bothun, G., Impey, C., and McGaugh, S. S. 1997, PASP, 109, 745
Dalcanton, J. J., Spergel, D. N., and Summers, F. J. 1997, ApJ, 482, 659
de Jong, R. S. 1996, A&A, 313, 45
Ellis, R. S. 1997, ARA&A, 35, 389
Ferguson, H. C. 1998, in The Hubble Deep Field, ed. M. Livio, S. M. Fall, and P. Madau (Cambridge, Cambridge University Press), p. 181
Ferguson, H. C. and Babul, A. 1998, MNRAS, in press
Ferguson, H. C. and McGaugh, S. S. 1995, ApJ, 440, 470
Fukugita, M., Yamashita, K., Takahara, F., and Yoshii, Y. 1990, ApJ, 361, L1
Lin, H., Kirshner, R. P., Schectman, S. A., Landy, S. D., Oemler, A., Tucker, D. L., and Schechter, P. L. 1996, ApJ, 464, 60
Loveday, J. 1997, ApJ, 489, 29
Loveday, J., Peterson, B. A., Efstathiou, G., and Maddox, S. 1992, ApJ, 390, 338
Marzke, R. O., Da Costa, L. N., Pellegrini, P. S., Willmer, C. N. A., and Geller, M. J. 1998, ApJ, 503, 617
McGaugh, S. 1996, MNRAS, 280, 337
McGaugh, S. S. 1994, Nature, 367, 538
McLeod, B. A. and Rieke, M. J. 1995, ApJ, 454, 611
Mo, H. J., Mao, S., and White, S. D. M. 1998, MNRAS, 295, 319
Roche, N., Ratnatunga, K., Griffiths, R. E., Im, M., and Neuschaefer, L. 1996, MNRAS, 282, 1247
Sawicki, M. J., Lin, H., and Yee, H. K. C. 1997, AJ, 113, 1
Schechter, P. 1976, ApJ, 203, 297
Schombert, J. M., Bothun, G. D., Schneider, S. E., and McGaugh, S. S. 1992, AJ, 103, 1107
Smith, R., Driver, S. P., and Phillipps, S. 1997, MNRAS, 287, 415
Tyson, J. A. 1988, AJ, 96, 1
Williams, R. E. et al. 1996, AJ, 112, 1335
Yoshii, Y. and Peterson, B. 1994, ApJ, 436, 551
Yoshii, Y. and Takahara, F. 1988, ApJ, 326, 1

Low Surface Brightness Dwarf Galaxies in the Bristol – Anglo-Australian Observatory Virgo Cluster Survey

J. B. Jones, S. Phillipps, J. M. Schwartzenberg

Astrophysics Group, Department of Physics, University of Bristol, Tyndall Avenue, Bristol, BS8 1TL, United Kingdom.

Q. A. Parker

Anglo-Australian Observatory, Siding Spring, Coonabarabran, New South Wales 2357, Australia.

Abstract. We describe a new, deep photographic survey of the Virgo Cluster which uses multiple exposures on Tech Pan film with the United Kingdom Schmidt Telescope to probe the dwarf population to fainter surface brightness limits than previous surveys. We have identified galaxies having sizes ($\geq$ 3 arcsec scale length) and surface brightnesses ($\leq$ 24.5 R mag arcsec^{-2}) characteristic of those expected for dwarf spheroidal galaxies in the cluster. The survey is providing substantial samples of extremely low luminosity galaxies outside the environment of the Local Group and nearby groups for the first time. An initial study of two small areas has found dwarf spheroidal candidates in large numbers (500 deg^{-2}) which indicate a steep, continuously rising luminosity function at low luminosities.

1. The Virgo Cluster

The Virgo Cluster, alongside its counterpart in Fornax, is the nearest sizeable galaxy cluster. It is close enough for detailed morphological studies to be possible even for low luminosity dwarf galaxies. It is an irregular, poor cluster of Bautz-Morgan type III (Abell 1975) and Abell richness class 0. As such, it allows detailed studies of the dwarf population in an environment substantially different from the Local Group and other nearby groups.

A seminal study was carried out by Binggeli, Sandage and Tammann, who generated the Virgo Cluster Catalog consisting of 1277 galaxies classified as certain members and a further 574 possible members over an area of 140 deg^2 (Binggeli et al. 1985). Membership was assigned by visual inspection, essentially based on the larger angular sizes of the cluster galaxies compared with the background population. Their dwarfs conformed to a moderately steep luminosity function (Sandage et al. 1985, Binggeli et al. 1988). Various detailed studies of cluster members have been performed subsequently, including the dwarf population (e.g. Ferguson & Sandage 1989, Binggeli & Cameron 1993, Durrell 1997, Young & Currie 1998).

Of particular note, Impey, Bothun & Malin (1988) performed a survey for large angular size low surface brightness galaxies in a single Schmidt field centred on the M87 cluster core, using a photographic stacking technique. They identified 137 galaxies having central surface brightnesses in the range 23 to 26 B mag arcsec^{-2}, of which 27 were new detections.

2. The Bristol – AAO Virgo Cluster Survey

The properties of the faint end of the galaxy luminosity function are poorly constrained outside the Local Group, both in terms of the numbers and characteristics of the galaxies. Some studies have found evidence for very steep luminosity functions in both cluster (e.g. De Propris et al. 1995, Smith, Driver & Phillipps 1997, Trentham 1997, 1998a) and field (Loveday 1997, Morgan, Smith & Phillipps 1998) environments. At the very lowest luminosities, dwarf spheroidal galaxies have been identified in nearby groups (e.g. Caldwell et al. 1998), extending the number of known dSphs and allowing a comparison of their properties with Local Group members. The importance of very low luminosity galaxies in understanding galaxy formation (e.g. Frenk et al. 1996, Kauffmann, Nusser & Steinmetz 1997) and evolution (Gallagher & Wyse 1994, Caldwell et al. 1998, Trentham 1998b) demands that progress is made in identifying and studying extremely low luminosity dwarfs in new environments (see Phillipps et al. 1998a, and references therein).

In order to extend surveys of galaxies to fainter surface brightnesses than previous surveys and over a full ten degree square region of the cluster, the Bristol–Anglo-Australian Observatory survey is using multiple exposures with the United Kingdom Schmidt Telescope (UKST) on Kodak Tech Pan emulsion through an R band filter (Schwartzenberg, Phillipps & Parker 1995, Schwartzenberg & Phillipps 1995, 1997, Schwartzenberg 1996, Phillipps et al. 1998b, Jones et al. 1998). Six individual exposures of each Virgo field of $1 - 1\frac{1}{2}$ hour duration are digitally stacked to give a total integration time of 7 hours. The Tech Pan emulsion combines a high efficiency (approaching 10%, Parker et al. 1998, Phillipps & Parker 1993) and fine grains (providing a high imaging resolution of 5 microns and a high uniformity).

The four UKST fields of the Virgo survey area are shown in Figure 1. Field coordinates have been selected so as to cover a 100 deg^2 region centred on the M87 cluster core ("Cluster A" of Binggeli, Sandage & Tammann 1987). The data therefore survey both the region of high density in the core and the lower density regions at the periphery. The survey area extends south as far as the M49 cluster ("Cluster B") and includes the "M Cloud" west of the M87 cluster. The central area in the grid has been imaged on all 24 films, in effect providing a further 1 mag depth gain in this region and offering the opportunity of a deeper survey in a restricted area. Table 1 presents details of the four fields.

The photographic study compares very favourably with any CCD surveys currently feasible. The very large solid angles of the Schmidt fields, and the very long integration times, overcome the modest telescope aperture (1.2m) and the low quantum efficiency of the emulsion compared with CCDs. It has been possible to perform a very deep survey of a 10 deg × 10 deg region in the central part of the Virgo Cluster in the equivalent of only four nights' observing on the

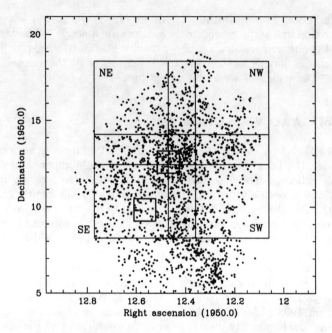

Figure 1. The four UKST fields and the Virgo Cluster. The boundaries of the four Schmidt fields are drawn superimposed on the distribution of cluster members given in the Virgo Cluster Catalog of Binggeli, Sandage & Tammann (1985). The full 6 deg × 6 deg extent of the fields are shown; in practice the scanned regions are slightly smaller. The grid of fields is centred on the M87 core of the cluster. The M49 condensation lies at the southern boundary of the surveyed region. The small squares show the fields used in the initial survey of Section 5.

Schmidt. An equivalent survey to the same depth using CCDs might reasonably take several weeks, dependent on the details of detector and telescope. To date, the photographic observations are 96% complete.

3. Data Reduction and Analysis

The photographic data are digitised using the SuperCOSMOS plate measuring machine at the Royal Observatory Edinburgh. Only the best quality films are used for stacking: adding poor quality films does not provide a useful gain in depth (Bland-Hawthorn, Shopbell & Malin, 1993). With 10 micron wide pixels, corresponding to 0.67 arcsec on the sky, data extents are very large, causing particular data processing difficulties.

Table 1. The four quadrants of the surveyed area.

Field	Central coordinates		No. exposures used	Total integ. time
	R.A. (1950.0)	Dec. (1950.0)		
Northwest	$12^h\ 16^m$	$+15.4°$	6	6.3 hr
Southwest	12 16	$+11.2$	6	6.0
Northeast	12 34	$+15.4$	6	6.3 [1]
Southeast	12 34	$+11.2$	6	7.1

[1] Estimated (one exposure still awaited).

The individual exposures are sky-subtracted using a pixel-by-pixel subtraction of a median-filtered version of the data. The use of a 3 arcmin square spatial filter maximises the ability to detect small, faint images but limits the survey, at least at the present time, to small galaxy images – for example, the larger galaxies in the survey of Impey, Bothun and Malin (1988) would be partially removed. All six films of a given part of sky have their intensities normalised to the same scale, thus correcting for factors such as differences in exposure time, atmospheric transmission or the details of the development process. Coaddition of the six exposures is accomplished using median stacking, which efficiently overcomes problems such as film defects, dust particle images or satellite trails that affect single films. Image detection is performed with a connected-pixel algorithm using a detection threshold $\mu_{lim} = 25.5$ R mag arcsec^{-2} and a minimum image area above this isophote of $A_{lim} = 11$ arcsec^2. These parameters ensure that each detection will have a signal-to-noise ratio of least 10 and magnitudes $R \leq 22$ mag.

4. Identifying Virgo Dwarf Spheroidals

Whilst the detection of galaxies in the direction of the Virgo Cluster may not present a particular challenge in itself, the identification of cluster members is significantly more difficult. Nearby clusters cover large angular scales and consequently the cluster population is swamped by the numerically dominant background population (see, for example, the review by Trentham 1998b). In the absence of redshift information (for example because of the practicalities of performing spectroscopy for very large samples of galaxies extending over wide areas, or because of the difficulties of obtaining spectra for low surface brightness galaxies), membership for the dwarf population must often be assigned on morphological grounds. Binggeli, Sandage & Tammann (1985) were able to assign membership for dwarf galaxies in the cluster on the basis of the galaxies' visual appearance: the dwarfs are characterised by their low surface brightnesses for their sizes.

In the present survey, even though the galaxies are more extreme than those of the Virgo Cluster Catalog, it is still possible to isolate samples of cluster galaxies likely to suffer only a small degree of contamination by the background population. A comparison of Virgo photographic data with deep CCD data from a South Galactic Pole field shows that the background contamination is as small as several percent for galaxies having central surface brightnesses in the range $\mu_0 = 22.0$ to 24.5 R mag arcsec^{-2} and exponential scale lengths $a \geq 3$ arcsec, even when allowance is made for differences in resolution between the two data sets. These issues are discussed in greater detail by Schwartzenberg (1996) and by Jones et al. (1998).

The galaxies selected have properties comparable to those that Local Group dwarf spheroidals would have if they were at the distance of the Virgo Cluster. They have exponential scale lengths $h \geq 260$ pc and absolute magnitudes $M_R = -11$ to -16 (for an assumed distance modulus of 31.3 mag, equivalent to $H_0 = 70$ km s^{-1} Mpc^{-1}). Local Group dwarf spheroidals have sizes between $h = 90$ and 400 pc and absolute magnitudes $M_R = -9$ to -14. The Virgo objects generally have fainter surface brightnesses than the galaxies of Binggeli, Sandage and Tammann (1985) and are smaller than those of Impey, Bothun and Malin (1988).

5. The Initial Survey

A preliminary study of two subfields in the Southeast quadrant of the full survey area has been carried out. The SuperCOSMOS scans of the quadrant were provided as nine sections, each 7680 × 7680 pixels in size, two of which were used for the initial survey (Schwartzenberg et al., 1995). Excluding their edges, both fields are 1.3 deg square, providing a total area of 3.2 deg^2. The fields, listed in Table 5., were selected to sample the core of the cluster and a region of lower density 3.1 deg to the southeast. The core field included M87, although due to the raised background light levels, the region immediately around M87 was excluded from the study. The data were further subdivided into 2280×2280 pixel subregions for data processing because of computer hardware limitations. Fuller details are given by Schwartzenberg (1996) and Phillipps et al. (1998b).

Table 2. Details of the fields studied in the initial survey.

Property	Core field	Outer field
Field area	1.58 deg^2	1.61 deg^2
Field centre:		
R.A.(1950)	12^h 28.2^m	12^h 33.9^m
Dec.(1950)	+12° 36′	+09° 49′

A total of 56 000 images were detected over the two fields. From these a subsample of galaxies was selected having central surface brightnesses in the range $\mu_0 = 22.0$ to 24.5 R mag arcsec^{-2} and scale lengths $a \geq 3$ arcsec after fitting exponential light profiles to the data. Through a detailed comparison of the numbers of Virgo galaxies with the South Galactic Pole field population at

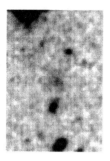 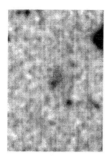 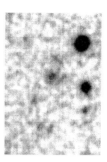

Figure 2. Examples of candidate Virgo Cluster dwarf spheroidal galaxies from the initial survey of 3.2 deg^2. The images have been produced by coadding six UKST exposures on Tech Pan film. None of the galaxies appears in either the Virgo Cluster Catalog or the sample of Impey, Bothun & Malin (1988). Each frame is 1.0 arcmin in width. North is at the top.

each point in the magnitude – surface brightness plane, it is possible to remove the background contamination. The overall contamination of the sample is expected to be 8 per cent. After this background subtraction, the sample contains 1570 galaxies across the two fields. A luminosity function can be constructed by binning the background-subtracted galaxy densities by magnitude. The luminosity function of this sample is found to be steep, with a formal faint end slope of $\alpha = -2.2$ (Phillipps et al. 1998b), and as such is comparable with the steep functions found in some more distant clusters (e.g. Driver et al. 1994, De Propris et al. 1995, Smith, Driver & Phillipps 1997, Wilson et al. 1997, Trentham 1997, 1998a). At the bright end, galaxy numbers are consistent with those of Sandage, Binggeli and Tammann (1985). The dwarf density in the cluster core field is actually smaller than that in the outer field (430 galaxies deg^{-2} against 560 deg^{-2}). The dense environment in the cluster core, and particularly the presence of M87, may have a direct effect on the low surface brightness galaxy population, either through the removal of these galaxies or suppressing their formation. A similar effect has been found in the core of the Coma Cluster by Thompson & Gregory (1993).

6. The Full Cluster Survey

Progress is underway on reducing the full 100 deg^2 Virgo survey area. To date, 23 of the required 24 exposures have been taken. It is intended that the full survey will measure the background population using data identical in format to those from Virgo, reducing systematic errors which might be introduced during the subtraction of the background galaxy numbers. Work is therefore progressing on the Virgo Northwest field and on a background field centred at right ascension 10h 40m, declination 0°00′ (1950). The photometric calibration has been accomplished using R band observations on the Anglo-Australian Telescope obtained for other projects.

For data reduction, each field is subdivided into 16 subregions, each 6800 × 6800 pixels. Data are stored as 4-byte numbers per pixel, and therefore each 1/16-th section is 185 Mbyte in size. New computer hardware resources provide sufficient memory to reduce each 6800 × 6800 pixel section as a whole, avoiding any need to break these sections into a large number of smaller regions as was done in the initial survey. This has the advantage of much simpler data handling than would be needed if large numbers of small sections were used. Similarly, establishing astrometric reference frames for a very large number of different sections would be prohibitive.

As in the case of the initial survey of 3.2 deg^2, emphasis is being put on searches for small angular size low surface brightness dwarfs. However, the feasibility of performing a parallel survey with a different sky background subtraction is being investigated, in order to allow searches for large angular size low surface brightness galaxies of the type studied by Impey, Bothun and Malin (1988), but over a wider area than was available for their study.

7. Conclusions

The new survey is already providing large samples of candidate Virgo Cluster dwarf spheroidal galaxies suitable for more detailed study. The galaxies have absolute magnitudes as faint as $M_R = -11$ to -16 (for $H_0 = 70$ km s^{-1} Mpc^{-1}). They will be important for addressing questions relating to the properties and evolution of the lowest luminosity galaxies in an environment very different from that in nearby galaxy groups. An initial survey of 3.2 deg^2 has found a high density of these galaxies and evidence for a steep luminosity function. The survey is continuing over an area of 100 deg^2. The very large numbers of dwarf spheroidals should reduce the errors in the faint end of the luminosity function and help to define its shape. They will enable clear differences to be found at these very faint limits between the cluster core and periphery to significantly better accuracy than has been possible for any other cluster before now. A parallel survey of the Fornax Cluster is planned for which several films are already available.

Acknowledgments. We wish to thank the UKST and SuperCOSMOS staff in providing their usual excellent service.

References

Abell, G. O. 1975, in 'Galaxies and the Universe', A. Sandage, M. Sandage & J. Kristian, University of Chicago Press, p. 601

Binggeli, B., & Cameron, L. M. 1993, A&AS, 98, 297

Binggeli, B., Sandage, A., & Tammann, G. A. 1985, AJ, 90, 1681

Binggeli, B., Sandage, A., & Tammann, G. A. 1988, ARA&A, 26, 509

Binggeli, B., Tammann, G. A., & Sandage, A. 1987, AJ, 94, 251
Bland-Hawthorn, J., Shopbell, P. L., & Malin, D. F. 1993, AJ, 106, 2154
Caldwell, N., Armandroff, T. E., Da Costa, G. S., & Seitzer, P. 1998, AJ, 115, 535
De Propris, R., Pritchett, C. J., Harris, W. E., & McClure, R. D. 1995, ApJ, 450, 534
Durrell, P. R. 1997, AJ, 113, 531
Frenk, C. S., Evrard, A. E., White, S. D. M., & Summers, F. J. 1996, ApJ, 472, 460
Ferguson, H. C., & Sandage, A. 1989, ApJ, 346, L53
Gallagher, J. S., Wyse, R. F. G. 1994, PASP, 106, 1225
Impey, C. D., Bothun, G. D., & Malin, D. F. 1988, ApJ, 330, 634
Jones, J. B., Phillipps, S., Schwartzenberg, J. M., & Parker, Q. A. 1998, in 'Dwarf Galaxies and Cosmology', T. X. Thuân et al., Gif-sur-Yvette: Editions Frontières, in press. astro-ph/9805287
Kauffmann, G., Nusser, A., & Steinmetz, M. 1997, MNRAS, 286, 795
Loveday, J. 1997, ApJ, 489, 29
Morgan, I., Smith, R. M., & Phillipps, S. 1997, MNRAS, 295, 99
Parker, Q. A., Malin, D. F., Canon, R. D., Phillipps, S., & Russell, K. S. 1998, MNRAS, submitted
Phillipps, S., Driver, S. P., Couch, W. J., & Smith, R. M. 1998a, ApJ, 498, L119
Phillipps, S., & Parker, Q. A. 1993, MNRAS, 265, 385
Phillipps, S., Parker, Q. A., Schwartzenberg, J. M., & Jones, J. B. 1998b, ApJ, 493, L59
Sandage, A., Binggeli, B., & Tammann, G. A. 1985, AJ, 90, 1759
Schwartzenberg, J. M., 1996, Ph.D. Thesis, University of Bristol
Schwartzenberg, J. M., & Phillipps, S. 1995, Anglo-Australian Observatory Newsletter, 72, 6
Schwartzenberg, J. M., & Phillipps, S. 1997, Astroph. Lett. & Comm., 36, 279
Schwartzenberg, J. M., Phillipps, S., & Parker, Q. A. 1995, A&A, 293, 332
Schwartzenberg, J. M., Phillipps, S., & Parker, Q. A. 1996, A&AS, 117, 179
Smith, R. M., Driver, S. P., & Phillipps, S. 1997, MNRAS, 287, 415
Thompson, L. A., & Gregory, S. A. 1993. AJ, 106, 2197
Trentham, N. 1997, MNRAS, 286, 133
Trentham, N. 1998a, MNRAS, 293, 71
Trentham, N. 1998b, in 'Dwarf Galaxies and Cosmology', T. X. Thuân et al., Gif-sur-Yvette: Editions Frontières, in press. astro-ph/9804013
Wilson, G., Smail, I., Ellis, R. S., & Couch, W. J. 1997, MNRAS, 284, 915
Young, C. K., & Currie, M. J. 1998, A&AS, 127, 367

A dichotomy between HSB and LSB galaxies

Marc Verheijen[1]

National Radio Astronomy Observatory, Socorro, New Mexico

Brent Tully

University of Hawaii, Institute for Astronomy, Honolulu, Hawaii

Abstract.
A complete sample of spiral galaxies in the Ursa Major cluster is imaged at various optical wavelengths and in the Near-Infrared K'-band. HI rotation curves were obtained for all gas rich systems. The Near-Infrared surface brightness distribution of disk galaxies turns out to be bimodal; galaxies avoid a domain around $\mu_0^i(K')\approx 18.5$ mag/arcsec^2. This bimodality is particularly striking when only the more isolated, non-interacting systems are considered. The Luminosity Function of the HSB family of galaxies is truncated well above the completion limit while the Luminosity Function of the LSB family is still sharply rising at our limiting magnitude. Near-Infrared mass-to-light ratios suggest that HSB galaxies are close to a kinematic maximum-disk situation while LSB galaxies are dark matter dominated at all radii. Assuming equal Near-Infrared mass-to-light ratios for both HSB and LSB systems, we find that the gap in the surface brightness distribution corresponds to a situation in which the baryonic mass is marginally self-gravitating. We finally conclude that the luminosity-line width relation is a fundamental correlation between the amount and distribution of dark matter mass and the total luminosity, regardless of how the luminous mass is distributed within the dark mater halo.

1. Introduction

The distribution of disk central surface brightnesses is determined with some confidence in the B-band. It shows a steep rise at the bright end and remains roughly flat towards fainter surface brightnesses (McGaugh *et al*, 1995; McGaugh, 1996; de Jong, 1996; Bothun *et al*, 1997 and references therein). In general, however, this distribution is determined using B-band photometry and the observed surface brightnesses are not or obscurely corrected for the effects of inclination and internal extinction. Furthermore, the effects of metallicity and age of a stellar population on the B-band surface brightness of a galaxy

---

[1] Jansky Fellow, Array Operations Center

can be considerable. It might not be surprising that the observed B-band surface brightness distribution is flat to some extent since large galaxy-to-galaxy variations in dust content, age and metallicity will smooth out any underlying, physically interesting distribution like the stellar surface density or angular momentum distribution. Therefore, "...it is possible that a great deal of our present understanding about the surface brightness distribution of galaxies is based on a parameter that is very insensitive to the actual physical characteristics of galaxies" (Davies, 1990).

In our talks we presented results from a Near-Infrared imaging survey of a complete volume limited sample of galaxies in the Ursa Major Cluster. In the Near-Infrared, internal extinction is negligible and the age and metallicity effects on the luminosity of a stellar population are minimal. We show that the distribution of Near-Infrared face-on disk central surface brightnesses is bimodal; there is a relative lack of spirals with $\mu_0^i(K')\approx 18.5^m$. Analyzing HI rotation curves shows that this gap in the Near-Infrared surface brightness distribution may correspond to a dynamical instability when a disk is marginally self-gravitating.

2. The Sample and Observations

The Ursa Major cluster, located in the Supergalactic Plane, consists of 80 galaxies in a $\sim$80 Mpc^3 volume at a distance of 15.5 Mpc. This volume is overdense by roughly a factor 10 compared to the average field. However, the population of galaxies is comparable to the field. It comprises mainly spiral galaxies, a dozen S0 and maybe one elliptical system. A complete sample of 62 galaxies, intrinsically brighter then the SMC, has been identified in the B-band. Since all galaxies are at the same distance, there is little question about their relative luminosities and disk scale lengths. This cluster is described in detail by Tully et al (1996).

Surface photometry in the B, R and I passbands was obtained for all galaxies in 11 observing runs between February 1984 and March 1996. Photometric K'-band images of 69 spiral galaxies in Ursa Major were obtained in May 1991, February 1992 and March 1993 using a 256^2 HgCdTe detector on the 24-inch and 88-inch telescopes of the University of Hawaii on Mauna Kea. Sixty galaxies form a nearly complete sample (only U6628 and U7129 have missing K'-band data) of spirals intrinsically brighter than the SMC. Images and luminosity profiles can be found in Tully et al (1996).

HI synthesis data for all galaxies with sufficient HI gas were obtained with the Westerbork Synthesis Radio Telescope between 1991 and 1996. From these data, HI rotation curves were derived by fitting tilted rings to the HI velocity fields. The rotation curves were corrected for the effects of beam smearing by overlaying them on position-velocity diagrams and adjusting them by eye. The radial surface density profiles of the gas were derived from the total HI maps by averaging the flux in ellipses following the adjusted tilted rings fits. The HI data are described in detail by Verheijen (1997).

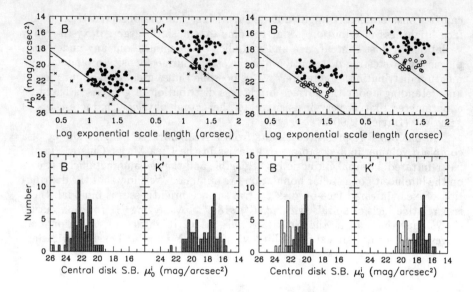

Figure 1. **Upper-left panels:** Plots of central disk surface brightness versus the log of the exponential disk scale length as measured in the B and K' bands. Galaxies are assumed to be transparent in both passbands ($C^{B,K'}=1$). Slanted lines indicate the completion limit for purely exponential disks. Crosses indicate galaxies fainter than the SMC. **Upper-right panels:** Complete sample only while the LSB family of galaxies is identified with open symbols. Here, $C^{K'}_{\rm HSB,LSB}=1$ while $C^{B}_{\rm HSB}=0.23$ to correct for internal extinction and $C^{B}_{\rm LSB}=1$. **Lower panels:** Histogram distributions of the central disk surface brightnesses corresponding to the upper panels.

3. Surface Brightness Distributions

From the calibrated optical and Near-Infrared images, luminosity profiles were extracted by averaging the flux density in concentric ellipses of equal ellipticity and position angle. Exponential disk fits were made by fitting straight lines to the quasi-linear part of the luminosity profiles, yielding disk scale lengths and central surface brightnesses. A possible upturn due to a bulge in the inner parts of the profiles was carefully excluded from the fit. The following discusssion of the photometric results will be restricted to the B- and K'-band data.

3.1. A Bimodal Distribution

The measured central surface brightnesses $\mu_0(\lambda)$ at a certain wavelength λ were corrected for inclination and internal extinction according to

$$\mu_0^i(\lambda) = \mu_0(\lambda) - 2.5\, C^\lambda \,\mathrm{Log}(b/a)$$

where C^λ accounts for internal extinction. If galaxies are transparent, $C^\lambda=1$ and $\mu_0(\lambda)$ will only be corrected for the path length through the inclined disk

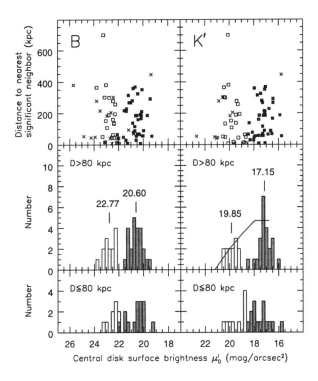

Figure 2. **Upper panels:** Disk central surface brightness in the B and K' passbands versus projected distance to nearest significant neighbor. Nearly all galaxies with intermediate surface brightness have a close companion. **Middle panels:** Surface brightness distributions for galaxies more isolated than 80 kpc from their nearest neighbor. The line in the K'-band panel roughly indicates the completion limit based on purely exponential disks. **Lower panels:** Surface brightness distributions for galaxies with close companions (in projection).

while $C^\lambda<1$ takes a correction for internal extinction into account. Values of μ_0^i are plotted against disk scale lengths in Figure 1. LSB galaxies are indicated with open symbols in the right panels in which a trend of surface brightness with total luminosity can be seen.

We assume that galaxies are transparent in the Near-Infrared and thus use $C^{K'}=1$. Doing so, we find an apparent lack of galaxies with $\mu_0^i(K')\approx 18.5$ mag/arcsec^2 (see K'-panels in Figure 1). This is a first hint to a possible bimodality in the Near-Infrared surface brightness distribution. We use the Near-Infrared data to define a low surface brightness galaxy in case its $\mu_0^i(K')>18.5$ mag/arcsec^2 and a high surface brightness galaxy if $\mu_0^i(K')<18.5$ mag/arcsec^2.

It should be pointed out that de Jong (1996) obtained K'-band images of 85 galaxies in his sample and derived a Near-Infrared surface brightness distribution without a hint of any bimodality. Unfortunately, his sample is dominated by HSB galaxies and the volume corrections for the LSB systems are quite uncertain.

The internal extinction in the B-band gives rise to much more uncertainty in any correction of $\mu_0(B)$ to face-on. Figure 1 shows the distribution of $\mu_0^i(B)$ assuming that either $C^B=1$ for all galaxies (left B-band panels) or $C^B=1$ only for LSB galaxies while $C^B=0.23$ for HSB galaxies (right B-band panels). The latter assumption is based on the notion that the overall fainter LSB galaxies in Ursa Major are in general not detected by IRAS, indicating a low dust content. Given this bimodal extinction correction in the B-band, the resulting bimodality of the B-band surface brightness distribution is largely artificial. It should be noted that the former assumption, $C^B=1$ for all galaxies, leads to a flat surface brightness distribution in the B-band.

3.2. A Near-Neighbor Effect

Investigating whether the LSB galaxies in our sample are more isolated than the HSB galaxies lead to an interesting result. In the upper panels of Figure 2 we plot the surface brightness of a galaxy versus the projected distance to its nearest significant neighbor; the nearest galaxy with an I-band luminosity of at least 10% of the galaxy considered. It turns out that galaxies of intermediate surface brightness, $\mu_0^i(K')\approx 18.5$ mag/arcsec^2, have near companions and are often involved in a tidal interaction. Considering the more isolated galaxies, the bimodality in the surface brightness distribution is striking (middle panels). The line in the middle K'-panel indicates how our completion limit maps onto the surface brightness distribution. Within our sample, the LSB galaxies are not more isolated than the HSB galaxies although the most isolated system in our sample is an LSB.

4. The Luminosity Functions.

This bimodality of HSB and LSB families is also reflected in the Luminosity Functions. Figure 3 shows a linear version of the Luminosity Functions in both the B and K$'$ bands including galaxies fainter than the completion limit. The filled part of the histogram corresponds to the HSB family of galaxies. The Luminosity Functions of the HSB galaxies clearly cut off well above the completion limit while the Luminosity Functions of the LSB galaxies keep rising toward the faint end until the completion limit is reached. In the K$'$-band, the HSB and LSB Luminosity Functions are more separated than in the B-band. This may result in a shallow dip in the Luminosity Functions around $M_{K'}^{b,i}=-21.5^m$ while the B-band Luminosity Function is quite flat.

5. Stellar Mass-to-Light Ratios

We derived extended rotation curves of 22 HI-rich, non-interacting and sufficiently inclined galaxies in the complete sample which allow us to investigate a possible relation between this Near-Infrared surface brightness bimodality and the mass-to-light ratios of the stellar populations and kinematics of the stellar disks. The HI rotation curves were decomposed using the K$'$ luminosity pro-

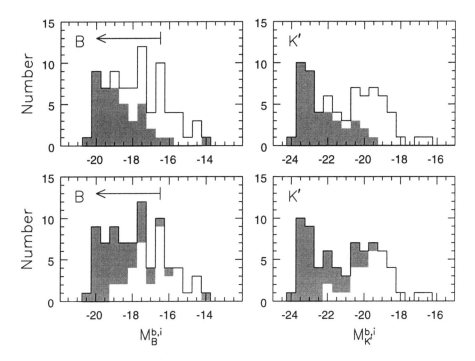

Figure 3. Luminosity Functions in the B and K′ passbands including all identified cluster members. The open part of the histograms correspond to the LSB family. The horizontal arrow in the B-band panels indicates the complete sample. The Luminosity Function of the HSB galaxies drops well above the completion limit. The sub-histograms are switched between the upper and lower panels.

files, the HI surface density profiles and an isothermal sphere model for the dark matter halo (Figure 5). Maximum-disk fits were made, avoiding a hollow halo core. The resulting stellar mass-to-light ratios in the K′-band are plotted versus B−K′ color in the upper panels of Figure 4.

First, let's consider the results for the HSB galaxies (filled symbols). In the K′-band, the scatter in the stellar mass-to-light ratios of the HSB galaxies is small (0.22) and the mean value of $<M_*/L_{K'}>_{HSB}=0.71$ (excluding NGC 3992) is in accordance with stellar population models (dashed lines) by Bruzual and Charlot (1993). Furthermore, for the HSB galaxies, there is a clear trend of the B-band mass-to-light ratios with B-K′ color as would be expected from stellar population models. From a stellar population point-of-view the maximum-disk hypothesis seems to make sense for HSB galaxies. Note that if HSB galaxies are systematically half-maximum-disk, the small K′-band scatter would remain but the average value would drop to $<M_*/L_{K'}>_{HSB}=0.18$ and the trend of (M_*/L_B) with B-K′ color would become too shallow.

The LSB galaxies on the other hand show a much larger scatter, have a significantly higher average mass-to-light ratio in the K′-band and do not

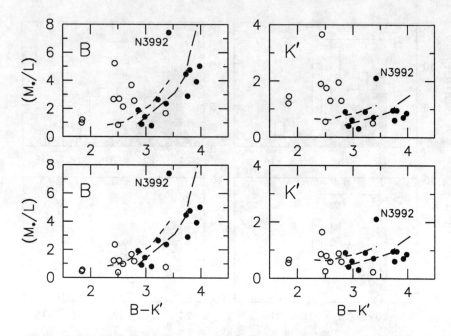

Figure 4. Stellar mass-to-light ratios in the B- and K'-band as a function of B-K' color for HSB (filled symbols) and LSB galaxies (open symbols). **Upper panels:** Mass-to-light ratios as derived from maximum-disk fits for both HSB and LSB galaxies. **Lower panels:** Same as upper panels for the HSB galaxies but the K'-band mass-to-light ratios of the LSB galaxies were scaled down such that $<M_*/L_{K'}>_{LSB} = <M_*/L_{K'}>_{HSB} = 0.71$. The dashed lines indicate expectations from the Bruzual and Charlot (1993) stellar population models with solar metallicities, a Salpeter IMF and ages of 2–17 Gyr. Long dashes: single burst, short dashes: exponential SFR.

show any trend of (M_*/L_B) with B-K' color. From this we conclude that we overestimate the mass-to-light ratios for LSB galaxies and, consequently, that it is unlikely that LSB galaxies in general are in a maximum-disk situation. This is also supported by the fact that the higher mass-to-light ratios for LSB galaxies can be understood easily from a geometrical point-of-view as explained by Zwaan et al (1995).

The lower panels of Figure 4 show the mass-to-light ratios in case the HSB galaxies have a maximum-disk (same as upper panels) but the K'-band mass-to-light ratios of the LSB galaxies are scaled down such that the average mass-to-light ratios of HSB and LSB galaxies are the same; $<M_*/L_{K'}>_{LSB} = <M_*/L_{K'}>_{HSB} = 0.71$. Although the scatter in $(M_*/L_{K'})$ of the LSBs is still somewhat larger than that of the HSB galaxies, the trend of (M_*/L_B) with B-K' color seems to be continued by the LSB galaxies in the lower left panel. The luminosities of both HSB and LSB galaxies were corrected for internal extinction in similar ways.

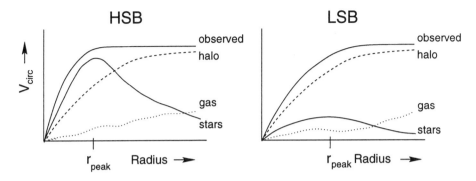

Figure 5. Schematic difference between the kinematics of an HSB and an LSB galaxy with a similar observed maximum rotational velocity.

6. A Dynamical Instability

Next, we investigate the relative dynamical importance of the dark matter halo, assuming equal stellar mass-to-light ratios in the K'-band for both HSB and LSB galaxies. Again, we decomposed the HI rotation curves but this time we assumed $(M_*/L_{K'})=0.6$ for all galaxies. This mass-to-light ratio is slightly less than the average in order not to violate the maximum-disk situation too badly for most systems. We calculated for each stellar disk the maximum rotational velocity V_{disk}^{max} which occurs at r_{peak} around 2.1 disk scale lengths and the rotational velocity $V_{halo}(r_{peak})$ induced by the dark matter halo, at the same radius where the rotation curve of the stellar disk peaks. If $V_{disk}^{max} > V_{halo}(r_{peak})$ then the stellar disk dominates the potential in the inner regions while the dark matter dominates if $V_{disk}^{max} < V_{halo}(r_{peak})$. The contributions by the gas component to the potential have been ignored. Figure 6 shows the measured ratios $V_{disk}^{max}/V_{halo}(r_{peak})$.

Again, we find a bimodal distribution along the lines of the HSB/LSB dichotomy; HSB disks are self-gravitating and close to a maximum-disk situation while the potentials of LSB galaxies are dominated by the dark matter halo at all radii. Furthermore, it seems that stellar disks avoid a situation in which the disk is marginally self-gravitating $V_{disk}^{max} \approx V_{halo}(r_{peak})$. This seems to hint at an instability, either in the dynamics of an established dissipationless stellar disk or during the formation process of a galaxy when it consisted mostly of dissipational gas. Testing the hypothesis of a dynamical instability in a marginally self-gravitating baryonic disk would require N-body (hydro-)dynamical simulations at a very high dynamic range.

7. Summary

Our findings and conclusions can be summarized as follows.

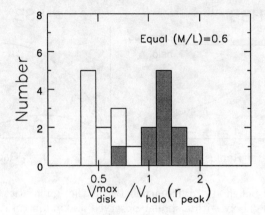

Figure 6. Ratio of the maximum rotational velocity induced by the stellar disk V_{disk}^{max} over the rotational velocity induced by the halo at r_{peak} where V_{disk}^{max} occurs. The filled histogram is for HSB galaxies, the open histogram for LSB galaxies. An equal stellar mass-to-light ratio in the K'-band of 0.6 is assumed for all galaxies. NGC 3992 is omitted.

- There is an avoidance of a domain ($\mu_0^i(K')\approx 18.5$ mag/arcsec^2) of Near-Infrared disk central surface brightness. For isolated galaxies, there is a distinct gap (factor 10 in density) between HSB and LSB families.

- The Luminosity Function of the HSB family is essentially truncated faintward of $M_B=-17^m$. The LSB family Luminosity Function is sharply increasing at our completion limit of $M_B=-16.5^m$. There might be a shallow dip in the K'-band Luminosity Function around $M_{K'}=-21.5^m$.

- Although HSB and LSB galaxies with similar (L,V_{max}) lie at the same position on the luminosity-line width relation, they may lie in very different domains in surface brightness-scale lengths plots.

- Maximum-disk decompositions of HSB rotation curves give reasonable Near-Infrared stellar mass-to-light ratios (average of 0.7 with rms scatter of only 0.2) and an obvious trend of B-band mass-to-light ratio with B-K' color. Maximum-disk decompositions of LSB galaxies give mass-to-light ratios with a higher average, a larger scatter and no trend with color. We infer that HSB disks are close to maximum-disk and self-gravitating inside ~2 scale lengths while LSB galaxies are well below maximum-disk and dark matter dominated at all radii.

- Assuming that both HSB and LSB galaxies have similar Near-Infrared mass-to-light ratios, we infer that galaxies avoid a situation in which their baryonic matter is marginally self-gravitating.

- The V^{max} that is relevant for the luminosity-line width relation is given by the mass distribution of the dark matter halo while the total luminosity of the dissipational matter is in fixed proportion to the halo mass. The exact distribution of the luminous matter within the dark matter halo is irrelevant for the luminosity-line width relation. In high angular momentum LSB systems, the dissipational matter reaches its rotational equilibrium at large radii and the mass distribution of dark matter remains dominant. In low angular momentum HSB systems, the dissipational matter collapses enough to become self-gravitating and might even give rise to a rotation curve which is declining in the inner parts before it becomes flat in the dark matter supported outer regions.

8. Future work

One might argue that our statistics are not overwhelming or that our findings are particular to the Ursa Major cluster. To tackle both issues we are currently analyzing the results of a blind VLA survey in the Perseus-Pisces ridge which yields an HI selected and volume limited sample of galaxies of both high and low surface brightness. The volume surveyed is similar to the entire Ursa Major volume. Near-Infrared surface photometry is planned for the fall of 1999 which should give us the Near-Infrared surface brightness distribution of this HI selected sample. In this way we hope to avoid any optical selection effects that might be strongly biased in favor of HSB systems. The combined data sets will yield a sample that is twice as large as the current Ursa Major sample itself.

Of course, we eagerly await the results of any other ongoing project aimed at measuring the *Near-Infrared* surface brightness distribution.

Acknowledgments. Mike Pierce, Jia-Sheng Huang and Richard Wainscoat participated in the collection of the data. The Westerbork Synthesis Radio Telescope is operated by the Netherlands Foundation for Research in Astronomy, with financial support by the Netherlands Organization for Scientific Research (NWO). This research has been supported by NATO Collaborative Research Grant 940271 and grants from the US National Science Foundation.

References

Bothun, G.D., Impey, C.D. and McGaugh, S.S., 1997, PASP, 109, 745
Bruzual, G.A. and Charlot, S., 1993, ApJ, 405, 538
Davies, J.I, 1990, MNRAS, 244, 8
de Jong, R.S., 1996, A&A, 313, 45
McGaugh, S.S., Bothun, G.D. and Schombert, J.M., 1995, AJ, 110, 573
McGaugh, S.S., 1996, MNRAS, 280, 337
Tully, R.B., Verheijen, M.A.W., Pierce, M.J., Huang, J.-S. and Wainscoat, R.J., 1996, AJ, 112,2471
Tully, R.B. and Verheijen, M.A.W., 1997, ApJ, 484, 145
Verheijen, M.A.W., 1997, PhD thesis, University of Groningen
Zwaan, M.A., van der Hulst, J.M., de Blok, W.J.G. and McGaugh, S.S., 1995, MNRAS, 237, L35

Searching for LSB - III

No sorry, it's here in my pocket.

Multi-Wavelength Surveys for Galaxies Hidden by the Milky Way

Renée C. Kraan-Korteweg

Departamento de Astronomía, Universidad de Guanajuato, Guanajuato GTO 36000, Mexico

Abstract. The systematic mapping of obscured and optically invisible galaxies behind the Milky Way through complementary surveys are important in arriving at the whole-sky distribution of complete galaxy samples and therewith for our understanding of the dynamics in the local Universe. In this paper, a status report is given of the various deep optical, near infrared (NIR), and systematic blind H I-surveys in the Zone of Avoidance, including a discussion on the limitations and selection effects inherent to the different multi-wavelength surveys and first results.

1. Introduction

Due to the foreground extinction of the Milky Way, galaxies become increasingly fainter, smaller and are of lower surface brightness as they approach the Galactic Equator. Although most of them are not intrinsically of low surface brightness, "whole-sky" mapping of galaxies is required (a) in explaining the origin of the peculiar velocity of the Local Group (LG) and the dipole in the Cosmic Microwave Background, (b) for our understanding of velocity flow fields such as the Great Attractor in the Zone of Avoidance (ZOA) with a predicted mass excess of a few times $10^{16} M_\odot$ at $(\ell, b, v) \sim (320°, 0°, 4500\,\mathrm{km\,s^{-1}}$, Kolatt et al. 1995), and (c) other suspected connections of nearby superclusters and voids behind the Milky Way.

This not only concerns large-scale structures. Nearby massive galaxies behind the obscuration layer of the Milky Way could significantly change our understanding of the internal dynamics and mass derivations of the LG.

Dedicated searches for galaxies in about 25% of the optically obscured extragalactic sky so far revealed a number of important features such as:

– the nearby bright spiral galaxy Dwingeloo 1, a neighbor to the LG (Kraan-Korteweg et al. 1994)

– the Puppis cluster at $(\ell, b, v) \sim (245°, 0°, 1500\,\mathrm{km\,s^{-1}})$ which may contribute at least $30\,\mathrm{km\,s^{-1}}$ to the motion of the LG *perpendicular* to the Supergalactic Plane (Lahav et al. 1993)

– the massive Coma-like cluster A3627 at $(\ell, b, v) \sim (325°, -7°, 4800\,\mathrm{km\,s^{-1}})$ which seems to constitute the previously unrecognized but predicted density peak at the bottom of the potential well of the Great Attractor (Kraan-Korteweg et al. 1996)

- the 3C129 cluster at $(\ell, b, v) \sim (160°, 0°, 5500\,{\rm km\,s^{-1}})$ connecting Perseus-Pisces and A569 across the Galactic Plane (Chamaraux et al. 1990, Pantoja et al. 1997)
- and the Ophiuchus (super-)cluster at $(\ell, b, v) \sim (0°, 8°, 8500\,{\rm km\,s^{-1}})$ behind the Galactic Center (Wakamatsu et al. 1994).

In the following, I will review the current status of deep optical searches behind the Milky Way, as well as the possibilities and results given with the recent near infrared surveys and blind H I-surveys.

2. Optical Surveys

Systematic optical galaxy catalogs are generally limited to the largest galaxies (typically with diameters D $\gtrsim$ 1', e.g., Lauberts 1982). These catalogs become, however, increasingly incomplete as the dust thickens, creating a "Zone of Avoidance" in the distribution of galaxies of roughly 25% of the sky. Systematic deeper searches for partially obscured galaxies – down to fainter magnitudes and smaller dimensions compared to existing catalogs – were performed with the aim to reduce this ZOA. These surveys are not biased with respect to any particular morphological type.

The various survey regions are displayed in Fig. 1 (cf. Woudt 1998, for an extensive overview). Further details and results on the uncovered galaxy distributions can be found in A: Aquila and Sagittarius (Roman et al. 1996), B: Sagittarius/Galactic (Roman & Saito 1997), C: Ophiuchus Supercluster (Wakamatsu et al. 1994), D: Galactic Center extension (Kraan-Korteweg, in progress), E: Crux and GA Region (Woudt & Kraan-Korteweg 1999, in prep.), F: Hydra/Antlia Supercluster (Kraan-Korteweg 1999, in prep.), G: Hydra to Puppis Region (Salem & Kraan-Korteweg, in progress) H & I : Puppis (Saito et al. 1990, 1991), I: Perseus-Pisces Supercluster (Pantoja 1997), J : northern crossing GP/SGP (Hau et al. 1996), K: northern ZOA (Seeberger et al. 1994).

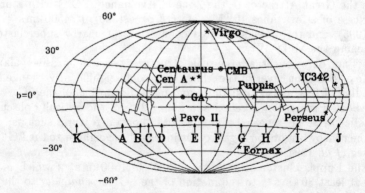

Figure 1. An overview of the different optical galaxy surveys in the ZOA centered on $\ell = 300°$. The labels identifying the search areas are explained in the text. Galaxy cluster positions (stars) and the CMB apex as well as the core of the GA are marked.

Meanwhile, as Fig. 1 illustrates, nearly the whole ZOA has been surveyed systematically. With the mapping of over 50000 previously unknown galaxies a considerable reduction of the ZOA was achieved. Analysing the galaxy density as a function of the galaxy size, magnitude and/or morphology in combination with the foreground extinction has led to the identification of various important large-scale structures. However, redshift follow-ups of well-defined samples are important in tracing the large-scale structures in detail. Such follow-up surveys have already revealed a number of dynamically important structures in the Zone of Avoidance (cf. Introduction).

Although the various optical surveys are based on different plate material and the criteria for inclusion in the respective surveys were not all identical, all surveys reveal the same dependence on extinction: for extinctions in the blue of $A_B \gtrsim 4 - 5^m$, the ZOA remains fully opaque (cf. top panel of Fig. 3), leaving a strip of about $\pm 5°$ devoid of galaxies.

3. NIR-Surveys

The extinction effects decrease with increasing wavelengths. In the NIR passbands I_c, J and K_s, the extinction compared to the blue is $A_{I_c} = 45\%$, $A_J = 21\%$, and $A_{K_s} = 9\%$. Moreover, NIR surveys are sensitive to early-type galaxies – tracers of massive groups and clusters missed in IRAS and H I surveys – and have little confusion with Galactic objects. Here, the recent near infrared surveys, 2MASS (Skrutskie et al. 1997) and DENIS (Epchtein 1997) might provide new insight at low Galactic latitudes.

In unobscured regions, the density of galaxies per square degree for the completion limit of $B_J \leq 19\overset{m}{.}0$ is 110 (Gardner et al. 1996). However, the number counts in the blue decrease rapidly with increasing obscuration as $N(A_B) \simeq 110 \times \text{dex}(0.6\,[-A_B])\,\text{deg}^{-2}$. In the NIR passbands I_c, J and K_s of the DENIS survey, the counts for the respective completeness limits of $I_{\text{lim}} = 16\overset{m}{.}0$, $J_{\text{lim}} = 14\overset{m}{.}0$, $K_{\text{lim}} = 12\overset{m}{.}2$ are considerably lower (30, 11, and 2, Mamon et al. 1997) but – as illustrated in Fig. 2 – the decrease in number counts as a function of "optical extinction" is considerably slower. The new cooling system for the focal instrument of DENIS installed in 1997 led to an increase in the K_s band counts of a factor of two for the completeness limit fainter by $\sim 0\overset{m}{.}5$.

Fig. 2 shows that the NIR becomes notably more efficient where the Milky Way becomes opaque in the optical ($A_B \geq 4 - 5^m$, cf. previous section). At an extinction of $A_B \simeq 3 - 4^m$, J becomes superior to I_c, while at $A_B \simeq 10^m$, K_s becomes superior to J. The cooled camera system will make the K_s passband competitive with J starting at $A_B \simeq 7^m$. These are very rough predictions and do not take into account any dependence on morphological type, surface brightness, orientation and crowding, which may lower the counts of actually detectable galaxies counts.

3.1. First Results from DENIS

To compare these predictions with real data, Schröder et al. (1997) and Kraan-Korteweg et al. (1998a) examined the efficiency of uncovering galaxies at high extinctions with DENIS images. The results are promising.

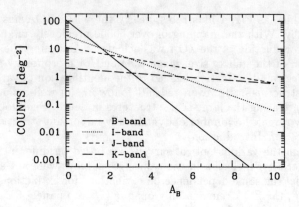

Figure 2. Predicted galaxy counts in B, I_c, J and K_s as a function of absorption in B, for highly complete and reliable DENIS galaxy samples and a $B_J \leq 19^m$ optical sample.

They found that down to intermediate latitudes and extinction ($|b| \gtrsim 5°$, $A_B \lesssim 4-5^m$), optical surveys remain superior for identifying galaxies. However, the NIR luminosities and colors together with extinction data from the NIR colors will prove invaluable in analysing the optical survey data and their distribution in redshift space, and in the final merging of these data with existing sky surveys. Despite the high extinction and the star crowding at these latitudes, I_c, J and K_s photometry from the survey data can be successfully performed at these low latitudes and led, for instance, to the preliminary I_c^o, J^o and K_s^o galaxy luminosity functions in A3627.

At low latitudes and high extinction ($|b| < 5°$ and $A_B \gtrsim 4-5^m$), the search for 'invisible' obscured galaxies on existing DENIS-images implicate that NIR-surveys can trace galaxies down to about $|b| \simeq 1°\!.5$. The J band was found to be optimal for identifying galaxies up to $A_B \simeq 7^m$, although this might change in favour of K_s with the new cooling system. NIR surveys can hence further reduce the width of the ZOA. Furthermore, this is the only tool that permits the mapping of early-type galaxies — tracers of density peaks — at high extinction.

The analysis of DENIS images behind the ZOA is being pursued in a more systematic way. Whether this will be performed by visual examination or whether galaxies can be successfully extracted using classical algorithms or artificial neural networks or a combination of both requires further exploration.

4. Blind H I surveys

In the regions of the highest obscuration and infrared confusion, the Galaxy is fully transparent to the 21-cm line radiation of neutral hydrogen. H I-rich galaxies can readily be found at lowest latitudes through detection of their redshifted 21-cm emission. Only low-velocity extragalactic sources (blue- and redshifted) within the strong Galactic H I emission will be missed, and – because of baseline ripple – galaxies close to radio continuum sources. Until recently, radio receivers were not sensitive and efficient enough to attempt large systematic surveys of the ZOA. In a pilot survey with the late 300-ft telescope of Green Bank, Kerr & Henning (1987) surveyed 1.5% of the ZOA and detected 16 new spiral galaxies.

4.1. The Northern Zone of Avoidance

Using the Dwingeloo 25m radio telescope, the whole northern Galactic ZOA ($|b| \lesssim 5°25$) is being surveyed in the 21cm line for galaxies out to 4000 km s^{-1}(see also Rivers et al., these proceedings). A shallow search (rms = 175 mJy) has been completed yielding five objects (Henning et al. 1998). This fast search for nearby massive galaxies uncovered no major unknown Andromeda-like galaxy. The most exciting discovery is the barred spiral galaxy Dwingeloo 1 (Kraan-Korteweg et al. 1994), a new neighbour of the Local Group with one third of the Galaxy's mass.

The deeper survey (rms=40 mJy) is 60% complete. 36 galaxies were detected of which 23 were previously unknown, the most surprising being the detection of a number of dwarfs at very low redshifts. They lie close to the Sdm galaxy NGC 6946 at v=48 km s^{-1}, suggesting a previously unrecognized nearby group or cloud of galaxies.

4.2. The Southern Zone of Avoidance

In March 1997, a systematic blind H I survey began in the the southern Milky Way ($|b| \leq 5°$) with the multibeam (MB) receiver (13 beams in the focal plane array) at the 64 m Parkes telescope. The survey covers the velocity range $-1200 \lesssim v \lesssim 12700$ km s^{-1} and will have a sensitivity of rms=6 mJy after Hanning smoothing.

So far, a shallow survey based on 2 out of the foreseen 25 driftscan passages has been analysed (cf. Henning et al. these proceedings, and Kraan-Korteweg et al. 1998b). 107 galaxies were catalogued with peak H I-flux densities of $\gtrsim 80$ Jy km s^{-1} (rms= 15 mJy after Hanning smoothing). Though galaxies up to 6500 km s^{-1} were identified, most of the galaxies (80%) are quite local (v< 3500 km s^{-1}) due to the (yet) low sensitivity.

Most detections are due to normal two-horned spiral galaxies. However, ATCA follow-up observations of three very extended (20' to $\gtrsim 1°$), nearby (v < 1500 km s^{-1}) sources revealed them to be interesting galaxies/complexes, with unprecedented low H I column densities (cf. Staveley-Smith et al. 1998).

As in the northern H I-survey, no Andromeda or other H I-rich Circinus galaxy has been found lurking undetected behind the extinction layer of the southern Milky Way. Both H I-surveys have, however, clearly proven the power of tracing spiral and H I-rich dwarf galaxies through the deepest extinction layer of the Milky Way (cf. Fig. 3 and 4, as well as Fig. 2 in Rivers et al., these proceedings).

5. Conclusions

Considerable progress has been made in mapping the galaxy distribution with various multi-wavelength approaches. The continuing surveys will lead to a much more complete picture of the galaxy distribution in the "former" ZOA.

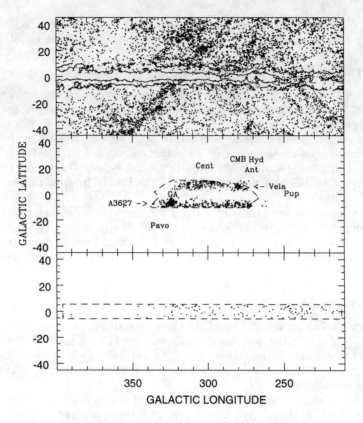

Figure 3. Galaxies with v<10000 km s^{-1}. Top panel: literature values (LEDA), superimposed are extinction levels $A_B \sim 1\overset{m}{.}5$ and 5^m; middle panel: follow-up redshifts (ESO, SAAO and Parkes) from deep optical ZOA survey with locations of clusters and dynamically important structures; bottom panel: redshifts from shallow MB-ZOA in H I with the Parkes radio telescope.

How complementary the various multi wavelength approaches are is illustrated in Fig. 3. The top panel shows the distribution of all known galaxies with velocities v $\leq$ 10000 km s^{-1} centered on the southern Milky Way. Although this constitutes an uncontrolled sample, it traces the main structures in the nearby Universe in a representative way. Note the near full lack of galaxy data for extinction levels $A_B \sim 1\overset{m}{.}5$ (outer contour).

The middle panel results from the follow-up observations of the optical galaxy search by Kraan-Korteweg and collaborators. Various new overdensities could be unveiled at intermediate extinction levels and low latitudes ($1\overset{m}{.}5 \lesssim A_B \lesssim 5^m, 5° \lesssim |b| \lesssim 10°$), but the innermost part of our Galaxy remains obscured ($A_B \gtrsim 4 - 5^m, |b| \lesssim 5°$). Here, the blind H I data finally provide the missing link for large-scale structure studies as indicated with the results from the detections in the shallow survey of the Parkes MB survey (lower panel).

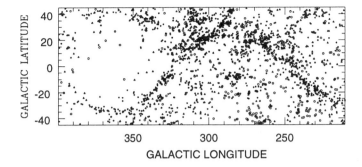

Figure 4. Redshift slices from data in Fig. 3 for the velocity range 500<v<3500 km s^{-1}. The open circles mark the nearest (500-1500), the triangles the medium (1500-2500) and the filled dots the most distant (2500-3500) slice.

In Fig. 4, the data from the various surveys displayed in Fig. 3 are combined for the redshift range 500<v<3500 km s^{-1}. The upper velocity limit reflects the depth achieved with the sensitivity of the shallow MB H I-survey. For the first time, structures are visible all the way across the ZOA. Note the continuity of the thin filamentary sine-wave-like structure that dominates the whole southern sky, and the prominence of the Local Void. With the full sensitivity of the MB-survey, we will be able to fill in the large-scale structures out to 10000 km s^{-1}.

The possibilities given for ZOA-research based on the currently ongoing NIR are very promising – and complementary in the sense that they finally allow the uncovering of early-type galaxies to low Galactic latitudes ($|b| \gtrsim 1 - 1°\!.5$). This is not the only addition NIR surveys provide. Schröder et al. (1997) and Kraan-Korteweg et al. (1998a) have shown that a fair fraction of heavily obscured spiral galaxies detected in blind H I surveys can be reidentified on DENIS images (cf., Fig. 5 in Kraan-Korteweg et al. 1998a). The combination of H I data with NIR data allow the study of the peculiar velocity field via the NIR Tully–Fisher relation "in the ZOA" compared to earlier interpolations of data adjacent to the ZOA (Schröder et al., in progress).

A difficult task still awaiting us in the future is the merging of ZOA data with catalogs outside the ZOA. This will have to be done with care to obtain 'unbiased' whole-sky surveys.

Acknowledgments. The collaborations with my colleagues in the various multi-wavelength surveys — P.A. Woudt with the optical surveys, A. Schröder and G.A. Mamon in the exploration of the DENIS survey, W.B. Burton, P.A. Henning, O. Lahav and A. Rivers in the northern ZOA H I-survey (DOGS) and the HIPASS ZOA team members R.D. Ekers, A.J. Green, R.F. Haynes, P.A. Henning, S. Juraszek, M. J. Kesteven, B. Koribalski, R.M. Price, E. Sadler and L. Staveley-Smith in the southern ZOA survey — are greatly appreciated.

References

Chamaraux P., Cayatte V., Balkowski C., Fontanelli P. 1990, A&A 229, 340
Epchtein N. 1997, in *The Impact of Large Scale Near-Infrared Surveys* p. 15, eds. F. Garzon *et al.*, Kluwer, Dordrecht
Gardner J.P., Sharples R.M., Carrasco B.E., Frenk C.S. 1996, MNRAS 282, L1
Hau G.K.T., Ferguson H.C., Lahav O. *et al.* 1996, MNRAS 277, 125
Henning P.A., Kraan-Korteweg R.C., Rivers A.J. *et al.* 1998, AJ 115, 584
Kerr F.J., Henning P.A. 1987, ApJ 320, L99
Kolatt T., Dekel A., Lahav O. 1995, MNRAS 275, 797
Kraan-Korteweg R.C., Loan A.J., Burton W.B. *et al.* 1994, Nature 372, 77
Kraan-Korteweg R.C., Woudt P.A., Cayatte V. *et al.* 1996, Nature 379, 519
Kraan-Korteweg R.C., Schröder A., Mamon G., Ruphy S. 1998a, in *The Impact of Near-Infrared Surveys on Galactic and Extragalactic Astronomy*, p.205, ed. N. Epchtein, Kluwer, Dordrecht
Kraan-Korteweg R.C., Koribalski B., Juraszek S. 1998b, in *Looking Deep in the Southern Sky*, eds. R. Morganti, W. Couch, Springer, in press
Lahav O., Yamada T., Scharf C.A., Kraan-Korteweg, R.C. 1993, MNRAS 262, 711
Lauberts, A. 1982, The ESO/Uppsala Survey of the ESO (B) Atlas, ESO, Garching
Mamon, G.A., Tricottet, M., Bonin, W., Banchet, V., 1997, in *Extragalactic Astronomy in the Infrared*, p. 369, eds. G. A. Mamon *et al.*, Frontières, Paris
Pantoja C.A., Altschuler D.R., Giovanardi C., Giovanelli R. 1997, AJ 113, 905
Roman A.T., Nakanishi K., Tomita A., Saito M. 1996, PASJ 48, 679
Roman A.T., Nakanishi K., Saito M. 1998, PASJ 50, 37
Saito M., Ohtani A., Asomuna A. *et al.* 1990, PASJ 42, 603
Saito M., Ohtani A., Baba A. *et al.* 1991, PASJ 43, 449
Schröder A., Kraan-Korteweg R.C., Mamon G.A. Ruphy S. 1997, in *Extragalactic Astronomy in the Infrared*, p. 381, eds. G. A. Mamon *et al.*, Frontières, Paris
Seeberger R., Saurer W., Weinberger R. *et al.* 1994, in *Unveiling Large-Scale Structures Behind the Milky Way*, p. 81, eds. C. Balkowski, R.C. Kraan-Korteweg, ASP Conf. Ser. 67
Skrutskie, M.F., *et al.* 1997, in *The Impact of Large Scale Near-Infrared Surveys*, p. 25, eds. F. Garzon *et al.*, Kluwer, Dordrecht
Staveley-Smith L., Juraszek S., Koribalski B.S. *et al.* 1998, AJ, in press
Tully R.B., Fisher J.R. 1977, A&A 54, 66
Wakamatsu K., Hasegawa T., Karoji H. *et al.* 1994, in *Unveiling Large-Scale Structures Behind the Milky Way*, p. 131, eds. C. Balkowski, R.C. Kraan-Korteweg, ASP Conf. Ser. 67
Woudt P.A. 1998, Ph.D. thesis, Univ. of Cape Town.

A Survey for Low Surface Brightness Dwarf Galaxies Around M31

Taft E. Armandroff
Kitt Peak National Observatory, National Optical Astronomy Observatories,[1] P.O. Box 26732, Tucson, AZ 85726

James E. Davies[2]
Department of Astronomy, University of Wisconsin, Madison, WI 53706

George H. Jacoby
Kitt Peak National Observatory, National Optical Astronomy Observatories,[1] P.O. Box 26732, Tucson, AZ 85726

Abstract. By applying a digital filtering technique to 1550 deg^2 of the POSS-II in the vicinity of M31, we found two previously unidentified very low surface brightness dwarf galaxies which we designate And V and VI. Follow-up imaging with the KPNO 4-m telescope resolved these into stars easily. The V- and I- band images of And V indicate a distance similar to that of M31, and $<$[Fe/H]$> \sim -1.5$. All evidence strongly supports its classification as a dwarf spheroidal companion to M31. Data for And VI are being analyzed, but preliminary indications support a similar conclusion. Our search for more dwarfs, including follow-up observations of numerous candidates found via digital filtering, is incomplete; thus, further identifications may be forthcoming.

1. Introduction

In order to understand many aspects of galaxy formation and galaxy evolution, a complete and unbiased census of the members of the Local Group is necessary. Some examples of the problems that can be addressed with such knowledge include: 1) Accurately defining the faint end of the galaxy luminosity function for the Local Group, allowing comparison with that in rich clusters of galaxies, for example. 2) Identifying additional dynamical probes within the Local Group in order to facilitate deciphering its dynamical history and total mass.

A survey for low surface brightness dwarf galaxies in the direction of M31 yields additional benefits. With a complete census of M31's dwarf companions

---

[1] The National Optical Astronomy Observatories are operated by the Association of Universities for Research in Astronomy, Inc., under cooperative agreement with the National Science Foundation.

[2] Currently at Dept. of Physics and Astronomy, Johns Hopkins University, Baltimore, MD 21218.

and knowledge of their properties, we can compare M31's satellite system with that of the Galaxy. Any differences or similarities between the two systems yield information on the effects of environment on the formation and evolution of dwarf galaxies. In addition, the properties of M31's dwarf companions tell us how much we can generalize from the Galaxy's dwarf satellite system.

The study of M31's dwarf spheroidal companions began with van den Bergh's (1972a, 1974) search of ∼700 square degrees around M31 using IIIaJ photographic plates taken with the Palomar 48-inch Schmidt telescope. By visually examining the plates, he identified three dwarf spheroidal candidates (And I, II & III). By taking deeper plates with the Palomar 5-m telescope, van den Bergh (1972b, 1974) found that And I, II & III resolve into stars at approximately the same magnitude as M31's companion NGC 185, strengthening the association of these galaxies with M31. Recent color–magnitude diagrams for And I, II & III from large terrestrial telescopes (Mould & Kristian 1990; Armandroff et al. 1993; Konig et al. 1993) and Hubble Space Telescope (Da Costa et al. 1996) have yielded distances that place And I, II & III in the outer halo of M31. These deep color–magnitude diagrams have also resulted in mean metallicity, metallicity dispersion, and age information. Surface photometry and structural parameters are available for And I, II & III from Caldwell et al. (1992). All available data suggest that And I, II & III closely resemble the Galactic dwarf spheroidals.

A number of arguments suggest that a new search of M31's environs for low surface brightness dwarfs may be worthwhile. First, the absolute magnitudes of And I, II & III do not extend as faint as the Galactic dwarf spheroidals (see Figure 2 of Armandroff 1994), suggesting that perhaps fainter dwarfs exist below the detection limit. Second, the area surveyed by van den Bergh (1972a, 1974) does not reach the projected distance from the center of M31 that one would expect based on the most distant Galactic dwarf companions (see Figure 6 of Armandroff 1994). Finally, why does M31 have three known dwarf spheroidal companions when the Galaxy has nine? Is this the result of incompleteness in the case of M31 or some intrinsic difference between the two systems?

Consequently, we have undertaken a new search for low surface brightness dwarf galaxies around M31. Our search uses digital filtering techniques applied to the Second Palomar Sky Survey (POSS-II). Our search is complementary to the nearby dwarf galaxy survey by Irwin (1994) that used star counts over much of the sky and revealed the Sextans dwarf spheroidal. Because dwarf galaxies at the M31 distance do not resolve into stars on survey plates, our search is not based on star counts. Since M31 is a northern object, our search is also complementary to that of Whiting et al. (1997) based on visual examination of the southern sky survey plates that called attention to the Antlia dwarf.

2. Search Strategy

The POSS-II (Reid et al. 1991) has higher resolution and extends substantially deeper than its first generation counterpart. The availability of the POSS-II in digital form (Lasker & Postman 1993) enables the use of digital processing techniques and allows full areal coverage around M31. Our survey program employs the POSS-II and seeks to: 1) search an area around M31 that is commensurate with the radial distribution of dwarf spheroidals around the Galaxy; 2) use digital processing techniques to enhance the detection of the faintest, lowest surface brightness dwarf spheroidals.

Our search methodology employs a matched filter and is described fully in Armandroff, Davies & Jacoby (1998). Briefly, the procedure includes: dividing each plate into overlapping 1 square degree regions; fitting a surface to remove the background; removing stars by clipping values $> 0.75\sigma$ above the median; applying a spatial median filter (77 × 77 arcsec; based on And II & III); visual examination of the processed and original images; selection of candidates that resemble And II & III on these images. Obvious bright-star ghosts and "optical cirrus" clouds are avoided in the selection process. We check the coordinates against SIMBAD and NED to eliminate known objects. To date, we have applied our detection procedure to 1550 square degrees of the POSS-II around M31.

As the next step in our survey, all the remaining candidates are imaged with the KPNO 0.9-m telescope. These CCD images eliminate any candidates that are plate-based false detections (e.g., remaining bright-star ghosts, unrecognized emulsion problems) and most astronomical misidentifications (e.g., distant galaxy clusters, background low surface brightness spirals, "optical cirrus" clouds). If a candidate shows incipient resolution into stars on the 0.9-m CCD images, then it is probably a nearby dwarf galaxy, so imaging with a 4-m-class telescope is undertaken. At the time that this paper was written (October 1998), we had not yet surveyed all the POSS-II plates in our desired search area, nor had we completed CCD imaging of our list of candidates from the POSS-II.

3. Andromeda V

Thus far, two candidates found by our survey have been resolved into stars (see Figure 1 for POSS-II images of these candidates). The first of these, called Andromeda V, is discussed extensively in Armandroff et al. (1998). We summarize our most important And V results below.

And V is located at the following coordinates: $\alpha = 1^h\ 10^m\ 17.1^s$, $\delta = +47°\ 37'\ 41''$ (J2000.0). And V was observed with the KPNO 4-m telescope and prime-focus CCD imager in the V and I filters, plus in Hα narrow-band and R. In the broad-band filters, And V resolves nicely into stars and exhibits a smooth stellar distribution (Figure 2). In these images, And V resembles the other M31 dwarf spheroidals. And V does not exhibit the features of classical dwarf irregulars, such as obvious regions of star formation or substantial asymmetries in its stellar distribution. In the And V continuum-subtracted Hα image, no diffuse Hα emission or H II regions are detected. The lack of Hα emission in And V reinforces the conclusion, based on And V's appearance on the broad-band images, that it is a dwarf spheroidal galaxy rather than a dwarf irregular.

And V is not detected in any of the IRAS far-infrared bands either. Because far-infrared emission traces warm dust, and because some Local Group dwarf irregular galaxies are detected by IRAS, And V's lack of far-infrared emission serves as additional, weaker evidence that it is a dwarf spheroidal. No information is currently available about the H I content of And V via 21 cm observations. Either an H I detection or a strict upper limit would be valuable.

We have measured And V's apparent central surface brightness via large-aperture photometry in its core: 25.7 mag/arcsec^2 in V. And V has a fainter

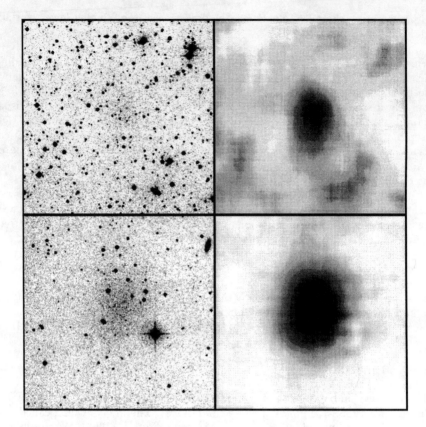

Figure 1. The left panels show images of And V & VI from the digitized POSS-II; the right panels show the results of applying our digital enhancement procedure. Each panel is 8.5 arcmin on a side.

apparent central surface brightness than And I, II & III (24.9, 24.8, and 25.3 mag/arcsec^2 in V, respectively; Caldwell et al. 1992). And V probably eluded detection until now due to its very dim apparent surface brightness.

We used our V and I images and photometric standard observations to construct a color–magnitude diagram for And V stars, in order to reveal And V's distance and stellar populations characteristics. Color–magnitude diagrams for the parts of the images dominated by And V stars reveal a red giant branch, which is absent in the outer regions of the images (see Figure 3). The tip of the red giant branch is well defined in the color–magnitude diagram and in the luminosity function. A distance has been derived for And V based on the I magnitude of the tip of the red giant branch (Da Costa & Armandroff 1990, Lee et al. 1993). On the distance scale of Lee et al. (1990), the resulting And V distance is 810 ± 45 kpc. How does this And V distance compare with that of M31? The most directly comparable distance determinations for M31, based on either red giant branch tip stars or RR Lyraes in the M31 halo or horizontal

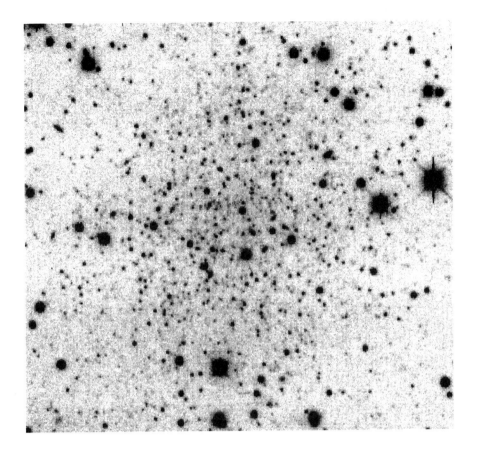

Figure 2. A V-band image of And V (three 900-second exposures) taken with the KPNO 4-m telescope. North is at the top, and east is to the right. The image covers 3×3 arcmin.

branch stars in M31 globular clusters, are 760 ± 45 kpc or 850 ± 20 kpc, both on the same scale (see Da Costa et al. 1996). Our distance for And V of 810 ± 45 kpc implies that And V is located at the same distance along the line of sight as M31 to within the uncertainties. And V's projected distance from the center of M31 is 112 kpc; And I, II & III have projected M31-centric distances of 46, 144 and 69 kpc, respectively. The above line-of-sight and projected distances strongly suggest that And V is indeed associated with M31.

In order to investigate the stellar populations in And V, we have compared its color–magnitude diagram with fiducials representing the red giant branches of Galactic globular clusters that span a range of metal abundance (Da Costa & Armandroff 1990; see Figure 3). Based on the position of the And V giant branch relative to the globular cluster fiducials, the mean metal abundance of And V is approximately −1.5. This metallicity is normal for a dwarf spheroidal (e.g., see Figure 9 of Armandroff et al. 1993). No bright blue stars are present

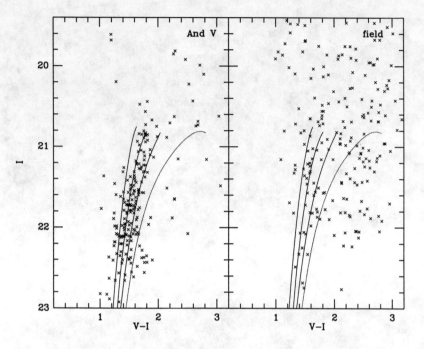

Figure 3. Color–magnitude diagrams. The left panel shows stars within a radius of 71 arcsec of the center of And V, where And V members outnumber field contamination. The right panel displays stars greater than 168 arcsec from the center of And V, where the contribution from And V stars is negligible. The right panel represents 5.5 times more area on the CCD than the left panel. Red giant branch fiducials for four Galactic globular clusters that span a range of metal abundance (Da Costa & Armandroff 1990), shifted to the distance modulus and reddening of And V, have been overplotted. From left to right, the red giant branch fiducials are M15 ([Fe/H] = −2.17), M2 (−1.58), NGC 1851 (−1.16), and 47 Tuc (−0.71).

in the And V color–magnitude diagram. Interpreting via isochrones, this lack of blue stars rules out any stars younger than 200 Myr in And V and is further evidence that And V is a dwarf spheroidal and not a dwarf irregular.

From the luminosities and numbers of upper asymptotic giant branch stars in a metal-poor stellar system, one can infer the age and strength of its intermediate age component (Renzini & Buzzoni 1986). Using the And V field-subtracted I luminosity function, there is no evidence for upper asymptotic giant branch stars that are more luminous than and redward of the red giant branch tip. Therefore, And V does not have a prominent intermediate age population; in this sense, it is similar to And I & III.

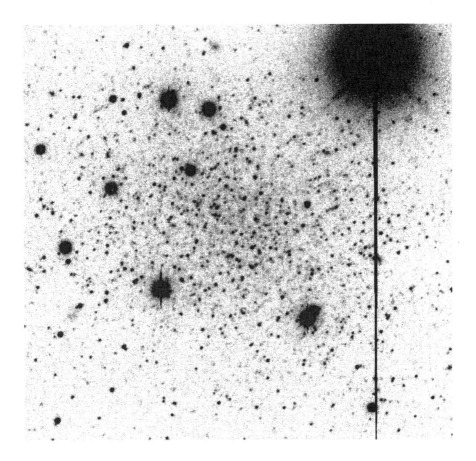

Figure 4. An R-band image of And VI (three 300-second exposures) taken with the KPNO 4-m telescope. North is to the left, and east is at the bottom. The image covers 3.9×3.7 arcmin.

4. Andromeda VI

Using our search methodology, we found a second candidate dwarf spheroidal galaxy, designated Andromeda VI, at the following celestial coordinates: $\alpha = 23^{\text{h}} 51^{\text{m}} 46.9^{\text{s}}$, $\delta = +24° 34' 45''$ (J2000.0). And VI was imaged with the KPNO 4-m telescope prime-focus CCD on 1998 January 23 in V for 300 seconds. And VI resolved nicely into stars in this short V image, suggesting that it is indeed a nearby dwarf galaxy. Several months after we had confirmed to ourselves that And VI is a bona fide nearby dwarf galaxy, we learned that Karachentsev & Karachentseva were reporting the discovery of this same galaxy, calling it the Pegasus Dwarf (see Brinks & Grebel 1998).

On 1998 July 15, And VI was imaged more deeply with the KPNO 4-m telescope through the R and Hα filters. The R image of And VI, displayed in

Figure 4, exhibits a smooth stellar distribution and a resemblance with the other M31 dwarf spheroidals. And VI does not look lumpy or show obvious regions of star formation that would suggest a dwarf irregular, as opposed to dwarf spheroidal, classification. In the And VI continuum-subtracted Hα image, no diffuse Hα emission or H II regions are detected. The lack of Hα emission rules out current high-mass star formation in And VI and serves as further evidence that And VI is a dwarf spheroidal. Like And V and the other M31 dwarf spheroidals, And VI is not detected in any of the IRAS far-infrared bands.

We are currently working with images of And VI from the WIYN 3.5-m telescope in B, V, and I. A deep color–magnitude diagram is being constructed which will reveal the distance and stellar populations characteristics of And VI. We will report all our findings on And VI in an upcoming journal paper.

5. Discussion

The discovery of the M31 dwarf spheroidal And V and the probable M31 dwarf spheroidal And VI increases the number of M31 dwarf spheroidals from three to certainly four, and probably five. It changes somewhat the properties of M31's satellite system, as discussed below. The most obvious change is that M31 is not as poor in dwarf spheroidals as previously thought.

Karachentsev (1996) discussed the spatial distribution of the companions to M31. The discovery of And V changes somewhat the spatial distribution of the M31 satellites. Curiously, And I, II & III are all located south of M31, while the three more luminous dwarf elliptical companions NGC 147, 185 & 205 are all positioned north of M31. Also, Karachentsev (1996) noted that there are more M31 companions overall south of M31 than north of M31. And V's location north of M31 lessens both of these asymmetries. If And VI proves to be a bona fide M31 companion, its location south of M31 would contribute to the asymmetry. With a projected radius from the center of M31 of 112 kpc, And V increases slightly the mean projected radius of the M31 dwarf spheroidals from 86 kpc to 93 kpc. If And VI is also an M31 dwarf spheroidal, with its projected distance from M31 of 271 kpc, the mean projected radius of the M31 dwarf spheroidals increases to 128 kpc.

The discovery of nearby dwarf galaxies like And V augments the faint end of the luminosity function of the Local Group. We do not yet have a reliable M_V value for And V, but it appears to be similar to that of And III ($M_V = -10.2$) since they have similar extinction-corrected central surface brightness. From a survey of nine clusters of galaxies, Trentham (1998) derived a composite luminosity function that is steeper at the faint end than that of the Local Group (see his Figure 2). He attributed the difference to poor counting statistics and/or incompleteness among the Local Group sample. The discovery of And V reduces slightly the discrepancy between the Local Group luminosity function and the extrapolation of Trentham's (1998) function. If And VI is a bona fide M31 dwarf spheroidal, it also reduces this discrepancy.

It is also of interest to learn whether dwarf spheroidals as faint as Draco and Ursa Minor ($M_V \approx -9$) exist around M31. By completing our survey and by understanding our completeness limits, we plan to address this question.

References

Armandroff, T. E. 1994, in ESO/OHP Workshop on Dwarf Galaxies, ed. G. Meylan & P. Prugniel (Garching: ESO), 211

Armandroff, T. E., Da Costa, G. S., Caldwell, N., & Seitzer, P. 1993, AJ, 106, 986

Armandroff, T. E., Davies, J. E., & Jacoby, G. H. 1998, AJ, 116, 2287

Brinks, E., & Grebel, E. 1998, Dwarf Tales Newsletter, Number 3

Caldwell, N., Armandroff, T. E., Seitzer, P., & Da Costa, G. S. 1992, AJ, 103, 840

Da Costa, G. S., & Armandroff, T. E. 1990, AJ, 100, 162

Da Costa, G. S., Armandroff, T. E., Caldwell, N., & Seitzer, P. 1996, AJ, 112, 2576

Irwin, M. J. 1994, in ESO/OHP Workshop on Dwarf Galaxies, ed. G. Meylan & P. Prugniel (Garching: ESO), 27

Karachentsev, I. 1996, A&A, 305, 33

Konig, C. H. B., Nemec, J. M., Mould, J. R., & Fahlman, G. G. 1993, AJ, 106, 1819

Lasker, B. M., & Postman, M. 1993, in ASP Conf. Ser. 43, Sky Surveys: Protostars to Protogalaxies, ed. B. T. Soifer (San Francisco: ASP), 131

Lee, M. G., Freedman, W. L., & Madore, B. F. 1993, ApJ, 417, 553

Lee, Y.-W., Demarque, P., & Zinn, R. 1990, ApJ, 350, 155

Mould, J., & Kristian, J. 1990, ApJ, 354, 438

Reid, I. N., Brewer, C., Brucato, R. J., McKinley, W. R., Maury, A., Mendenhall, D., Mould, J. R., Mueller, J., Neugebauer, G., Phinney, J., Sargent, W. L. W., Schombert, J., & Thicksten, R. 1991, PASP, 103, 661

Renzini, A., & Buzzoni, A. 1986, in Spectral Evolution of Galaxies, ed. C. Chiosi & A. Renzini (Dordrecht: Reidel), 195

Trentham, N. 1998, MNRAS, 294, 193

van den Bergh, S. 1972a, ApJ, 171, L31

van den Bergh, S. 1972b, ApJ, 178, L99

van den Bergh, S. 1974, ApJ, 191, 271

Whiting, A. B., Irwin, M. J., & Hau, G. T. 1997, AJ, 114, 996

The Fornax Spectroscopic Survey — Low Surface Brightness Galaxies in Fornax

M.J. Drinkwater

Physics, University of New South Wales, Sydney 2052, Australia

S. Phillipps, J.B. Jones

Physics, University of Bristol, Tyndall Avenue, Bristol BS8 1TL, UK

Abstract. *The Fornax Spectroscopic Survey* is a large optical spectroscopic survey of *all* 14 000 objects with $16.5 < B_J < 19.7$ in a 12 deg^2 area of sky centered on the Fornax Cluster. We are using the 400-fibre Two Degree Field spectrograph on the Anglo-Australian Telescope: the multiplex advantage of this system allows us to observe objects conventionally classified as "stars" as well as "galaxies". This is the only way to minimise selection effects caused by image classification or assessing cluster membership.

In this paper we present the first measurements of low surface brightness (LSB) galaxies we have detected both in the Fornax Cluster and among the background field galaxies. The new cluster members include some very low luminosity ($M_B \approx -11.5$ mag) dwarf ellipticals, whereas the background LSB galaxies are luminous ($-19.6 < M_B < -17.0$ mag) disk-like galaxies.

1. Introduction: The Fornax Spectroscopic Survey

Several remarks have already been made at this Colloquium about the difficulty of performing optical redshift surveys of low surface brightness (LSB) galaxies, in particular using fibre-fed spectroscopy, despite the pressing need for redshifts for these objects. It is hard to obtain optical spectra of LSB galaxies at the best of times. The limited apertures of fibre spectrographs and problems with sky subtraction would normally be thought to make matters even worse. However by using a system with a very large number of fibres like the 400-fibre Two Degree Field (2dF) on the Anglo-Australian Telescope (AAT) these limitations are outweighed by the multiplex advantage of the system.

The Fornax Spectroscopic Survey (FSS) is designed to sample the largest possible range in surface brightness, including high surface brightness (HSB) as well as LSB galaxies. It does this by targeting *all* objects in a region of sky centred on the Fornax Cluster. No morphological information is used in the target selection, so objects conventionally classified as "stars" are included as well as "galaxies". Such a complete survey of all objects is the only way to minimise selection effects caused by image classification or assessing cluster membership. Previous attempts at "all-object" surveys have been limited to small areas. Morton, Krug & Tritton (1985) obtained spectra of all 606 star-like objects brighter than $B = 20$ in an area of 0.31 deg^2 and Colless et al. (1991) extended their galaxy survey by measuring spectra of 117 compact objects with $21 < B_J < 23.5$ in a 0.1 deg^2 area. *The Fornax Spectroscopic Survey* will cover four 2dF fields (totaling 12 deg^2) centered on the Fornax Cluster. It will include all 14,000 objects in the magnitude range of $16.5 < B_J < 19.7$ (and somewhat deeper for unresolved images).

We selected our targets from an APM (Irwin et al. 1994) scan of the blue and red sky survey plates which provided accurate positions, image classifications and photographic B_J and R magnitudes (optimised for stellar profiles). The galaxy B_J magnitudes were taken from Davies et al. (1988), as were estimates of the surface brightness and image scale length.

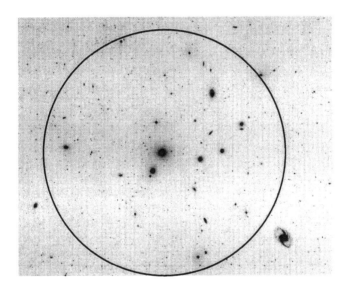

Figure 1. The first 2dF Fornax field observed. We measured spectra of 2500 "stars" and "galaxies" with $B_J < 19.7$ inside the 2 degree diameter circle shown. The field is centred at R.A. 03^h 38^m 29^s, dec. $-35°$ 27' 01" (J2000).

In this paper we present some preliminary results from the FSS based on the first season of observing in which we have almost completed our first field, shown in Figure 1. Here we concentrate on the LSB galaxies and in a companion paper (Drinkwater et al., this volume) we present the detection of HSB galaxies from the survey.

2. Observations

The spectroscopic data were obtained with the 2dF in 1996 and 1997. Full details of the observations and the other objects observed are given in Drinkwater et al. (1999). In the limited time available during these initial observations, we successfully observed 1041 (80%) of the resolved objects to a limit of $B_J = 19.7$ and a total of 1123 (45%) of the stellar objects to the deeper limit of $B_J = 20.3$. Most of the observations were 2 hour exposures for the galaxies and about 45 minutes for the stars.

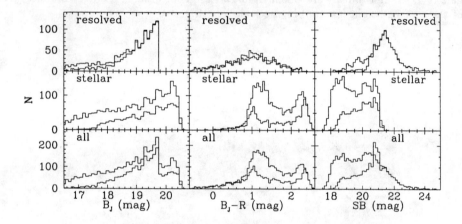

Figure 2. The completness of our observations in the first 2dF field as functions of magnitude, colour and central surface brightness for resolved images, stellar images and all images combined. In each panel the upper histogram indicates the total sample and the lower histogram represents the numbers of objects observed. Note that the R magnitudes are only reliable for stellar images, so the colours of the resolved objects are only indicative. Similarly the surface brightness estimates are not reliable for the stellar images, being determined by seeing and saturation effects.

The numbers of objects observed are shown in Figure 2 as functions of their magnitude, colour and surface brightness for the resolved and stellar image classes and for the whole sample combined. The star–galaxy classification is taken from the APM catalogue. The colours $(B_J - R_F)$ were derived using the R magnitudes from the APM catalogue data. These have not been independently calibrated and are optimized for stellar images so the galaxy colours are only indicative. However the stellar objects show the normal colours with the classic bimodal distribution of blue halo and red disk populations (Kron 1980).

The selection functions in Figure 2 show that, as discussed above, we observed the stellar images to a fainter magnitude limit than the resolved objects, with some bias not to observe the brightest stars. The selection by colour was fairly uniform for resolved objects, but the stellar objects were chosen with a bias to extreme colours so as preferentially to include unusual objects. There is a slight bias to faint surface brightness for all images classes. We will complete this first field during out next observing run to remove these biases.

3. Analysis

We reduced the data using standard IRAF routines for multi-fibre spectroscopy, although an automated pipeline reduction package is now also available at the AAT. We then cleaned the spectra by interpolating across the strongest sky line residuals and correcting for atmospheric absorption. We analysed all the spectra with the RVSAO (Kurtz & Mink 1998) cross-correlation package to identify the objects and measure their radial velocities. Instead of the normal selection of galaxy templates for the cross-correlations, we used a set of ten stellar templates from the Jacoby, Hunter & Christian (1984) library plus one emission line galaxy template. These templates were capable of identifying and measuring galaxy redshifts as well as a set of galaxy templates, but had the advantage, when applied to Galactic stars, of giving a good first estimate of the stellar type. The object identifications were accepted if the Tonry & Davis (1979) "R" coefficient was greater than 3, although all the identifications were checked by visual inspection. If an object was not identified and there were any signs of broad emission lines in the spectrum, it was subsequently tested against a QSO template spectrum (Francis et al. 1991). We used the same process to analyse all the spectra regardless of the image morphology, to give object identifications based on the spectra alone.

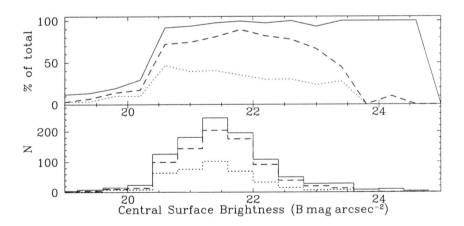

Figure 3. The galaxies observed (solid), with measured redshifts (dashed), and with strong emission lines (dotted) plotted as a fraction of the total sample (upper panel) and as histograms of the actual numbers (lower panel), both as a function of surface brightness.

Figure 3 indicates the success rate of the identifications for galaxies in the sample (i.e. fraction of objects with measured redshifts) as a function of surface brightness. This is fairly constant at about 90% except for the extremely LSB objects: the success rate drops below 50% fainter than 23 B mag arcsec^{-2}. We plan to extend our survey to fainter surface brightness limits in future observations with one long (order 6 hour) exposure in each field. This should permit us to get a reasonable completeness to 23.5 Bmag arcsec^{-2}. The Figure also shows that about half the identified galaxies have strong emission lines. This fraction increases to lower surface brightness, demonstrating the bias against identifying absorption line spectra at low signal-to-noise ratios.

4. Results

In this section we summarise the initial results of the survey, concentrating on the low surface brightness galaxies, but briefly reviewing the other objects. Most of the galaxy results are summarised in Figure 4 which shows the surface brightness plotted against redshift for all the galaxies measured so far. Note that we use the spectra to define which objects are galaxies: those with redshifts greater than $700\,\mathrm{km\,s^{-1}}$, irrespective of morphology.

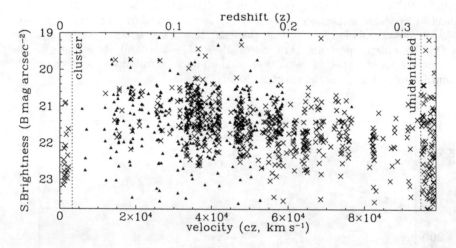

Figure 4. The distribution in surface brightness and redshift of all the galaxies observed so far. The triangles represent objects measured with strong emission lines; all others are indicated by crosses. Note that at redshifts above $55{,}000\,\mathrm{km\,s^{-1}}$ the Hα emission line is shifted out of the 2dF spectra, so very few galaxies are flagged for emission. Galaxies for which we could not measure a redshift are plotted at the right of the figure.

4.1. Galactic stars

In addition to the galaxies shown in Figure 4, we also have a large sample of stars. These will be used for studies of Galactic structure in later papers.

4.2. Cluster galaxies

Our galaxy observations were limited in both magnitude and surface brightness, so only 20 galaxies classified as cluster members by Ferguson (1989, FCC) were observed. Of these, 4 were actually found to be background galaxies, the remaining 16 being confirmed as members.

We detected 5 more members of the Fornax Cluster that were not listed as cluster members in the FCC. Two of these new members were listed in the FCC as background galaxies because they had relatively high surface brightness (reported by Drinkwater & Gregg, 1998). Examples of these galaxies are shown in the top row of Figure 6. All the cluster members observed, most of which have central surface brightness $\mu_0 > 22.5 B$ mag arcsec^{-2}, were identified by absorption line spectra, having no strong emission lines. The remaining 3 new members were not listed at all in the FCC, presumably because they were too faint. These new galaxies have red colours ($1.1 < B_J - R < 1.7$ mag) and appear to be very small (scale lengths less than 200 pc), low luminosity dwarf ellipticals ($-11.9 < M_B < -11.3$ mag). We use 15.4 Mpc as the Fornax cluster distance and 30.9 mag as the distance modulus (Bureau et al. 1996). Some of the brighter new cluster dwarf galaxies are shown in the middle row of Figure 6.

The detection of new cluster members like these weakens the correlation between magnitude and surface brightness proposed by Ferguson & Sandage (1988) and refuted by Irwin et al. (1990). We have replotted the data for this correlation in Figure 5 with the new cluster members plotted. The scatter in the relation is significantly increased, but the relation may still exist. We hope to resolve this with further observations.

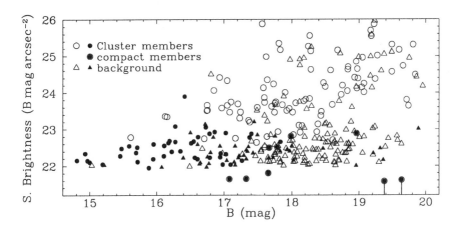

Figure 5. Magnitude–surface brightness relation for galaxies in the Fornax Cluster. The B band central surface brightness is plotted against B magnitude. The membership classifications are taken from the FCC (open symbols), except where they are now spectroscopically confirmed (closed symbols). The two new compact galaxies indicated at the lower right have surface brightnesses of 20.9.

4.3. Field galaxies

We denote as "field galaxies" all those galaxies beyond the Fornax cluster (i.e. $cz > 3500\,\mathrm{km\,s^{-1}}$ but excluding those identified as QSOs by having very broad emission lines. The field galaxies include several high surface brightness compact emission line galaxies which we have identified among the unresolved "stellar" objects. These are discussed in more detail by Drinkwater et al. (this volume).

It is apparent from Figure 4 that there is a tail of LSB galaxies extending below the main galaxy population at all redshifts. If we select the galaxies with central surface brightness $\mu_0 > 22.5\,\mathrm{B\,mag\,arcsec^{-2}}$ and redshift $cz < 50,000\,\mathrm{km\,s^{-1}}$ (so Hα is detectable), we note that they *all* have emission line spectra and have bluer colours ($-0.4 < B_J - R < 0.7\,\mathrm{mag}$) than the cluster LSBs. These field LSBs are luminous ($-19.6 < M_B < -17.0\,\mathrm{mag}$ for $H_0 = 75\,\mathrm{km\,s^{-1}Mpc^{-1}}$) and have relatively large scale lengths (2–5 kpc) making them more like local spirals (de Jong 1996) than dI galaxies. A selection of these LSB galaxies is shown in the bottom row of Figure 6.

We also note from Figure 4 that there remains a large number of LSB galaxies not yet identified (at the right edge of the Figure). These galaxies are very interesting: if cluster members they represent a significant new addition to the cluster luminosity function. On the other hand if they are background field galaxies they must include some very large galaxies.

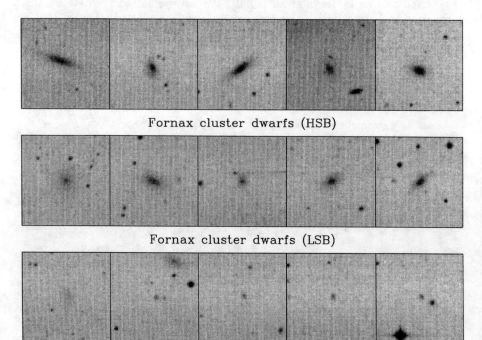

Figure 6. B_J-band images of the new galaxies. Each is $2'$ across.

4.4. QSOs

We have so far identified 52 QSOs in the first field to a limit of $B_J = 20.3$, generally consistent with published QSO number counts. We will eventually generate a large QSO sample relatively free of selection bias compared to other QSO surveys: for instance the current sample covers a redshift range of $0.3 < z < 3.1$.

5. Conclusions

In this paper we have presented results from the first 2500 spectra measured for *The Fornax Spectroscopic Survey*. We are continuing this project with more observations scheduled in 1998 November when we plan to extend the observed sample to fainter limits of surface brightness. When complete the survey will comprise a unique sample of the spectra of all 14,000 objects with $16.5 < B_J < 19.7$ in this $12\,\mathrm{deg}^2$ area of the Fornax Cluster. The ability to record spectra through 400 fibres simultaneously allows long exposures to be used while still sampling large numbers of galaxies. It is therefore possible to measure redshifts for the many low surface brightness galaxies observed in the direction of the Fornax Cluster, including both cluster members and background objects. Already new cluster members have been found, while other objects have been shown to be large field LSBGs in the background.

References

Bureau, M., Mould, J.R., Staveley-Smith, L., 1996, ApJ, 463, 60

Colless, M., Ellis, R.S., Taylor, K., Shaw, G. 1991, MNRAS, 253, 686

Davies, J.I., Phillipps, S., Cawson, M.G.M. , Disney, M.J., Kibblewhite, E.J. 1988, MNRAS, 232, 239

Drinkwater, M. J., Gregg, M. D. 1998, MNRAS, 296, L15

Drinkwater, M. J., Phillipps, S., Davies, J. I., Gregg, M. D., Jones, J. B., Parker, Q. A., Sadler, E. M., Smith, R. M. 1999. In preparation.

Ferguson, H. C., 1989, AJ, 98, 367 (FCC)

Ferguson, H. C., Sandage, A. 1988, AJ, 95, 1520

Francis, P.J., Hewett, P.C., Foltz, C.B., Chaffee, F.H., Weymann, R.J., Morris, S.L. 1991, ApJ, 373, 465

Irwin, M.J., Davies, J.I., Disney, M.J., Phillipps, S., 1990, MNRAS, 245, 289

Irwin, M., Maddox, S., McMahon, R. 1994, Spectrum, 2, 14

Jacoby, G.H., Hunter, D.A., Christian, C.A. 1984, ApJS, 65, 257

de Jong, R.S. 1996, A&A, 313, 45

Kron, R.G. 1980, ApJS, 43, 305

Kurtz, M.J., Mink, D.L., 1998, PASP, 110, 934

Morton, D.C., Krug, P.A., Tritton, K.P. 1985, MNRAS, 212, 325

Tonry, J., & Davies, M. 1979, AJ, 84, 1511

Found: High Surface Brightness Compact Galaxies

M.J. Drinkwater

Physics, University of New South Wales, Sydney 2052, Australia

S. Phillipps, J.B. Jones

Physics, University of Bristol, Tyndall Avenue, Bristol BS8 1TL, UK

M.D. Gregg

IGPP, Lawrence Livermore Lab., L-413, Livermore, CA 94550, USA

Q.A. Parker

Anglo-Australian Observatory, Coonabarabran, NSW 2357, Australia

R.M. Smith, J.I. Davies

Physics & Astronomy, University of Wales, Cardiff CF2 3YB, UK

E.M. Sadler

School of Physics, University of Sydney, NSW 2006, Australia

Abstract. We are using the 2dF spectrograph to make a survey of all objects ("stars" and "galaxies") in a 12 deg^2 region towards the Fornax cluster. We have discovered a population of compact emission-line galaxies unresolved on photographic sky survey plates and therefore missing in most galaxy surveys based on such material. These galaxies are as luminous as normal field galaxies. Using Hα to estimate star formation they contribute at least an additional 5% to the local star formation rate.

1. Introduction

Most galaxy surveys only detect galaxies with a limited range of surface brightness (SB): the difficulty of detecting low SB (LSB) galaxies is well-accepted (Impey, Bothun & Malin 1988, Ferguson & McGaugh 1995) but at the other extreme, it has been argued that there is no selection against high SB (HSB) galaxies (Allen & Shu 1979). This was based on a small sample of bright galaxies, but the conclusion has since been applied to nearly all galaxy samples based on photographic sky surveys. Several groups have in fact detected compact galaxies unresolved on photographic survey plates, often termed compact emission-line galaxies (CELGs). Many have been found in optical QSO surveys (Guzman et al. 1998, Stobie et al. 1997, Boyle et al. 1998). Similar galaxies have been found

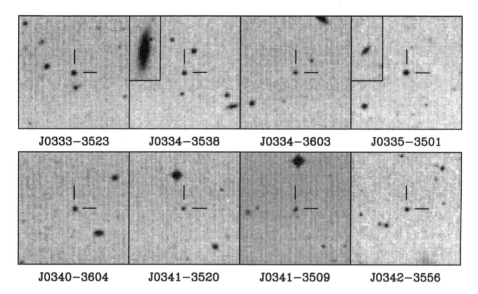

Figure 1. B_J-band images of the new compact field galaxies. Each is $2'$ across. The insets in the images of J0334−3538 and J0335−3501 show two normal (M^*) galaxies at the same distances.

in the University of Michigan emission line galaxy survey (Salzer et al. 1989) and among HII galaxies (Terlevich et al. 1991). Unfortunately, none of these are from simple flux- or size-limited samples so they cannot be used to determine the true extent of the compact galaxy population.

In this paper we present the detection of a new population of compact galaxies based on initial results of *The Fornax Spectroscopic Survey* (see Drinkwater, Phillipps & Jones, this volume), a complete survey of *all* objects in a region of sky centred on the Fornax cluster, being carried out with the 2dF multi-object spectrograph on the AAT. Of 2947 "stars" and 1290 "galaxies" in the central 2dF field down to $B_J = 19.7$, we successfully observed 1249 (42%) and 1250 (97%) respectively. Among the "stars" we found a number of objects with HII region-type spectra at redshifts of 10-50,000 km/s making them field galaxies well beyond the Fornax cluster (1500 km/s). These results are presented in more detail by Drinkwater et al. (1998).

2. Properties of the New Compact Galaxies Galaxies

We present images of the new compact galaxies in Figure 1. The new CELGs have similar absolute magnitudes and distances to "normal" field galaxies, but were classified as stars both by eye and in the APM catalogue. For a typical seeing in survey plates (2") this is equivalent to scale sizes less than about 1", or physical scale sizes 0.6–2.2 kpc ($H_0 = 75\,\mathrm{km\,s^{-1}\,Mpc^{-1}}$). These upper limits are less than is typical of the local spiral galaxy population (de Jong 1996). However, these are not dwarfs in terms of their luminosities which are within a factor 10 or so of L*; indeed two of them exceed L*. This is clearly because of their high surface brightnesses. If we take the limit of 1" to be the scale size of an exponential profile, then the magnitudes imply central surface brightnesses 19.5–21.8 Bmag/arcsec^2, from 1 to 5 times brighter than "normal" spirals or irregulars (van der Kruit 1987, de Jong 1996). Interestingly, if we extrapolate to larger distances, the CELGs have similar properties to the small galaxies in the Hubble Deep Field. For instance, one of our CELGs placed at redshifts z=0.2–1 would have very similar scale lengths (0.1–0.2") and surface brightness to those measured by Jones, Disney & Phillipps (1998) for disk-like galaxies in the HDF.

3. Significance of the New Population

The discovery of these CELGs poses several questions, notably what fraction of all galaxies has been missed? At redshifts less than 55000 km/s we observed 600 resolved objects (galaxies): about 300 of these exhibited strong emission line features. Within the same distance and magnitude limits we found seven new CELGs. These therefore constitute about 1% of the observed local galaxy population, but have been missed by existing surveys. All the new compacts have strong Hα emission, so they represent twice this fraction of emission line galaxies (2%). We can go one step further and consider their contribution to the local star formation rate as Hα emission is a measure of recent star formation (Tresse & Maddox 1998). Our spectra are not flux-calibrated and the 2" apertures only partially sample the flux from resolved objects. However the equivalent width of the emission lines is proportional to the absolute flux in the emission lines of the different galaxies to first order if we assume that the galaxies all come from the same distribution of intrinsic magnitudes and distances as shown above. On this basis the seven new compact galaxies contribute an extra 5% of star formation rate over that in the whole sample of 600. As we have surveyed less than half of the unresolved objects, these numbers are likely to double in the final analysis.

References

Allen R.J., Shu F.H., 1979, ApJ, 227, 67
Boyle B.J., et al., 1998, Phil. Trans. R. Soc. Lond. A, in press
Drinkwater, M.J., et al., ApJ, submitted
Ferguson H.C., McGaugh S.S., 1995, AJ, 440, 470
Guzman R., Jangren A., Koo D., Bershady M., Simard L. 1998, ApJ, 495, L13
Impey C., Bothun G., Malin D., 1988, ApJ, 330, 634
Jones J.B., Disney M.J., Phillipps S., 1998, MNRAS, submitted
de Jong, R.S. 1996, A&A, 313, 45
van der Kruit P.C., 1987, A&A, 173, 59
Salzer J.J., MacAlpine G.M., Boroson T.A., 1989, ApJS, 70, 479
Stobie et al., 1997, MNRAS, 287, 848
Terlevich R., Melnick J., Masegosa J., Moles M., Copetti, 1991, A&AS, 91, 285
Tresse L., Maddox S.J., 1998, ApJ, 495, 691

Counting the Ghosts: Optical Field Surveys for Low Surface Brightness Galaxies

Julianne J. Dalcanton

Observatories of the Carnegie Institution of Washington, 813 Santa Barbara Street, Pasadena CA 91101 and University of Washington, Department of Astronomy, Box 351580 Seattle WA 98195

Abstract. Given the difficulties in detecting even a single LSB, surveying statistically significant numbers of LSBs presents a daunting task. Large, systematic surveys with well understood selection criteria are necessary for assessing the full cosmological significance of the LSB population. Here I briefly review some of the progress which has been made in the last ten years, and suggest strategies which may prove fruitful in the future.

1. Introduction

Now that the community of low surface brightness astronomers has grown to the point of supporting an IAU meeting, it is clear that what we have learned about low surface brightness galaxies (LSBs) is of general interest to the community of astronomers at large. The size of these proceedings demonstrates the astounding progress which has been made towards understanding these peculiar objects, and towards revealing their links with better known populations of galaxies. The question still arises, however, as to their importance to larger questions in cosmology. Can the matter associated with previously hidden LSBs provide sufficient mass to close the universe, or to challenge the predictions of big bang nucleosynthesis? Can LSBs play a role in explaining observations of the universe at higher redshifts? Can their distribution affect measurements of large scale structure, and does their absence from nearby surveys affect measurements of the power spectrum? How do they relate to well-studied normal galaxies, in properties, numbers, and distribution?

As with normal galaxies, such questions are best answered through large systematic surveys. At the minimum, any survey can provide a database of galaxies to study, and given the paucity of information on LSBs until recently, simply providing the community with objects to observe has been a tremendous boon. Few of the articles in this volume would have been possible without the existence of large catalogs of LSBs from which to draw. At their best, however, large surveys can provide much more, by placing the population of galaxies in a broader, well-quantified cosmological context. Such surveys are critical for considering low surface brightness galaxies in continuity with normal galaxies, and elevating them above being freakshow oddballs, or solely laboratories for testing interesting physics.

2. Optical Field Surveys to Date

Over the past 20 years, long, arduous work by many astronomers has given us a number of incredibly useful catalogs of low surface brightness galaxies. While much work has been done on low surface brightness galaxies in clusters, or on low surface brightness dwarf galaxies in particular, for this review I will be concentrating on the surveys of the general field population. These field surveys have typically covered hundreds of square degrees, and have largely relied on Schmidt plates to map these large areas.

The classic of such surveys is the Uppsala General Catalog (UGC; Nilson 1973), which catalogs all galaxies northward of $\delta = -2°30'$ with angular diameters greater than $1'$ on the original Palomar Sky Survey plates. While the UGC was not selected to be a low surface brightness galaxy catalog per se, its galaxies were chosen to be above a fixed diameter limit, without including a more traditional magnitude cut. This criteria admitted a large fraction of low surface brightness galaxies, which could be then studied either as a subset of the larger catalog (e.g. Romanishin et al. 1983, Knezek 1993), or in continuity with the normal galaxies in the catalog (e.g. de Jong & van der Kruit 1994).

When the second generation Palomar Sky Survey plates became available, Schombert and collaborators (Schombert & Bothun 1988, Schombert et al. 1992) repeated the same search criteria as the UGC, but on the deeper plates, for declinations between $0°$ and $+25°$. Their survey revealed several hundred new galaxies, and decreased the limiting average surface brightness by an additional magnitude per square arcsecond beyond the surface brightness limit of the UGC. Most of the recent work on LSB colors, dynamics, and gas content have used galaxies drawn from both this catalog and the UGC.

More recently, Impey and collaborators (Impey et al. 1996) have used a combination of visual searches and APM machine scans to identify nearly 700 galaxies from 24 equatorial fields of the United Kingdom Schmidt Telescope (UKST) survey plates; 500 of these galaxies were previously uncataloged. As the newest large survey, the APM catalog has not yet been exploited to the degree of the earlier POSS-II LSB survey, but with its large size, and copious redshifts and HI observations, it should prove as fertile a field in the years to come.

These large field surveys have been complemented by many smaller field surveys which either push to fainter limiting surface brightnesses than can be reached in large plate surveys (for which $\mu_0 < 24\, B\, mag/\square''$, typically), or which use different methods for galaxy selection. Most of these surveys sacrifice area in exchange for the increased sensitivity and linearity of CCDs (e.g. Dalcanton et al. 1996, O'Neil et al. 1997), and/or use stacked plates and/or novel search techniques to reach fainter surface brightnesses (e.g. Schwartzenberg et al. 1994, Davies et al. 1994). While lacking the numbers and generality of the large surveys, such smaller surveys can push into previously unexplored territory, revealing extremely low surface brightness galaxies, or LSBs with unusual colors (e.g., see contributions by O'Neil in this volume). These mini-surveys can be extremely useful complements to the largest LSB catalogs, and will be important input for designing the next generation surveys.

2.1. The Density of LSBs

These large optical surveys have built a tremendous base for examining the properties of LSB galaxies. The galaxies within them are all relatively nearby (typically with 1000 km/s< $V_\odot$ < 10000 km/s), and thus they can be studied in a fair bit of detail. The surveys have also given us the means to start placing LSBs into their full cosmological context, by assessing the size of the LSB population as a whole.

For example, using several large field surveys, McGaugh (1996) has used visibility corrections to reconstruct a *relative* surface brightness distribution for galaxies, and shown that the distribution is largely flat, with an exponential cutoff at bright surface brightness. For angular diameter or magnitude limited surveys, these reconstructions do not require that the redshifts of the galaxies are known, and thus they can be instantly applied to any large field survey. However, the methods applied by McGaugh (1996) rely upon assuming that the distributions of surface brightness and of scale length are disjoint and independent, which unfortunately they do not seem to be. Thus, these results can only be taken as as a rough indication that the population of LSBs is not negligible, at least in numbers.

The best progress towards measuring the full bivariate luminosity function (i.e. the number density of galaxies as a function of luminosity and surface brightness) has been made by Sprayberry et al. (1997), using the APM LSB survey of Impey et al. (1995). Using a subset of CCD observations to calibrate the plates, and after measuring the selection function of the survey with fake galaxy tests (Sprayberry et al. 1996), Sprayberry et al. (1997) derived a steeply rising luminosity function for the APM LSB sample, which encompassed surface brightnesses between $22 < \mu_0 < 25\, B\, mag/\square''$. Their derived luminosity function is similar in form and normalization to the luminosity function of the irregular galaxies from the CfA survey, derived by Marzke et al. (1994), both of which suggest that LSBs make up only a small fraction of bright L_* galaxies, but that their numbers overtake normally cataloged galaxies at magnitudes fainter than roughly -15 in B. However, because of the paucity of bright LSBs, Sprayberry's results suggest that the luminosity density in uncataloged LSBs contributes less than 1/3 of the luminosity density known to be in normal galaxies. These results will need to be confirmed as full redshift information for their sample becomes available; while most of the galaxies in the APM sample have a central surface brightness of $\mu_0 \sim 24\, B\, mag/\square''$, the redshifts at this surface brightness are only 20% complete. A significantly smaller survey with complete redshift information by my collaborators and I (Dalcanton et al. 1997) derived similar upper limits to the luminosity density of lower surface brightness LSBs ($23 < \mu_0 < 25\, mag/\square''$ in V), suggesting that the APM result will not change drastically with complete redshift information.

3. How Can We Improve?

While these nearby, wide-field surveys have provided us with samples of galaxies which are near enough to study in detail, for the seemingly simple task of "counting" LSBs, these surveys are not the easiest approach. To date, existing analyses of the number density of LSBs have suffered from some combination of very incomplete redshift information, woefully small sample sizes, and limited range of surface brightness. These limitations are nearly unavoidable consequences of focusing our attentions on the nearest LSBs, a focus which has arisen from our need to identify galaxies with large diameter limits (typically $> 30''$).

The difficulties with nearby LSB surveys are several. First, field surveys of nearby galaxies require both deep exposures and large areas to uncover large enough samples to be statistically meaningful. The volume of a survey is proportional to the mean distance cubed, and thus when restricting a search to nearby galaxies, astronomers must cover prodigious areas to find even a few hundred galaxies. For LSB galaxies, which have additional selection biases restricting their mean distance (see contributions by McGaugh in this volume), these volume limits are even more stringent. The second difficulty is that distances to nearby LSBs must be determined one by one, as the galaxies are near enough that there are a scant handful per square degree. This inability to multiplex makes follow-up observations incredibly time-consuming, testing the willpower of both the astronomer and the Time Allocation Committee. Finally, the large survey areas needed make it difficult to maintain a consistent selection efficiency across the survey data, which in turn makes an accurate analysis of the sample challenging and subject to large systematic errors.

I believe that many of these problems could be avoided by instead targeting LSB surveys towards slightly higher redshifts ($z \sim 0.1$, comparable to the LCRS (Shectman et al. 1996) redshift survey). At these larger distances (which are still close enough to be considered a reasonable sample of the nearby universe), individual galaxies will require more telescope time to be observed with the same signal-to-noise as they would be seen with nearby. However, the number of galaxies per unit area will go up dramatically, allowing hundreds to thousands of LSBs to be revealed in a single image. With multi-slit and multi-fiber spectrographs available on most telescopes, the spectroscopic follow-up required for a survey becomes much more efficient as well. This capability is vital, given the hideously long exposure times needed to measure redshifts of such low surface brightness objects. A move to higher redshifts also has the added benefit of allowing one to survey both normal and LSB galaxies simultaneously, giving us not only the bivariate luminosity function, but also the correlation function of LSBs as well. The resulting sample can also be used to study galaxy properties continuously over surface brightness and luminosity, instead of treating LSBs as a disjoint class of galaxies.

To demonstrate the gains possible with such an approach, in Figure 1 I have estimated the time which would be needed to image a constant number of LSBs with constant signal-to-noise, as a function of the typical LSB redshift in a survey. Including the effects of seeing and cosmological $(1 + z)^{-4}$ dimming, the middle panel of Figure 1 shows that much longer exposure times are needed to reach LSBs at distances greater than $z > 0.1$. However, the bottom panel, which also considers the increasing volume which is visible at larger redshifts, shows that the total survey time actually *decreases* out to $z \sim 0.25$, because less area is needed to detect the same number of LSBs. This exercise suggests that surveys of LSBs at $z \sim 0.1$ could be hundreds of times more efficient than surveys of LSBs at $z \sim 0.01$. The increase in efficiency will be similar for the spectroscopic follow-up as well.

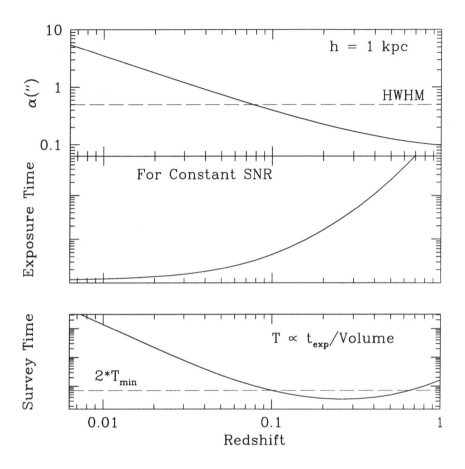

Figure 1. LSB survey properties as a function of typical redshift: The top panel shows the apparent exponential scale length as a function of redshift, assuming a physical scale length of $h = 1$ kpc; the HWHM of a 1″ seeing disk is marked as the horizontal dashed line. The middle panel shows the exposure time needed to reach the same signal-to-noise, as an exponential disk galaxy is moved to higher redshifts. The effects of seeing and cosmological dimming are included in the calculation. The bottom panel shows how the total needed survey time would vary with redshift. The plotted curve is the exposure time calculated in the middle panel, divided by the volume of the universe seen out to each redshift. The minimum survey time T_{min} is reached near $z \sim 0.25$; a survey optimized for $z \sim 0.1$ would take only twice as long as T_{min}.

While moderate redshift surveys for LSBs clearly have a great advantage in observational efficiency, they are likely to require more complicated analysis. For example, because of atmospheric seeing, assigning an accurate surface brightness or scale length to each galaxy becomes increasingly difficult at large redshift. By $z \sim 0.1$, a typical galaxy's scale length will be comparable to the HWHM of the seeing disk (see the top panel of Figure 1), and thus its apparent surface brightness profile will be strongly affected by the PSF. Uncertainty in a galaxy's surface brightness can also cause great difficulty in reconstructing the galaxy's V_{max}, and thus can lead to highly biased derivations of the luminosity function (e.g. Dalcanton 1998). However, the effect of the seeing can easily be modeled, and correcting for such effects may be no more difficult than modeling the selection effects in large area surveys. In fact, it may be somewhat easier, given that the selection function will be much more uniform over the survey area.

I am optimistic that astronomers will find it worthwhile to solve the problems associated with $z \sim 0.1$ LSB surveys. Most of us are in the lucky (or unlucky?) position of having more brains than telescope time, and have to make the best of both. As we have now entered an epoch where we have learned so much about internal LSB properties, I believe we can now sacrifice the detail revealed in nearby LSBs, to gain knowledge about the many unanswered cosmological questions which remain.

Acknowledgments. It is a great pleasure to thank the LSB community of Cardiff for organizing such a delightful meeting, and for hosting the conference in a location with so much good beer. I am also grateful for financial support from the IAU and from Carnegie Observatories, which made my attendence at this meeting possible.

References

Dalcanton, J. J., Spergel, D. N., Gunn, J. E., Schmidt, M., & Schneider, D. P. 1997 AJ, 114, 635

Dalcanton, J. J. 1998, ApJ, 495, 251.

Davies, J. I., Disney, M. J., Phillipps, S., Boyle, B. J., & Couch, W. J. 1994, MNRAS, 269, 349

de Jong, R. S., & van der Kruit 1994, A&AS, 106, 451

Impey, C. D., Sprayberry, D., Irwin, M., & Bothun, G. D. 1996, ApJS, 105, 209

Knezek, P. 1993, Ph.D. Thesis, University of Massachusetts

McGaugh, S. S. 1996, MNRAS, 280, 337

Nilson, P. 1973, Uppsala General Catalog of Galaxies, Uppsala Astr. Obs. Ann., Vol.6.

O'Neil, K, Bothun, G. D., & Cornell, M. E. 1997, AJ, 113, 1212

Romanishin, W., Strom, K. M., & Strom, S. E. 1983, ApJS, 53, 105.

Sprayberry, D., Impey, C. D., Irwin, M. J., & Bothun, G. D. 1997, ApJ, 482, 104

Schombert, J. M., & Bothun, G. D. 1988, AJ, 95, 1389

Schombert, J. M., Bothun, G. D., Schneider, S. E., & McGaugh, S. S. 1992, AJ, 103, 1107

Schwartzenberg, J. M., Phillipps, S., Smith, R. M., Couch, W. J., & Boyle, B. J. MNRAS, 275, 121

Shectman, S. A., Landy, S. D., Oemler, A., Tucker, D. L, Lin, H., Kirshner, R. P., & Schechter, P. L. 1996, ApJ, 470, 172

Sprayberry, D., Impey, C. D., & Irwin, M.J. 1996, ApJ, 463, 535

The radial extent of the Fornax cluster Low Surface Brightness galaxy population

J. I. Davies, A. Kambas, Z. Morshidi-Esslinger and R. Smith

Department of Physics and Astronomy, University of Cardiff, Cardiff, UK

Abstract. In this paper we compare the radial distribution of Fornax cluster galaxies selected at different surface brightness values. The bright galaxy (RC3) sample surface brightness distribution peaks at a $\mu_x^B = 21.5$, the Low Surface Brightness (LSB) sample at $\mu_x^B = 23.7$ and the very Low Surface Brigthness (VLSB) sample at $\mu_x^B = 25.7$. The bright galaxy surface density decreases with an exponential scale length of $\alpha = 0.5°$, the LSB galaxies with $\alpha = 1.3°$ and the VLSB galaxies with $\alpha = 2.0°$. Thus the LSB and VLSB populations define a mass scale size some 2.5 and 4 times larger than the bright galaxies respectively. The contribution of the VLSB population to the total cluster luminosity is comparable to that of the brighter galaxies. If their mass to light ratios are significantly larger than than that of the bright galaxies they will dominate the mass and possibly account for the missing mass in clusters.

1. Introduction

An important issue in observational cosmology is the characteristic scales over which mass is distributed. Recently there has been a surge in interest in trying to relate the locally observed mass distribution to that determined from the fluctuations in the microwave background. The approach in the main has been to use numerical models to simulate the expected distribution of galaxies given the early distribution of matter defined by the CMB fluctuations. From this one can try to infer the dominant physical processes at work and the important cosmological parameters. The interpretation of these simulations rely on measurements of the local mass distribution determined from observations of relatively nearby galaxies. In many cases a biasing parameter is used to relate the models to observations because the bright galaxies are thought to be "biased" tracers of the mass, in the sense that they trace out smaller scale structure than the underlying "true" distribution of mass (see Jenkins *et al.*, (1998). In this paper we ask the question "are LSB galaxies more accurate tracers of the underlying mass distribution in the Universe ?". We have identified three samples of Fornax cluster galaxies selected by their surface brightness and we compare the length scales over which they are distributed. We also assess their contribution to the total luminosity (mass ?) of the cluster.

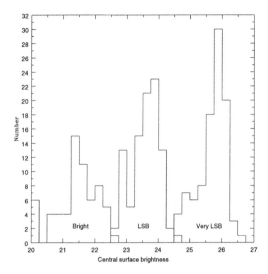

Figure 1. The B band surface brightness distributions of the three samples. Numbers in the VLSB sample have been scaled (reduced) by a factor of 20 so that all the samples can be easily displayed on the same plot.

2. The bright galaxy sample

We have taken the bright galaxy sample from the paper by Jones and Jones (1980). They list 64 Fornax cluster galaxies with $m_B \leq 16$. We have previously measured the surface brightness distribution of these galaxies (Disney et al. 1990) and we reproduce this in fig 1. This sample illustrates the familiar Freeman result. The distribution of surface brightness is sharply peaked at a value of ≈ 21.5 Bμ with an rms width of ≈ 0.4 Bμ. These galaxies will be used to define the mass distribution as delineated by the bright galaxies.

3. The LSB sample

For this sample we use data from an APM scan of UKST photographic plate (f358). The APM machine scans an area of $5.8° \times 5.8°$ of a $6° \times 6°$ UKSTU photographic plate following a 2 pass procedure. First, it determines the sky background as a function of position on the plate. Secondly, it detects images which are in the form of connected pixels above a predetermined threshold above sky.

In addition, the processing computer assigns a flag number to each image it picks up. This is done by looking at the distribution of various image parameters of all detected objects; stars, galaxies, plate defects, other noise, etc. (Maddox 1988). A star which falls along a well defined stellar ridge line is tagged as -1. Other objects are plotted in the same manner and their separations from the stellar ridge line are measured. Images which are more than 2.0σ away are tagged with 1 as non-stellar objects (galaxies). The remaining objects with parameters which are highly unlikely for real stars and galaxies are tagged with 0 as noise. Each image is also assigned 16 scan parameters, these include the isophotal intensity, position and the image area at 8 isophotal levels (the 'areal' profile).

Only objects which are classified by the APM as galaxies are included in our sample. All of the objects we are interested in are assumed to have exponential light profiles (Davies, 1989), so we have fitted the APM areal profiles to an exponential (i.e. $I = I_0 e^{-(\frac{r}{\alpha})}$) and calculated μ_0, α and m_B.

The calibration of the Fornax Plate was carried out using data from Davies et al. (1988). The detection isophote was set to $\mu_L^B = 25.6$. We made several searches for LSB galaxies in the Fornax Cluster and adjusted our selection criteria to optimise our detection technique, i.e. to minimise the number of background contaminating galaxies and to maximise the number of detected cluster LSB galaxies. Checking against Ferguson (1989) and the Digitized Sky Survey, test samples of Fornax galaxies with scale lengths smaller than 3.0 arcseconds or with central surface brightnesses brighter than 22.5 B mag/arcsec^2 suffer greater background contamination and larger numbers of spurious detections (see also Scwartzenberg 1996). Based on the results of our trial runs, the search technique was optimised using the following selection criteria: μ_0 fainter than 22.5 B mag/arcsec^2 and α larger than 3.0 arcseconds.

Checking each object we detected on the plate which satisfy the selection criteria against the Digitized Sky Survey we found that 94% of the detected objects were genuine LSB galaxies while the remaining objects were either stellar halos, edges of bright galaxies or plate defects. Having removed the spurious detections from our test sample we compared the results with those of Davies et al. (1988), Caldwell (1987) and Ferguson (1989) and Irwin et al. 1990.

Using a cross-reference against Ferguson (1989) we found that 70 of our detections were listed as cluster galaxies, 19 were background galaxies while the remaining 3 were not classified at all. Out of 92 detections, 89 appeared in Ferguson's catalogue and he classified 72 of these as dE galaxies while the rest were either Spirals, dS0 galaxies or appear as groups of galaxies. Caldwell (1987) detected 145 dwarf ellipticals galaxies in the Fornax cluster. Checking against his catalogue, we found that 69 of our APM sample were listed as dE galaxies. Davies et al. (1988) (see also Irwin et al. (1990)) used a similar APM data set with slightly different selection criteria. They obtained a catalogue of 134 LSB galaxies. Checking against this 67 galaxies in our sample were found to have been detected by them.

We have also obtained the redshifts of about 20 galaxies and carried out a numerical simulation to predict the numbers of background galaxies that might be contaminating our sample. In both cases we agree with Ferguson's classification which indicated about 20 background galaxies in our sample. We conclude that this sample consists predominately of Fornax cluster galaxies (full details of the above are given in Morshidi-Esslinger et al., 1999a and b).

The distribution of surface brightnesses for this sample is shown in Fig. 1. The peak occurs at about 23.5 Bμ. This is a result of the high surface brightness cut-off at 22.5 Bμ and the incompleteness beyond 24 Bμ and is not indicative

4. The VLSB sample

To obtain data for a sample of lower surface brightness galaxies we have utilized the large format large field of view CCD camera on the Curtis Schmidt Telescope. The 2048 × 2048 pixel CCD provides a 1.7 × 1.7 $sq\ deg$ field of view with large 2.3 $arc\ sec$ pixels. The poor response in the blue meant that it was most efficient to use the R filter. We have adjusted all R band surface brightness and magnitudes to the B band calibration of the other samples using (B-R)=1.5 (Evans et al. (1990)). We have 12 fields extending outwards from the cluster centre to a distance of 9.5 deg. The data have been reduced in the standard way and calibrated against Landolt standards (see Kambas et al., 1999).

We have used the image detection package SExtractor to select a sample of Fornax galaxies. SExtractor detects objects with isophotal areas above a specified size. In this case we used a one sigma isophote that corresponds to $\mu_L^B = 27.5$, almost a factor of 10 fainter than that used for the LSB sample. In the same way as the previous sample we select all objects with exponential scale lengths greater than 3 $arc\ sec$, but we set a minimum surface brightness of $\mu_o^B = 24.5$. We have detected 3998 galaxies over the 13.8 $sq\ deg$ of the survey. Thus we have a VLSB sample with a surface brightness distribution disjoint from the LSB sample (fig 1). The sample extends down to $M_B \approx -10.5$ at the distance of the Fornax custer. This is just about entering the realm of the local group dSph galaxies. We have no way of knowing the contamination of this sample by background galaxies, but we believe that by selecting at such a large scale size we will, in a similar way to the LSB sample, predominantly select nearby galaxies. In the next section we will show that the surface density of galaxies increases rapidly as the cluster centre is approached and so one has strong reasons for believing that in the main these are cluster galaxies. Further details of this work will be given in Kambas et al., (1999).

5. The radial distribution of galaxies of different surface brightness

Using the three data sets described above we can measure the decrease in surface number density from the cluster centre, for galaxies of different surface brightnesses, this is shown in fig. 2. Fitting an exponential to the data we find scale sizes of 0.5, 1.3 and 2.0 deg for the bright, LSB and VLSB samples respectively. This corresponds to 0.1, 0.3 and 0.5 Mpc for a Fornax distance of 15 Mpc. The LSB and VLSB samples are some 3 and 4 times more spatially extended than the bright galaxy sample. The result for the LSB sample has also been shown to be true using a much larger sample of galaxies over a much larger area of sky that includes other nearby galaxy groups (see Morshidi-Essliger et al. 1999a and b). The central galaxy number densities are 40, 10 and 1000 $per\ sq\ deg$ respectively. This implies a sharp rise in the luminosity function at the faint end (slope of $\alpha \approx -2$) as the VLSB galaxies also have the faintest magnitudes.

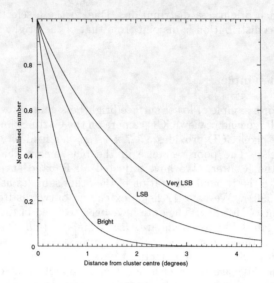

Figure 2. The normalised surface number density of galaxies as a function of distance from the centre of the Fornax cluster.

6. The total luminosity (mass ?) of each sample

Given the exponential fits to the number density distributions we can use the mean luminosities of each sample to estimate the ratio of total luminosities produced at each surface brightness. These are 1:0.02:0.5 for the bright:LSB:VLSB samples respectively. The LSB galaxies contribute little to the light of the cluster, but the VLSB galaxies contribute about half that of the brighter galaxies, but this is extended over approximately 16 times the area. The VLSB galaxies also make a VLSB contribution to the total cluster light. We estimate that the intra-cluster light due to the VLSB galaxies is at most 31 Bμ at the centre of the cluster. The origin of this VLSB population could be the galaxy harrassment mechanism of Moore (this volume). Inter-cluster stars have previously been identified in Fornax by Theuns and Warren (1997) and they state that these stars could account for 40% of the total cluster light. The mass to light ratios of LSB dwarf galaxies in the local group are large (($M/L)_\odot$=10-200 (Irwin and Hatzidimitriou, 1995)), so the VLSB population could contribute the majority of the mass of the cluster and account for the missing mass in clusters.

7. The extent of the VLSB population

The Curtis Schmidt data extends to some 9 *deg* away from the cluster centre and terminates with one field centred on the peculiar lenticular galaxy NGC1291, at a velocity of 802 $km\ s^{-1}$ (de Vaucouleurs 1975). A visual inspection of the fields around NGC1291 confirms a large number of LSB companions. In fig. 3 we show the normalised surface density of galaxies as a function of distance from

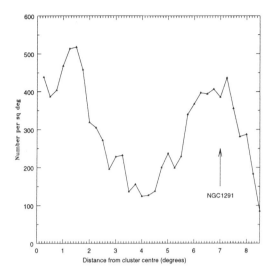

Figure 3. The normalised surface number density of galaxies as a function of distance from the centre of the Fornax cluster with extention to include NGC1291.

the centre of Fornax extending to 9 *deg*. The most striking feature is the initial decline and then large increase in galaxy numbers as NGC1291 is approached. NGC1291 appears to be at the centre of a cluster of LSB galaxies !

8. Conclusions

Although our work is at a preliminary stage we believe that it is quite possible that a significant fraction of the mass (but not light) of the Universe resides in small LSB dwarf galaxies that are dark matter dominated. This mass is distributed over scales far larger than that delineated by the bright galaxies.

References

Abraham R.G., et al., 1996, ApJS, 471, 694
Baugh C.M., Cole S., Frenk C.S., 1996, MNRAS, 283, 1361
Caldwell N., 1983, AJ, 88, 804
Davies J., Phillipps S., Cawson M., disney M. and Kibblewhite E., 1988, MNRAS, 232, 239
de Vaucouleurs G., 1975, AJSS, 284, 29
Disney M., Phillipps S., Davies J., Phillipps S., Cawson M. and Kibblewhite E., 1990, MNRAS, 245, 175
Evans R., Davies J.and Phillipps S., 1990, MNRAS, 245, 164

Ferguson H.C., 1989, AJ, 98, 367
Irwin M., Davies J., Disney M. and Phillipps S., 1990, MNRAS, 245, 289
Irwin M. and Hadzidimitriou D., MNRAS, 277, 1354
Jenkins *et al.*, 1998, ApJ, 499, 20
Jones J. and Jones B., 1980, MNRAS, 191, 685
Kambas A., Davies J. and Smith R., 1999, MNRAS, in preparation
Maddox S., 1988, PhD thesis, University of cambridge
Morshidi-Esslinger Z., Davies J. and Smith R., 1999a, MNRAS, in press
Morshidi-Esslinger Z., Davies J. and Smith R., 1999b, MNRAS, in press
Schwartzenberg J., 1996, PhD thesis, University of Bristol
Theuns T. and Warren S., 1997, MNRAS, 284, L11

The Discovery of Red Low Surface Brightness Galaxies

Karen O'Neil
Arecibo Observatory, PO Box 995, Arecibo, PR 00613
koneil@naic.edu

Abstract. We have performed a digital survey for Low Surface Brightness (LSB) galaxies in the spiral-rich Cancer and Pegasus clusters as well as the low density regime defined by the Great Wall. A total of 127 galaxies were found with $\mu_B(0) > 22.0$ mag arcsec^{-2}, 119 of which were previously unidentified.

Consistent with other surveys, we find a significant number of galaxies with $\mu_B(0) > 23.0$ mag arcsec^{-2} which suggests that the space density of galaxies as a function of $\mu_B(0)$ is not strongly peaked. To more rigorously test this hypothesis we compare the actual surface brightness distribution from our survey with that from two different types of Monte-Carlo based sky images and show that it is not possible to distinguish between the flat distribution and the Gaussian one as the proper description of the underlying surface brightness distribution for this survey beyond 24.0 mag arcsec^{-2}.

The colors of the survey galaxies range continuously from very blue to the first discovery of very red LSB galaxies. It also includes a group of old galaxies which show evidence for recent star formation. This continuous range of colors clearly shows that LSB galaxies at the present epoch define a wide range of evolutionary states.

1. The Data

All the data for this survey was taken on the University of Texas McDonald Observatory 0.8m telescope from 1993 – 1996 using a Loral-Fairchild 2048x2048 pixel CCD camera (1.32"/pixel resolution). Approximately fifty-four 45'x45' fields were imaged during that time, resulting in a total of 27 degrees^2 of sky being imaged. The fields covered both the Pegasus and Cancer sky regions as well as the area around various (known) galaxies distributed in the Great Wall (Geller & Huchra 1989; Dell Antonio, Geller, & Bothun 1996). Six to eight flat field images were taken each night for each filter and were (median) combined to obtain the flat field images for that night's data. The typical seeing disk was 1.5-2.0" during these observations, though we emphasize that in conditions of relatively good seeing the data are undersampled.

When conditions were photometric magnitudes and colors, corrected for atmospheric extinction, were obtained from both standard stars (Landolt 1973,1983) and by using the brighter galaxies in Bothun et al (1985) and Dell'Antonio, Geller, & Bothun (1996) as photometric standards. The precision of these pho-

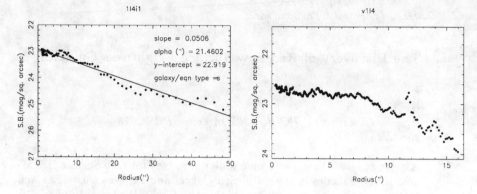

Figure 1. Surface brightness profiles of V1L4. The profile on the left (a) is from the Las Campanas 1.5-m telescope (B band), while on the right the profile (b) is from a HST WFPC-2 F814W (I band) image.

tometric standards is the lowest in the I-band filter and, in general, the signal-to-noise of the galaxies are also lowest in this filter (due to the increased noise and brightness in the sky background at I-band wavelengths). Typical precision for the I band photometric standards is 0.04 mag, while the secondary calibrators are good to 0.03 mag. Typically, three images were taken through both the B (exp. time 600s) and U (exp. time 900s) filters, and one was taken in V (300s) and I (120s). Multiple images were averaged to eliminate extraneous noise.

The average sky brightnesses through the B and I filters were $\langle B \rangle = 21.8 \pm 0.3$ mag arcsec^{-2} and $\langle I \rangle = 18.5 \pm 0.5$ mag arcsec^{-2}. The sky was often brighter than would have been ideal due to the presence of thin cirrus clouds during many of our observations (hence requiring the use of previously observed galaxies in the field to act as photometric standards). On average we were able to detect LSB galaxies down to $\mu_B(0) = 25.5$ mag arcsec^{-2}, or 3% of the sky background. Surface brightness profiles could accurately (error ≤ 0.25 mag arcsec^{-2}) be measured to $27.5\mu_B$, or 0.5% of the sky background. Since images of LSB galaxies are difficult to print, we refer the reader to http://guernsey.uoregon.edu/~karen where all the digital data is reproduced.

Further details about the data reduction, zeropoint determination, and galaxy identification can be found in O'Neil, Bothun, & Cornell (1997) and O'Neil, et al (1997) (OBC and OBSCI, respectively).

2. Central Surface Brightness

Once galaxies were identified, the galaxy's peak intensity was found and ellipses were fit around that point to obtain the intensity in each annulus using the GASP software package (Cawson 1983). In the cases where no obvious peak intensity existed in the galaxy (the clumpier LSB galaxies) the physical center, estimated by centroiding with respect to the outer isophotes, of the galaxy was chosen. Because of the coarse pixel size of many of the galaxies in this survey, our accuracy in locating galactic centers was only good to within 2". Surface brightness profiles and other data cannot be trusted inside that boundary, which also corresponds well with the typical size of the seeing disk. The average sky-subtracted intensity within each (annular) ellipse was found and calibrated with the determined photometric zeropoint.

Surface brightness profiles were then plotted against the major axis (in arcsec) and a best fit line was found. The majority of the galaxies in this survey (80%, or 102 galaxies) being well fit by an exponential profile:

$$\mu(r) = \mu(0) + (\frac{1.086}{\alpha})r \qquad (1)$$

where $\mu(0)$ is the central surface brightness in (mag arcsec^{-2}) and α is the scale length in arcsec. In 17% of the galaxies, though, the exponential fit was clearly not representative of the light profile. In these cases we found that the surface brightness profiles were much better fit (in the reduced χ^2 sense) with King's (1965) model of star clusters, approximated by:

$$\mu(r) = \mu(0) + 2.5\log(1 + (\frac{r}{R_E})^n) \qquad (2)$$

where $\mu(0)$ again is the central surface brightness in (mag arcsec^{-2}), R_E is the core radius, and n is a free parameter. Finally, it should be noted that 3% of the galaxies (4) could not be fit by any curve, while *none* of the galaxies were fit by a $r^{1/4}$-type (elliptical) profile.

All that will be noted herein about those galaxies best fit by a King profile is simply that the galaxies' surface brightness profiles are similar to those found in high resolution imaging of known LSB galaxy cores. Figure 1(a) shows a typical exponential profile of a nearby LSB galaxy (V1L4) taken with the Las Campanas 2.5-m telescope (Impey, Bothun, & Malin 1988). To its right, Figure 1(b) shows the surface brightness profile of the same galaxy as imaged with the HST Wide Field Planetary Camera-2 (0.0996"/pixel resolution) (O'Neil, Bothun, & Impey 1998). As all four LSB galaxies imaged by O'Neil, Bothun, & Impey (1998) show both an overall exponential profile and this flat inner core (see Figure 1 b), it seems likely that the presence of a flat inner core is not uncommon in LSB galaxies. This leads to the hypothesis that the galaxies we categorized as 'king profile' are galaxies whose size is large enough and surface brightness profile faint enough that only the inner core of the galaxies could readily be discerned.

The central surface brightness distribution of the exponential profile (disk-like) galaxies is given in Figure 2, which shows a flat profile from $\mu_B(0) = 22.0$ mag arcsec^{-2} through 24.0 mag arcsec^{-2} and a sharp drop-off on either side. The drop-off at $\mu_B(0) = 22.0$ mag arcsec^{-2} is purely a factor of our selection criteria and should be ignored as such. The drop-off at 24.0 mag arcsec^{-2}, though, is worthy of further investigation as it occurs well before the detection threshold of $\mu_B(0) \sim 26.5$ mag arcsec^{-2}.

Three different tests were run in an attempt to determine whether the central surface brightness drop-off at 24.0 mag arcsec^{-2} is a real effect (possibly due to the dense cluster environment) or simply further evidence of our selection biases. The first two tests were based off the assumed knowledge of the cluster

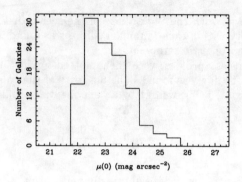

Figure 2. Distribution of $\mu_B(0)$ for the galaxies found in this survey.

densities surrounding each galaxy.[1] The first test utilized the 2-dimensional galaxy density in each field to look for trends between this density and the average central surface brightness and scale length in the fields. A definite trend toward brighter average central surface brightness and smaller average scale lengths with increased density was found, indicating cluster densities do affect the central surface brightness and scale lengths of galaxies. The second test simply separated the galaxies into three groups based on their cluster (or lack thereof) identity and looked for trends between these three groups (Pegasus, Cancer, and Great Wall galaxies). In this case, no difference between the three groups' central surface brightness or scale length distribution was found. Thus, the second test led to the conclusion that the cluster environment plays no role in determining a galaxy's morphology, directly in contradiction with the results of the first test.

Finally, in an attempt to understand whether or not we would be able to detect a difference in the underlying central surface brightness distribution were it to (a) fall off as a Gaussian after $\mu_B(0) = 23.0$ mag arcsec^{-2} or (b) be completely flat from 22.0 mag arcsec^{-2} through 28.0 mag arcsec^{-2}, a series of Monte Carlo simulations were run. The simulations were based off these two different underlying surface brightness distributions and continued the same sky brightness, noise, number of stars, etc. as the original Texas data (see OBSCI for details). Twelve fields of each distribution were created and analyzed, with no detectable difference between the galaxies detected with Model A (the Gaussian distribution), the galaxies detected with Model B (the flat distribution), and the galaxies detected from the Texas survey. The final result, then, is that there is no way to determine whether the observed central surface brightness cut-off at $\mu_B(0) = 24.0$ mag arcsec^{-2} is due to the effect a cluster environment has

---

[1] It should be noted for both these tests that the redshifts of the individual galaxies were not known. Instead, an assumed distance was given to the galaxies based on the distances to other known galaxies within each field. Redshifts were estimated due to the unfortunate timing of the galaxies' discovery with the closing of the Arecibo 100-m telescope for refurbishment. As the refurbishment is now complete redshifts of these galaxies have been obtained and re-analysis of the data is underway. See O'Neil, et al (1999) for further details.

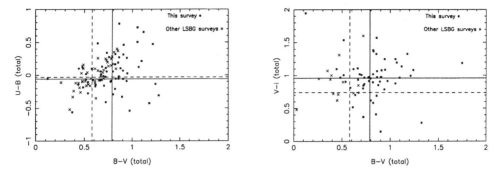

Figure 3. U−B, B−V, and V−I colors both for all the galaxies found in this survey (squares) as well as for numerous other LSB galaxy surveys (de Blok, van der Hulst, & Bothun 1995; Sprayberry, et al 1995; McGaugh 1992; Bothun, et al 1991; Impey, Bothun, & Malin 1988; Caldwell & Bothun 1987; Romanishin, Strom, & Strom 1983)

on galaxy morphologies, or due to the same selection effects which prevented significant detection of LSB galaxies for many decades.

3. Color

The use of multiple filters (Johnson/Kron-Cousins U,B,V, & I) for the sky images in this survey allowed for the determination of broadband colors for most of the galaxies discovered. Figure 3 shows the color distribution for both the galaxies discussed herein as well as for most other LSB galaxies with known colors. In an attempt to describe these colors, the galaxies will be separated into three categories – the very blue, the very red, and everything else.

3.1. The Very Blue

The galaxies in this category are those with colors of U−B< −0.2, B−V< 0.6, V−I<1.0. The existence of very blue LSB galaxies has been previously documented (i.e. McGaugh & Bothun 1994; de Blok, van der Hulst, & Bothun 1995; Impey, et al 1996) and is important primarily because of the restrictions they put on the star formation history of LSB galaxies. The simplistic explanation – that LSB galaxies are merely faded remnants of HSB galaxies – is quickly eliminated by the existence of these very blue LSB galaxies. The average colors of E0/S0 galaxies (good examples of old stellar populations) are U−B=0.54, B−V=0.96 (Tinsley 1978), while late type spiral galaxies have colors similar to these very blue LSB galaxies. Sc or later spiral galaxies with active star formation typically have colors in the range $0.35 \leq$ B−V ≤ 0.65, $-0.2 \leq$ U−B ≤ -0.4 (Huchra 1977; Bothun 1982), with mean values $\langle$B−V$\rangle \approx 0.50$, $\langle$U−B$\rangle \approx -0.20$ (Bothun 1982) and $\langle$ V−I $\rangle \approx 1.0$ (Han 1992). These colors encompass most of the very blue galaxies in our survey, and come close to matching the LSB galaxy colors found by McGaugh & Bothun (1994) (median colors of $\langle$ B−V $\rangle = 0.44$, $\langle$ U−B $\rangle = -0.17$, and $\langle$ V−I $\rangle = 0.89$) and Romanishin, Strom, & Strom

(1983) (average colors of $\langle$ B–V $\rangle$ = 0.43± 0.04, $\langle$ V-R $\rangle$ = 0.60± 0.02). Of course it should be noted that LSB disk galaxies, on average, have factors of ten lower star formation rates (SFRs) than Sc galaxies (i.e. McGaugh & Bothun 1994; McGaugh & de Blok 1997).

The issue here is how to best explain these very blue colors in galaxies with very low SFRs. One explanation for the very blue LSB galaxies could be their low metallicity. McGaugh (1994) has shown that LSB galaxies typically have metallicities approximately 1/3 the solar metallicity. This could lead to a bluing of the galaxies' colors. A comparison with the colors of globular clusters ($\langle$ U–B$\rangle$ ≈0.1, $\langle$ B–V$\rangle$ ≈0.65, $\langle$ V–I$\rangle$ ≈0.95 (Reed 1985)), however, shows that these very blue LSB galaxies are actually *bluer* than very metal poor globular clusters. Thus, although low metallicity may play a role in the color of very blue LSB galaxies, it can only account for part of the reason (see McGaugh & Bothun 1994; de Blok, van der Hulst, & Bothun 1995). For these and many other reasons (i.e. O'Neil, Bothun, & Schombert 1998; Bothun, Impey, & McGaugh 1997; de Blok, van der Hulst, & Bothun 1995; McGaugh 1992), the simplest surviving theory to explain the very blue LSB galaxies is that the galaxies have a young mean age.

3.2. The Very Red

A new and significant discovery in this survey are very red LSB galaxies. As seen in Figure 3 , before this study few LSB galaxies found had U−B > 0.3, B−V > 0.8, and no previous survey had found LSB galaxies with U−B > 0.4 and B−V > 0.9. Detailed studies (OBSCI) show these red colors are unlikely to be the result of photometric errors and/or the presence of strong bulges. The red colors are also unlikely to be the manifestation of large internal reddening, as there is little evidence of dust in LSB disks in general (i.e. Schombert et al 1992; McGaugh 1994; OBSCI). This may reflect both the relatively unevolved state of LSB disks and/or a deficiency of massive stars born in each, sporadic star formation event.

This lack of detected red LSB galaxies has been a concern in the past (i.e. McGaugh & Bothun 1994) because without them it appears LSB galaxies have colors equivalent only to Sc and later galaxies and do not cover the full spectrum of HSB galaxy colors – that is, it appears *no* faded LSB galaxies exist at the present epoch. Since at some future epoch when star formation ceases, all galaxies will fade into LSB objects, the lack of this process occurring now, given the low fractional gas contents of many galaxies, is a bit disconcerting. Indeed, McGaugh (1994) has argued that the photographic selection process has prevented red LSBs disks from being detected previously so perhaps the CCD selection criteria, defined by simultaneous detections in the B, and V filters, has effectively corrected for this selection effect. Possibly, as argued below, LSB disk galaxy evolution is accelerated in clusters and hence clusters contain more red LSB disks than blue LSB disks, although in general this is not the case for HSB disks in spiral-rich clusters (*e. g.* Bothun et al 1982) nor really for dwarf galaxies in the Fornax cluster (*e. g.* Evans, Davis, & Phillips 1990; Bothun et al 1991).

To test if the selection through multiple filters influenced our finding of red LSB galaxies, we compared the (B−V) color distribution for those galaxies within 1σ of the average $\langle\,\mu_B(0)\,\rangle$ against the the (B−V) colors of the galaxies within 1σ of $\langle \mu_V(0) \rangle$. Both distributions had $\langle B-V \rangle = 0.72 \pm 0.1$, and a mean value of B−V = 0.69. Thus, had images only been available in one of the two filters (B or V), the same overall color distribution, including the very red LSB galaxies, would still have been found.

The average colors of E0/S0 galaxies are $\langle U-B \rangle = 0.54$, $\langle B-V \rangle = 0.96$ (Tinsley 1978). The galaxies in the upper right corner of Figure 3(a) & (b) include colors from $0.8 \leq B-V \leq 1.2$, $0.4 \leq U-B \leq 0.8$, and $1.0 \leq V-I \leq 1.7$, showing that the colors of the very red LSB galaxies in our survey are similar to the colors of some of the oldest galaxies. Larson and Tinsley (1978) approximate these colors with their model of a galaxy which formed stars for only the first 10^7 years, and has been quiescent for it's remaining 20 Gyrs. (For this model they obtain U−B=0.74, B−V=1.02). It is quite likely, then, that these galaxies are also the result of a bursting SFR. Unlike their blue counterparts, though, these galaxies underwent starburst early in their existence, consuming most of the galaxies' gas and leaving them with only an old stellar population. In fact, this is the expected route to low disk galaxy surface brightness but until now, it had not been observationally detected.

3.3. The Rest

Naturally, the rest of the galaxies in our survey fall somewhere between the last two extremes – neither consisting of only a young, blue population nor containing only old, red stars. These galaxies fill in the gap between the very blue LSB galaxies and the very red LSB galaxies, showing that we are not merely seeing two distinct galaxy groups – a group of very young, LSB galaxies and a group of very old, faded galaxies – but instead that LSB galaxies cover the entire HSB color spectrum.

One significant group of galaxies have colors U−B< −0.2, B−V> 0.8, and V−I< 1.0, corresponding to primarily young to 'middle-aged' (A0-G8) stars and lack a large population of very old, red stars. The colors of these galaxies are reproduced by Guiderdoni & Rocca-Volmerange (1987,1988) (GR from now on) for young Sc/Sd galaxies. GR obtains the colors U−B=−0.18 − −0.27, B−V=0.75 − 0.88, and V−I=0.91 − 1.00 for Sc/Sd galaxies 11.5 − 11.8 Gyrs old (using GR's values of h_0=0.5 and Ω=0.1). GR assumed the galaxies would have a heavy element fraction of Z=0.02, an unrealistic assumption for most LSB galaxies. McGaugh (1994) gives LSB galaxies a heavy element fraction of Z< 30% $Z_\odot$, or Z< 0.01. Reducing the metallicity of the galaxies in GR would primarily reduce their values for V−I, which follows the trend of the galaxies in our survey quite nicely. As the actual age of these galaxies is presumably similar to the age of all the other galaxies in the various clusters (i.e. Pegasus, Cancer), linking them with galaxies only 11.6 Gyr old merely indicates they have a lower overall SFR than their red counterparts.

This group of galaxies is well represented by galaxies which have already experienced their primary phase of starburst but have not completely left their star formation era. These galaxies most likely represent the next evolutionary phase for the very blue LSB galaxies mentioned earlier.

The other interesting galaxy group consists of galaxies whose B−V and V−I colors indicate a large, old stellar population, yet whose U−B colors expose a population of young stars. That is, they appear to be old galaxies currently undergoing a small yet significant burst of star formation. In all cases the very red V−I or B−I colors can only be obtained by a significant population of K0 and later stellar types, yet the U−B colors in this group are similar to a Sc or earlier galaxy (Huchra 1977).

Results from the models of O'Neil, Bothun, & Schombert (1998) shows that if 15% of a galaxy undergoes a starburst, the galaxy's U−B color will change by −0.6 mag while it's B−V and V−I colors change by only −0.3 and −0.15 mag, respectively. If the starburst is evenly distributed throughout the galaxy the central surface brightness will increase by roughly 1 mag arcsec^{-2}. Realistically, though, it is likely tidal interactions will result in local compressions in the LSB galaxy's gas (Mihos, McGaugh, & de Blok 1997). In this case, the color changes would occur in outer annuli and the galaxy's central surface brightness would show a smaller increase. Note that rotation curve data obtained by de Blok, McGaugh, & van der Hulst (1996) indicates that LSB disks are dark matter dominated at all radii and hence are likely to be quite stable to tidal disruption so repeated encounters in the cluster environment are possible. In addition, most LSB disks are in an environment that would promote only weak tidal interactions (Bothun et al 1993).

4. Summary

Our survey resulted in the detection of 127 galaxies in the local universe with $\mu_B(0) > 22.0$ mag arcsec^{-2}, 119 of which were previously undetected. The central surface brightness distribution of the discovered galaxies is flat from 22.0 B mag arcsec^{-2} (the artificial survey cut-off) through 24.0 mag arcsec^{-2}, at which point there is a sharp drop-off. Through a series of Monte Carlo simulations, though, we've shown that there is no way of knowing whether this cut-off is real or due to selection effects.

Additionally, this survey found LSB galaxy colors covering the entire HSB galaxy color spectrum, including the first discovery of a population of very red LSB galaxies. All of the galaxy colors in this survey can be explained through a low overall star formation rate with sporadic bursts of star formation, and through modeling we determined that these bursts of star formation will not significantly increase the galaxies' central surface brightness.

Acknowledgments. Thanks to Greg Bothun, Mark Cornell & Jim Schombert for help with this work. This work was done while at the University of Oregon.

References

Bothun, G.D., Impey, C., & McGaugh, S. 1997 PASP 109, 745
Bothun, G., et al 1993, AJ, 106, 530
Bothun, G., et al 1991, ApJ, 376, 404
Bothun, G., et al 1985, ApJS, 57, 523
Bothun, G., et al 1982, AJ, 87, 725
Bothun, G. 1982, PASP, 94, 774
Caldwell, N. & Bothun, G. 1987, AJ, 94, 1126

Cawson, M. 1983, Ph.D. thesis, University of Cambridge
de Blok, W., McGaugh, S. & van der Hulst, J. 1996, MNRAS, 283, 18
de Blok, W., van der Hulst, J., & Bothun. G. 1995, MNRAS, 274, 235
Dell'Antonio, I., Geller, M., & Bothun, G. 1996, AJ, 112, 1780
Evans, R., Davies, J., and Phillipps, S. 1990 MNRAS, 245, 164
Geller, M., & Huchra, J. 1989, Sci, 246, 897
Guiderdoni, B. & Rocca-Volmerange, B. 1988, A&AS, 74, 185 (GR)
Guiderdoni, B. & Rocca-Volmerange, B. 1987, A&A, 186, 1 (GR)
Han, M. 1992, ApJS, 81, 35
Huchra, J. 1977, ApJS, 35, 171
Impey, C., Sprayberry, D., Irwin, M., & Bothun, G., 1996, ApJS, 105, 209
Impey, C., Bothun, G., & Malin, D. 1988, ApJ, 330, 634
King, I. 1965, AJ, 71, 64
Landolt, A. 1983, AJ, 88, 439
Landolt, A. 1973, AJ, 78, 959
Larson, R., & Tinsley, B. 1978, AJ, 219, 46
McGaugh, S., de Blok, W. J. 1997, ApJ, 481, 689
McGaugh, S. 1994, ApJ, 426, 135
McGaugh, S. & Bothun, G. 1994, AJ, 107, 530
McGaugh, S. 1992, Ph.D. thesis, University of Michigan
Mihos, C., McGaugh, S., & de Blok, W. 1997, ApJ, preprint
O'Neil, et al 1999 in preparation
O'Neil, K., Bothun, G., & Impey, C. 1998 submitted to AJ
O'Neil, K., Bothun, G., Schombert, J. 1998 AJ, preprint
O'Neil, K., et al 1998 AJ, 116, 657
O'Neil, K., et al 1997 AJ, 114, 2448 (OBSCI)
O'Neil, K., Bothun, G., & Cornell, M. 1997 AJ, 113, 1212 (OBC)
Reed, B. 1985, PASP, 97 120
Romanishin, W., Strom, K., & Strom, S. 1983, ApJS, 53, 105
Schombert, J., Bothun, G., Schneider, S., & McGaugh, S. 1992, AJ, 103 1107
Sprayberry, D., Impey, C., Bothun, G., & Irwin, M. 1995, AJ, 109, 558
Tinsley, B. 1978, ApJ, 222, 14

Dwarf Galaxies in Nearby Groups

T. Bremnes and B. Binggeli

Astronomical Institute, University of Basel, Switzerland

P. Prugniel

Observatoire de Lyon, France

Abstract. We present preliminary results from two observing campaigns where global photometric data for most dwarf galaxies in the M81 and M101 groups as well as some field dwarfs were obtained. The galaxies in the denser M81 group are more often of dwarf elliptical type and are redder and fainter than those of the M101 group and surrounding field, which are mostly of the dwarf irregular types. But both types follow the same total magnitude – central surface brightness relation, so there might be an evolutionary connection between the two classes.

1. Introduction

Our work is aimed at filling the gap of data and knowledge between LG and cluster dwarfs. We have until now determined global photometric properties of members of the M81 and M101 groups (Bremnes et al. 1998a, Lesaffre et al. 1998, Bremnes et al. 1998b) as well as possible M101 group members and nearby field galaxies. These observations are part of a long term project aimed at doing systematic imaging of dwarf galaxies in nearby groups and the general field, based on the "10 Mpc Catalogue" of galaxies by Kraan-Korteweg and Tammann (1979), updated by Schmidt and Boller (1992). Our goal is to compare these galaxies with existing data on the dwarf galaxy populations of the LG and clusters such as Virgo and Fornax. Total magnitudes, effective radii, effective surface brightnesses and galaxy diameters at various isophotal levels have been determined in the Cousins B and R bands as well as best-fitting exponential parameters and colour gradients.

2. The ABC of dEs dIs BCDs ...

There is evidence pointing towards the existence of an evolutionary link between dwarf irregulars, BCDs and dwarf ellipticals (Evans et al. 1990, Patterson & Thuan 1996, Moore et al. 1998). Their structural parameters are quite similar. They form continuous sequences in different parameter spaces, like for example the central surface brightness – absolute magnitude plane. The existence of a density – morphology (Dressler 1980, Binggeli et al. 1987, Ferguson and Sandage 1988, Binggeli et al. 1990) or density – dwarf population (Phillipps et al. 1998) relation also points towards the possibility of an evolutionary link between the different types. Dwarf ellipticals are often associated with large galaxies, dense groups and clusters, whereas the irregulars are mostly present in the field and loose groups. There is also a clear difference in the distribution of the different galaxy types in the CM diagram: the early galaxy types populate the fainter and redder part, whereas the late-type galaxies are bluer and brighter. Note that in *dense* clusters, the trend is opposite, the fainter dEs being bluer and fainter than the brighter dEs (Secker et al. 1997), but see also Evans et al. (1990). The general picture could be one where the dEs are the remnants of spirals or dIs which have been "harassed" by their cluster/group environment (Moore et al. 1998). The M81 group dwarfs are on average redder and fainter than those of the less dense M101 group and the field dwarfs, which are still bluer and brighter, see Fig. 1. This is to be expected if for example the dEs were dIs that were "harassed" by their environment to produce stars earlier and with a larger SFR, whereas the field dIs simply float around, forming stars in a "quiescent" manner over a much longer timescale (Van Zee et al. 1997).

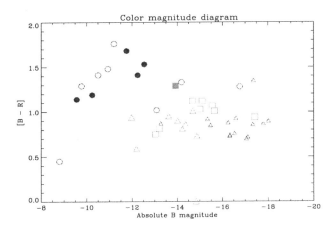

Figure 1. CM diagram for dwarf galaxies in the M81 (circles) and M101 (squares) groups as well as field dwarfs (large triangles) in the projected vicinity of M101 and "quiescent" dwarf irregulars (small triangles) from Van Zee et al. (1997). Filled symbols represent early-type galaxies, open ones late-type galaxies.

References

Binggeli, B., Tammann, G. A. and Sandage, A. 1987, AJ, 94, 251
Binggeli B., Tarenghi, M. and Sandage, A. 1990, A&A, 228, 42
Bremnes, T., Binggeli, B. and Prugniel, P. 1998a, A&AS, 129, 313
Bremnes, T., Binggeli, B. and Prugniel, P. 1998b, A&AS, in preparation
Dressler, A. 1980, ApJ, 236, 351

Evans, R., Davies, J. I. and Phillipps, S. 1990, MNRAS, 245, 164
Ferguson, H. C. and Sandage, A. 1988, AJ, 96, 1520
Kraan-Korteweg, R. C. and Tammann, G. A. 1979, AN, 300, 181
Lesaffre, P., Prugniel, P., Binggeli, B. and Bremnes, T. 1998, A&AS, in preparation
Moore, B., Lake, G. and Katz, N. 1998, ApJ, 495, 139
Patterson, R. J. and Thuan, T. X. 1996, ApJS, 107, 103
Phillipps, S., Driver, S. P., Couch, W. J. and Smith, R. M. 1998, ApJ, 498, 119
Schmidt, K. H. and Boller, T. 1992, AN, 313, 189
Secker , J., Harris, W. E. and Plummer, J. D. 1997, PASP, 109, 1377
Van Zee, L., Haynes, M. P. and Salzer, J. J. 1997, AJ, 114, 2479

Low Surface Brightness Dwarf Galaxies in Nearby Clusters

Kristin Chiboucas and Mario Mateo
Department of Astronomy, University of Michigan, Ann Arbor, MI 48104, USA

Abstract. We present initial results of a study of low surface brightness dwarf galaxies within galaxy clusters at z $\leq$ .03 as part of our program to determine the clustering properties, luminosity functions, and morphologies of dwarf galaxies in a wider range of cluster environments. In addition to deep V-band images covering up to 1 deg^2 in each of 13 different clusters, we have obtained velocities from fiber spectroscopy for 235 galaxies in A3526. In A3526, we find a drop in cluster galaxy counts at intermediate magnitudes which is supported by our spectroscopic results.

1. Introduction

Dwarf galaxies are the most common type of galaxy thus far discovered and may be substantial contributors of mass to the universe. They also provide important clues to understanding galaxy structure, formation, and evolution, and are ideal candidates for probing environmental effects in galaxy clusters.

We are deriving galaxy luminosity functions (LFs) for 13 nearby clusters. Results of previous studies find widely varying values for the faint-end LF slope (Bernstein et al. 1995, Trentham et al. 1998, de Propris et al. 1995) which may be due to different treatments of the selection effects inherent in detecting these low surface brightness systems. Different cluster environments, however, may play a significant role in determining the abundance and morphologies of member galaxies causing intrinsic cluster-to-cluster variations in the LF. With our large data set, and utilizing the same method for each cluster, we will be able to ascertain whether this is the case. A thorough understanding of the faint-end slope will determine whether dwarfs (and their associated dark matter halos) can account for a substantial fraction of the cluster dark matter content.

We are also studying environmental influences on dwarf properties. Morphology, luminosity, structure (such as existence of a nucleus), and surface brightness may all be affected by environmental factors. With our large areal coverage, we shall look at dwarf / giant ratios and the dwarf properties with respect to their spatial distributions within each cluster and in comparison between different clusters to determine which factors most influence dwarf populations.

2. Observations and Methods

Observations of 13 galaxy clusters were acquired between 1993 and 1997 on the LCO 1m, CTIO 1.5m and 4m, and MDM 1.3m telescopes. Deep ($\sim$ 1 hour)

V-band images were taken for each cluster in a mosaic pattern achieving total coverage for each of $\geq .5$ deg^2. A spectroscopic follow-up was made of the brightest 500 galaxies in A3526 using the LCO 2.5m with a 2D Frutti + fibers reaching cluster galaxies as faint as $M_V = -14$ (for $H_0 = 75$ km/s/Mpc). We used two software packages, FOCAS (Valdez 1989) and SExtractor (Bertin & Arnouts 1996), to detect and classify objects. Control fields, which were observed 3 degrees away from each cluster, were reduced in exactly the same manner and were used to determine the contribution of background galaxies. We are addressing incompleteness and selection effects through false galaxy analysis.

3. Luminosity Function of A3526

Fig. 1 shows the LF for A3526 after subtraction of background galaxy counts. These background counts were determined both from our control fields and from our spectroscopic data. We have corrected the field size for the areal coverage of large galaxies, but have yet to correct the counts for incompleteness. It is readily apparent that incompleteness sets in by $m_V = 22$ ($M_V = -11$). More importantly, between $19.5 < m_V < 21.5$ there is an excess of control field counts over the cluster counts.

We investigated several possibilities for this anomalous behavior. One possibility is that our control fields landed on regions of higher background density. Field counts can vary up to 25%, even in nearby regions (Picard 1991), so by lowering our control field counts 25%, we get an upper limit for the faint-end slope of the cluster LF of $\alpha = -1.14$. Other possibilites for this apparent truncation of the LF include selection effects which cause us to miss a particular population of LSB galaxies, or that FOCAS and SExtractor are splitting up the larger LSB galaxies. We are addressing these concerns with false galaxy tests.

To estimate the LF from our spectral results, we determined the fraction of members / non-members in each magnitude bin. These were corrected to include as members those galaxies with spectra from which velocities were not extracted but which fell within a surface brightness - size region of other proven members. Using these fractions, along with the total number of galaxies detected in the cluster fields, we calculated the expected number of members in each magnitude bin. In fig. 1 it can be seen that these results match those from statistical subtraction of background galaxies quite well, and thus supports a genuine drop-off in membership at $m_V = 18.5$.

In other recent spectroscopic studies of cluster dwarfs by Adami et al. (1998) and Secker et al. (1998) Coma dwarf candidate spectra turned up far fewer dwarf members than was expected based on the cluster luminosity function as determined by Bernstein et al. (1995) and Secker. Between the two studies, of the program candidates which yielded velocities, only 20-45% of the expected number of members turned out to be members placing concerns as to the validity of control field statistical subtraction. While our photometric and spectroscopic results do agree, these 3 studies finding a flattening of the LF may contradict work finding steeper slopes at intermediate magnitudes.

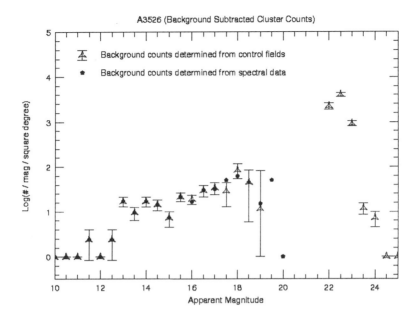

Figure 1. Open triangles show cluster member counts after subtraction of the control field counts while solid symbols represent counts determined from spectroscopic membership results. For clarity, error bars are only supplied for the former method. Neither spectra nor control field counts find background galaxies with $m_V < 16$, and the spectra were only obtained up to $m_V = 20$. The two member populations agree remarkably well. Both show a drop in member counts at $m_V \sim 18.5$ supporting a truncation of the LF (assuming selection effects are not to blame).

References

Adami, C, Nichol, R.C., Mazure, A., Durret, F., Holden, B., & Lobo, C. 1998, A&A, 334, 765.
Bernstein, 1995, AJ, 110, 1507.
Bertin & Arnouts 1996, A&A, 117, 393.
De Propris, R., Pritchet, C., Harris, W.E., McClure, R.D. 1995, ApJ, 450, 534.
Picard, Alain 1991, AJ, 102, 445.
Secker, J., Harris, W.E., Cote, P., Oke, J.B. 1998, Untangling Coma Berenices, Proceedings of the meeting held in Marseilles, Eds.: Mazure, A., Casoli, F., Durret, F., Gerbal, D., Word Scientific Publishing Co Pte Ltd, p 115.
Trentham, Neil, 1998, MNRAS, 294, 193.

The Interstellar Medium in LSB Galaxies

W.J.G. de Blok
School of Physics, University of Melbourne, Parkville VIC 3052, Australia

Abstract. I describe the properties of the interstellar medium in LSB galaxies, and conclude that apart from low density the other important factor determining the slow evolution of LSB galaxies is the metallicity, and (due to inefficient cooling) the corresponding lack of large amounts of molecular gas.

1. Introduction

Deep surveys of the night-sky have uncovered a large population of disk galaxies with properties quite different from those of the extensively studied "normal" high surface brightness (HSB) galaxies. These so-called Low Surface Brightness (LSB) galaxies, which I will be discussing here, are generally dominated by an exponential disk, with scale lengths of a few kpc. Morphologically they form an extension of the Hubble sequence towards very late-type galaxies.

As the evolutionary rate of a galaxy may in fact be reflected in its surface brightness, LSB galaxies are interesting in that they may be a local example of unevolved galaxies. For example, the gas fraction ($M_{\rm gas}/M_{\rm gas+stars}$) increases systematically with decreasing surface brightness, from a few percent for early type spirals to values approaching unity for late type LSB galaxies (McGaugh & de Blok 1996). In many LSB galaxies the gas mass exceeds the stellar mass.

It is still unclear what the physical driver is for the difference between HSB and LSB galaxies. Investigations of their dynamics, using H I observations (de Blok et al. 1996), suggest that LSB galaxies are low-density galaxies. This is one of the favoured explanations for the low evolution rate of LSB galaxies (see e.g. van der Hulst et al. 1987), as this implies a large dynamical time-scale.

In this paper I will briefly discuss what is known about the interstellar medium (ISM) in LSB galaxies, and show that apart from density, metallicity is one of the main causes for the HSB/LSB difference. A fuller treatment can be found in Gerritsen & de Blok (1998) and de Blok & van der Hulst (1998b).

2. Properties of "normal" LSB galaxies

Measurements of the colours of LSB galaxies by McGaugh (1992), van der Hulst et al. (1993), McGaugh & Bothun (1994), de Blok et al. (1995) all showed that the colours of LSB galaxies are among the bluest known for non-interacting disc galaxies. Two possible causes for these blue colours are low metallicity and recent star formation. Both are found in LSB galaxies. Hα imaging (McGaugh

1992) shows that a few regions of star formation are usually present in LSB galaxies. The low surface brightness of the underlying disc ensures that only a small amount of star formation is needed to significantly influence the colours (de Blok et al. 1995). Measurements of the oxygen abundances in H II regions in LSB galaxies (McGaugh 1994, de Blok & van der Hulst 1998) show that the metallicity is on average 0.2-0.5 solar.

In summary, LSB galaxies are extended, low density galaxies, that are still in an early stage of galaxy evolution. But is it just density (that is gravity) that causes these long time-scales, or does the state of the ISM itself also play a role. This was investigated using N-body simulations. These will be discussed in the next section.

3. N-body simulations

A first attempt was made at modelling the ISM in LSB galaxies to test the density hypothesis by using a hybrid N-body/hydrodynamics code (TREESPH; Hernquist & Katz 1989). An extensive description is of the model, as well as the recipe for transforming gas into stars and the method for supplying feedback onto the gas given in Gerritsen & Icke (1997, 1998). The recipe works well for normal HSB galaxies, with the energy budget of the ISM as prime driver for the star formation. The simulations allow for a multi-phase ISM with temperature between $10 < T < 10^7$ K. One can therefore consider cold $T < 10^3$ K regions as places for star formation (Giant Molecular Clouds in real life).

There are many ways to construct model galaxies. Here a galaxy model is built after an existing galaxy with well-determined properties. LSB galaxy F563-1 is a late-type LSB galaxy, representative of the field LSB galaxies. The optical properties of this galaxy are described in de Blok et al. (1995); measurements of metallicities in H II regions are described in de Blok & van der Hulst (1998a); a neutral hydrogen map and rotation curve are given in de Blok et al. (1996). Parameters such as stellar velocity dispersion, which cannot be measured directly, are set in comparison with values measured locally in the Galaxy. The current star formation rate was deduced from Hα imaging. For convenience these data are summarised in Table 1.

Table 1. Parameters for F563-1 ($H_0 = 75$ km s^{-1} Mpc^{-1}).

L_B	$1.35 \times 10^9 \, L_\odot$	
h_*	2.8 kpc	center ($R < 5$ kpc)
h_*	5.0 kpc	outside ($R > 5$ kpc)
SFR	0.05 $M_\odot$/yr	
$M_{\rm H\,I}$	$2.75 \times 10^9 \, M_\odot$	
$v_{\rm max}$	113 km/s	
$\rho_0^{\rm halo}$	0.0751 $M_\odot$/pc^3	
$R_c^{\rm halo}$	1.776 kpc	

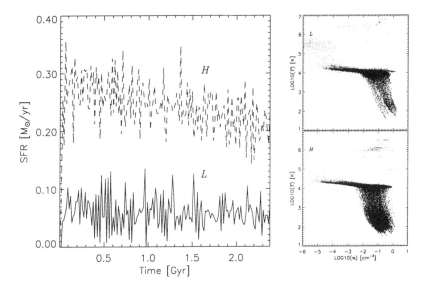

Figure 1.   **Left:** The evolution of the SFR versus time. The lower line shows the SFR from simulation L, the upper line represents simulation H. **Right:** Phase diagrams (temperature versus number density) for the simulations. Top panel shows simulation L, bottom panel shows simulation H; each dot represents an SPH particle. The individual particles at the top of each diagram are hot SN particles. Simulation H clearly shows a two-phase structure, while in simulation L almost all gas is in the warm ($T \approx 10^4$ K) phase (85%)

The most difficult problem faced in constructing a model is converting the measured luminosity to a stellar disk mass. This is one of the most persistent problems in analysing the dynamics of galaxies, and, unfortunately, the present observations do not provide a unique answer for this stellar disk mass-to-light ratio $(M/L)_*$. Rather than using the so-called "maximum disk" value $(M/L)_* = 9$, which is an upper limit to the possible values of $(M/L)_*$, a value was adopted based on colours and velocity dispersions of $(M/L)_* = 1.75$. An extensive motivation for this choice is given by de Blok & McGaugh (1997).

An isothermal halo is included in the calculations as a rigid potential. This is justified since the galaxy model evolves in isolation. Any contraction of the halo under the influence of the disk potential will thus be ignored. This is not expected to be important anyway as the mass of the disk (assuming $(M/L)_* = 1.75$) is only 4 per cent of the measured halo mass. For the simulations 40,000 SPH particles and 80,000 star particles are used initially. This corresponds to an SPH particle mass of 9.6×10^4 $M_\odot$ and a star particle mass of 3.0×10^4 $M_\odot$. The low density as found in LSB galaxies, by itself is not sufficient to reproduce the low observed SFRs of LSB galaxies. *Low metallicity gas is required to explain the properties of LSB galaxies.*

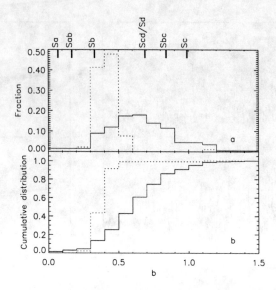

Figure 2. (a) Distribution of the birthrate parameter b, the ratio of the current SFR to the average past SFR. Solid line shows the b values derived for simulation L, dotted line shows the b distribution for a simulation of an Sc galaxy (Gerritsen & Icke 1998). On the top are average b values for different types of galaxies (Kennicutt et al. 1994). The bottom panel shows the cumulative distribution for the two simulations. Less than 20% of the b values for the LSB galaxy are below $b = 0.4$, hence one expects at most 20% of LSB galaxies to be "red"

To show this two model galaxies were constructed using the structural parameters of F563-1. Model H represents an LSB galaxy with a solar metallicity gas. Although the structural parameters relevant for F563-1 were used, the model is in effect a model HSB galaxy, which is "stretched out" to give the low (surface) densities found in LSB galaxies. This model therefore tests the low-density hypothesis. The other model, L, has the same structural parameters as H, but in addition the cooling efficiency of the gas below 10^4 K was lowered by a factor of seven. Cooling below 10^4 K is dominated by metals, so lowering the efficiency is equivalent to lowering the metallicity by an equal amount. Model L thus most closely approximates what is currently known observationally about LSB galaxies. Direct observational support for a low metallicity ISM in LSB galaxies comes from oxygen abundance measurements of H II regions in LSB galaxies. Those studies yield metallicities of approximately 0.5 times solar metallicity (McGaugh 1994). Measurements of the oxygen abundance in F563-1 (de Blok & van der Hulst 1998a) give an average oxygen abundance of 0.15 $Z_\odot$ (compare with the difference in metallicity between models H and L).

For both simulations the SFR varies on time scales of a few tens of Myr and the amplitude of these variations can exceed 0.1 $M_\odot$/yr. However, only simulation L yields a simulated average SFR that comes close to the observed value. The SFR of model H is a factor of ~ 5 too high. The rapid variability in the SFR is due to the discrete nature of star formation in the simulation. New star particles have a mass of approximately 5×10^4 $M_\odot$ and thus represent (large) stellar clusters. A small number of clusters will give the impression of a rapidly varying SFR (as in LSB galaxies). When the SFR increases, the large numbers of fluctuations going on at the same time will give the impression of a smoothly varying SFR. It is therefore not so much the absolute value of the average SFR which determines the colours, but the contrast of any SF fluctuation with respect to the average SFR. For LSB galaxies the large contrast leads to blue colours.

Figure 1 shows the temperature versus density for all gas particles in both simulations. The top panel shows simulation L; the bottom panel shows the phase diagram for simulation H. Most particles have a temperature of 10^4 K. In simulation H a large fraction of the particles has a temperature of about 100 K. This simulation shows a two-phase structure, and resembles the ISM in simulations of HSB galaxies (Gerritsen & Icke 1998). Quantitatively, the cold gas fraction ($T < 1000$ K) makes up 37% of the total gas mass in simulation H and only 4% in simulation L. This reflects the cooling properties of the gas: in simulation L seven times less heat input is required to keep the gas at 10^4 K as in simulation H. In practice simulation L only contains a warm, one-phase ISM. The simulated absence of metals prevents the ISM from cooling efficiently. The essential information to retain from this phase diagram is that we need a different ISM for LSB galaxies, where the bulk of the gas is not directly available for star formation, as it is too warm (of order 10^4 K). The simulations do not include phase transitions from neutral to molecular gas, but as an estimate for the H_2 mass one can consider all star forming gas ($T \lesssim 300$ K) to be molecular. This gas represents less than 2% of the total gas mass. One thus expects the disks of LSB galaxies to contain only negligible amounts of (cold) molecular gas. This is consistent with observations by Schombert et al. (1990) and de Blok & van der Hulst (1998b) which will be discussed below.

Fluctuations in the SFR as shown in Fig. 3, may very well explain why most of the LSB galaxies detected in surveys are blue. As McGaugh (1996) argues, the selection effects against finding red ($B - V > 1$) LSB galaxies on the blue sensitive plates on which surveys have been carried out are quite severe. Assuming that blue LSB galaxies are currently undergoing a period of enhanced star formation, implies that there exists a population of *red, non-bursting, quiescent* LSB galaxies. The fraction of red LSB galaxies can be estimated by calculating the distribution of the birthrate parameter b for simulation L. The b parameter is the ratio of the present SFR over the average past SFR. Birthrate parameters have been determined for a large sample of spiral galaxies by Kennicutt et al. (1994). The trend is that early type galaxies have small values for b, while late type and irregular galaxies have large values for b, often exceeding 1, indicating that those galaxy are still actively forming stars. Here this analysis is applied to simulation L in order to estimate the fractions of blue and red LSB galaxies.

In Fig. 2a I plot the distribution of b values for simulation L (solid line), where I have followed the value of b over the duration of the simulation in steps of 15 Myr. Thus if the SFR peaks in a particular time interval, the corresponding value of b will be high. If the SFR is low in this time interval b is also low. In total there are 200 b values. Also shown in Fig. 2a are the b distribution for a simulation of an HSB Sc galaxy (Gerritsen & Icke 1998, dotted line) and the mean values for different galaxy types (from Kennicutt et al. 1994). Due to the low average SFR the distribution for the LSB simulation is much broader than the distribution for the HSB simulation, and the average b value is larger. The LSB galaxy has b values larger than the average for early type galaxies for most of the time. "Classical" LSB galaxies are blue compared to HSB galaxies, they thus have an excess of recent star formation or equivalently a higher b value. I now define a LSB galaxy to be "blue" if its birthrate parameter b exceeds the average value of b for a HSB late-type galaxy (see Kennicutt et al. 1994 for relations between birthrate parameter and colour). Fig. 2a shows that this requires that $b_{LSB} > \langle b_{HSB} \rangle \approx 0.4$.

LSB galaxies that do not meet this requirement are "red": non-bursting, but nevertheless still gas-rich. From Fig. 2b (which shows the cumulative b distribution) once can see that over 80 percent of the fluctuations result in blue LSB galaxies. Less than 20 percent of the fluctuations therefore results in red LSB galaxies. Using population synthesis models and the burst strengths found in simulation L one can derive that the red LBS galaxies must have $B - V \sim 1$, $R - I \sim 0.6$ and $\mu_0(B) = 24.5$ for the red population.

A recent CCD survey (O'Neill et al. 1997) has picked up a class of LSB galaxies which have $\mu_0(B) \simeq 24$ and $B - V \simeq 0.8$. If some of these galaxies are indeed the non-bursting counterparts of the blue LSB galaxies, they should be metal-poor and gas-rich, and share many of the properties of the modelled galaxies.

In summary, if the blue colours found in LSB galaxies are the result of fluctuations in the star formation rate, then this implies that the red gas-rich LSB galaxies constitute less than 20% of the gas-rich LSB disk galaxies. This does not rule out the existence of a population of red, gas-poor LSB galaxies. These must however have had an evolutionary history quite different from those discussed here and possibly have consumed or expelled all their gas quite early in their life.

4. Molecular gas

Three galaxies from the sample of LSB galaxies in de Blok et al. (1996) were observed with the 15-m James Clerk Maxwell Telescope at Mauna Kea, Hawaii, in the ^{12}CO (J= 2 − 1) line at 230 GHz rest-frequency (see de Blok & van der Hulst 1998b). No CO emission was detected at any of the positions after on-source integration times of ~ 1.5 hours per position. Typical RMS-noises at 500-kHz-resolution were $T_A^* \sim 6$ mK.

The non-detections of CO can be taken at face-value to suggest that LSB galaxies are poor in H_2, thus confirming the models. There are, however, several factors which complicate this naive interpretation. The most important of these is the conversion factor X which is used to convert the measured CO brightness temperature into an H_2 mass. The value of X is uncertain and is inferred to have a large range.

Wilson (1995) demonstrated that Maloney & Black's (1988) ideas concerning a variation in X with metallicity is borne out in observations of galaxies in the

Local Group. Based on measurements of the CO luminosity and determination of the virial masses of individual clouds, Wilson finds that the conversion factor increases as the metallicity decreases. Israel (1997) investigated the metallicity dependence of X in a different way using the FIR surface brightness and H I column density to estimate the column density of H_2 and found an even steeper relationship.

The average oxygen abundance for the observed LSB galaxies is $12 + \log(O/H) \sim 8.4$. Using the above results this would lead to conversion factor values X of 2 to 6 times the Galactic value.

The star formation rates and H I column densities in the galaxies used to derive the dependence of X on metallicity are appreciably higher than those commonly found in LSB galaxies. The lower star formation rate implies a lower energy density of the radiation field and consequently lower dissociation of the CO and H_2. The result will be that X probably is not as large as in the extreme case of the SMC, so some care should be exercised in using these results for estimating the H_2 mass limits for LSB galaxies. One effect of the low metallicities is a less efficient cooling of the ISM, which leads to *higher* cloud temperatures, making it difficult for a cold molecular phase to exist. Bearing these effects in mind I estimate that X will be ~ 4 times the Galactic value in LSB galaxies. In other words, LSB galaxies should contain 4 times more H_2 than the Galactic value suggests. The upper limits then imply that LSB galaxies roughly have (less than) 25% of their gas mass in the form of H_2. This is still lower than is found in HSB galaxies.

The conclusion then is that there are no large amounts of H_2 hidden in LSB galaxies. The low star formation rates measured in LSB galaxies can thus be explained by the lessened importance of a molecular component. A detailed comparison between the properties of star forming regions in LSB and HSB galaxies may be a good way to put more constraints on the way stars form in environments that lack a cold component.

5. Conclusions

What then causes the low evolution rate for LSB galaxies? The low density has often been invoked to explain this, since the dynamical time scales with $1/\sqrt{\rho}$. This scenario is exactly what is tested in simulation H. The result is striking: adopting "standard" values for the star formation process results in a SFR identical to the rates of HSB galaxies.

Thus the low density in itself seems not capable of doing the job, and we have to rely on a scarcity of heavy elements to reproduce a true LSB galaxy. This fits in logically with the notion that stars are the producers of these elements; the low star formation activity prevents metal enrichment of the ISM. It implies that the SFR has been low throughout the evolution of LSB galaxies, and that these galaxies are "trapped" in their current evolutionary state: low density prevents rapid star formation, which prevents enrichment of the ISM, which prevents cooling, resulting in a warm one-phase ISM. So although the lack of metals is directly responsible for the low SFR, the low density may ultimately determine the fate of LSB galaxies.

Due to the fluctuations in the SFR in LSB galaxies and their large contrast with the average SFR, the spread in colours among LSB galaxies will be larger than among HSB galaxies. From the distribution of birthrate parameters in the simulations one deduces that, if the currently known blue *gas-rich* LSB galaxies are the most actively star forming LSB galaxies, they constitute over 80 percent of the total population of *gas-rich* field LSB disk galaxies. This implies that there is at most an additional 20 percent of quiescent, gas-rich LSB disk galaxies. This does not preclude the existence of an additional red, gas-poor population. However this population must have an evolutionary history quite different from that described in this work.

Acknowledgments. I thank Jeroen Gerritsen and Thijs van der Hulst for their invaluable contributions to this work.

References

de Blok W. J. G., van der Hulst J. M., 1998a, A&A 335, 412
de Blok W. J. G., van der Hulst J. M., 1998b, A&A 336, 49
de Blok W. J. G., McGaugh S. S., 1997, MNRAS, 290, 533
de Blok W. J. G., McGaugh S. S., van der Hulst J. M., 1996, MNRAS 283, 18
de Blok W. J. G., van der Hulst J. M., Bothun G. D., 1995, MNRAS 274, 235
Gerritsen J.P.E., de Blok, W.J.G., A&A, in press
Gerritsen J.P.E., Icke V., 1997, A&A, 325, 972
Gerritsen J.P.E., Icke V., 1998, in press
Hernquist L., Katz N., 1989, ApJS 70, 419
Israel F.P., 1997, A&A 328, 471
Kennicutt R.C., Tamblyn P., Congdon C. W., 1994, ApJ 435, 22
Maloney P., Black J.H., 1988, ApJ 325, 389
McGaugh S.S., 1992, PhD Thesis, University of Michigan
McGaugh S. S., 1996, MNRAS 280, 337
McGaugh S. S., Bothun G. D., 1994, AJ 107, 530
McGaugh S. S., de Blok W. J. G., 1996, 481, 689
O'Neill K., Bothun G. D., Schombert J., Cornell M.E., Impey C.D., 1997, AJ 114, 2448
Schombert J. S., Bothun G. D., Impey D. D., Mundy L. G., 1990, AJ 100, 1523
van der Hulst J. M., Skillman E. D., Kennicutt R. C., Bothun G.D., 1987, A&A 117, 63
van der Hulst J. M., Skillman E. D., Smith T.R. et al., 1993, AJ 106, 548
Wilson C.D., 1995, ApJL 448, L97

Dwarf Galaxies as Low Surface Brightness Galaxies

Evan D. Skillman
Astronomy Department, University of Minnesota, Minneapolis, MN 55455

Abstract. Because of the weak correlation between surface brightness and luminosity, as galaxy surveys reach to lower surface brightnesses, increasing numbers of dwarf galaxies are found. Thus, studies of dwarf galaxies are relevant to our understanding of the low surface brightness universe. In particular, studies of nearby LSB dwarfs may help to answer questions concerning: (1) star formation in low abundance and low mass surface density environments, (2) star formation histories of galaxies and the relative youth of LSB galaxies, and (3) the chemical evolution of LSB galaxies. HST WFPC2 photometry of the resolved stars in nearby dwarf galaxies is particularly useful for all three questions.

H II region abundances remain the best probe of current ISM abundances. Current evidence favors total mass (or luminosity) as the primary factor for determining the present day chemical abundance in both LSB and HSB galaxies. It remains possible that mass surface density (or surface brightness) may be a second controlling parameter.

1. THE IMPORTANCE OF NEARBY DWARF IRREGULARS

I started to prepare for this talk by reading the Impey & Bothun (1997; IB97) Annual Review article on LSBs. Their Figures 1 and 2 compare the surface brightness/luminosity (μ/L) distribution of galaxies from the RC3 with that of galaxies from the APM survey (Impey et al. 1996). As IB97 state, there is a weak correlation between between μ and L for the newly discovered galaxies, such that as these surveys probe to fainter and fainter μ, the catalogs begin to be dominated by dwarfs (absolute magnitudes ≥ -18). Thus, it makes sense to pay attention to what has been learned about dwarf galaxies when attempting to understand the LSB galaxy population.

In other words, for important questions like estimating the fraction of Ω(baryons) or Ω(mass) that resides in LSB galaxies, large statistical samples are required. However, for certain other questions, detailed studies of prototypes are useful. Obviously, for these prototype studies, the nearest prototype galaxies are the best targets. Studies of these nearby galaxies have the potential to yield insights into the "LSB Universe."

Luckily, there is a great wealth and variety of nearby LSB dwarfs. Mateo (1998) has assembled a marvelous overview of all of the Local Group members. This compilation will be invaluable for future studies and I will take advantage of it right now. Figure 1 shows the μ, L distribution (in V band) for the Local

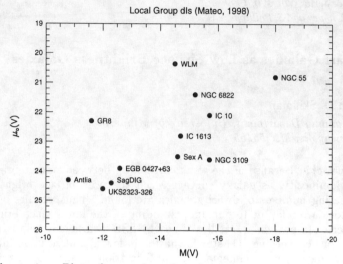

Figure 1. Plot of central surface brightness versus absolute magnitude for several Local Group dwarf irregular galaxies (from Mateo 1998). Note the relative abundance of LSB ($\mu \geq 23$) systems.

Group dwarf irregular galaxies. Here we see a large diversity in the currently star forming dwarfs, with many galaxies with central surface brightnesses below 23 (the nominal definition for a LSB galaxy). Note that not all Local Group dIs are included (some are still lacking reliable surface brightness measurements!).

These galaxies, and other nearby galaxies like them can be used to answer many questions about LSB galaxies. Here I concentrate on three topics:

• Is star formation different in low surface density and/or low chemical abundance environments? Theoretical calculations indicate that both mass surface density and abundance provide feed back to the star formation process - is there observational evidence for this?

• What are the star formation histories (SFHs) of the LSB dwarfs? Do they experience delayed formation or do their evolution clocks run more slowly (or both)? Is it possible to have galaxy formation occurring in the present epoch?

• Is the chemical evolution of LSB galaxies different from that of HSB galaxies? Is mass surface density a second parameter in determining the present day chemical abundance of galaxies (where total mass is the first parameter)?

2. STAR FORMATION IN LOW SURFACE DENSITY AND LOW CHEMICAL ABUNDANCE ENVIRONMENTS

2.1. Current Star Formation Rates and Thresholds

Theoretical calculations (e.g., Spaans & Norman, 1997; Spaans & Carollo, 1998) and observational evidence (e.g., Israel 1997; Taylor et al. 1998) support the idea that the character of the ISM changes both as a function of decreasing mass surface density and decreasing chemical abundance.

In the last decade, one idea which has gained considerable attention is that galaxies have star formation thresholds. In his now famous paper, Kennicutt (1989) found that the Toomre disk stability criterion for gravitational instability (Toomre 1964) can be used to predict critical values of the surface gas density (Σ_c). These values show a remarkable ability to predict the limits of star formation in the (predominantly high surface brightness) spiral galaxies studied by Kennicutt.

In the last few years, the focus of related research has been to test the theory in different environments, with a particular emphasis on LSB and dwarf galaxies. The following studies all support the threshold theory: (1) van der Hulst et al. (1993) observed a small collection of large LSB spirals. Their HI surface densities were found to lie at or below the critical densities calculated from the Toomre criterion across the extent of their disks. (2) Taylor et al. (1994) observed five HII galaxies. In all cases, the central HI surface densities exceeded Σ_c, and the radius where the HI surface density fell below Σ_c roughly corresponded to the optical diameter of the galaxy. (This result has been supported by new observations of blue compact galaxies by van Zee et al. 1998). (3) van Zee et al. (1997a) observed ten isolated dwarf galaxies which were divided into LSB dwarf galaxies and "normal" gas-rich dwarf galaxies. Their HI surface densities were found to lie at or below the critical densities. Interestingly, *local* peaks in the HI surface densities approach the critical value, and these are associated with sites of active star formation as determined from Hα observations.

2.2. Recent Star Formation Histories

A tremendous advance in this field has come about due to the fantastic imaging abilities of the Hubble Space Telescope. This has resulted in great improvements in the accuracy of the photometry of resolved stars, and this allows us to produce very detailed recent SFHs for nearby galaxies. As an example, Figure 2 shows two color magnitude diagrams (CMDs) constructed from HST WFPC2 observations. A very important advance is represented by the separation of the "blue plume" into main sequence and blue core helium burning ("blue loop" or HeB) stars.

The blue HeB stars provide a parallel track to the MS in which to observe star formation events. From the number of blue HeB stars, we can calculate the SFR for the age corresponding to this phase of evolution. There are two advantages to using the blue HeB as an indicator of SFH: (1) The blue HeB stars are about 2 magnitudes brighter than the MS turnoff stars of the same age (e.g., Bertelli et al. 1994). This allows us to probe the recent SFH further back in time (for the same photometric limits); (2) There is little confusion from overlapping generations. All of the blue HeB stars of a certain magnitude come from the same generation of stars. In practice, the blue HeB stars can probe the SFH back to almost 1 Gyr. At older ages, they blend with the red clump and horizontal branch, becoming degenerate in time.

Figure 3 shows the SFHs for four Local Group dIs constructed from the blue HeB stars. Note the unprecedented time resolution in these SFHs. For this sample, it turns out that the average recent star formation rate correlates with the gas mass content of the galaxies (see Dohm-Palmer et al. 1998).

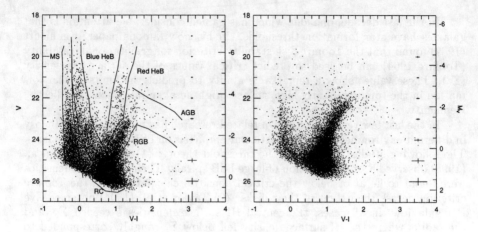

Figure 2. The CMDs in V and I for the Local Group dIs Sextans A (left, from Dohm-Palmer et al. 1997) and the Pegasus dI (right, from Gallagher et al. 1998). Note the paucity of the MS and HeB populations in Pegasus compared to Sextans A.

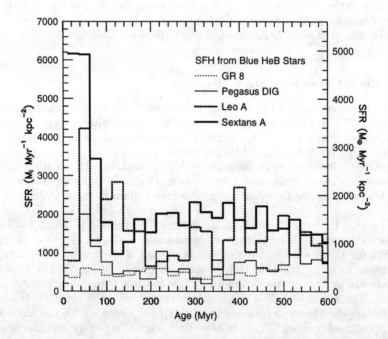

Figure 3. The star formation history of four nearby dI galaxies based on the blue HeB luminosity function. The bins are 25 Myr. Sextans A consistently has the highest SFR/area, followed by Leo A. Pegasus and GR 8 have a very similar SFR/area, which is consistently lower than in the other two galaxies. From Dohm-Palmer et al. 1998.

3. THE STAR FORMATION HISTORIES OF NEARBY DWARF IRREGULARS – THE YOUNG GALAXY HYPOTHESIS

McGaugh & Bothun (1994) pointed out that the blue colors of LSB galaxies meant that they were inconsistent with "fading" scenarios and hypothesized that LSB galaxies probably formed relatively late and evolved slowly. While more recent surveys (e.g., O'Neil et al. 1997) have shown that LSBs show a large range in color (and therefore evolutionary state), the hypothesis that LSB galaxies may form late and evolve slowly is still of interest (cf. McGaugh & de Blok 1998). Determining detailed star formation histories (SFH) of galaxies has always been a long-term goal of extragalactic observational astronomy (e.g., Hodge 1989), and measures of the resolved stars *in nearby galaxies* offer the best hope for very detailed SFHs (e.g., Smecker-Hane et al. 1994). Nonetheless, while §2.2 emphasized the successes in modeling the recent SFHs of dIs, constraining the old SFHs of galaxies (≥ 1 Gyr) is a very difficult problem.

For very close galaxies, the SFH can be reconstructed from the MS luminosity function (e.g., Butcher 1977). However, few galaxies are close enough and the old SFHs must come from an understanding of the distribution of stars on the red giant branch. This field has been pioneered with ground-based observations (e.g., Tosi et al. 1991; Greggio et al. 1993; Gallart et al. 1996; Tolstoy & Saha 1996). Imaging with the WFPC2 on the HST allows vast improvements in the fidelity of the photometry, yielding similar improvements in the observational constraints on the SFHs (e.g., Aparicio et al. 1996).

Recently it has been suggested that the stellar population in the Local Group dwarf irregular Leo A is predominantly young (i.e., stellar ages less than 2 Gyr; Tolstoy et al. 1998). The main diagnostic is the ratio of "red clump" stars (core helium burning stars which overlap the red giant branch in the HR diagram) to red giant branch stars (cf. Bertelli et al. 1992; Gallagher et al. 1998; Tolstoy et al. 1998). Since these stars probe the oldest ages and since the lifetime in the red clump is roughly constant while the lifetime in the red giant branch is a strong function of the mass of the star, this ratio is a relatively robust measure of the average age of the stellar population.

The study of Leo A was based on two orbits of HST observations. Clearly there is potential in this type of study to characterize the SFHs for a large number of nearby galaxies. As the sample increases it will be possible to study SFH (or average stellar age) as a function of both intrinsic galaxy properties (mass, luminosity, surface brightness) and extrinsic galaxy properties (environment, local density). This will allow us to answer the question whether surface mass density and surface brightness correlate with mean stellar age.

4. ISM ABUNDANCES OF LSB DWARFS

Surveys of the abundances in a number of H II regions in irregular galaxies show a clear correlation of oxygen abundance with galaxy luminosity (Skillman et al. 1989 and reference therein). (This trend is also seen in spiral galaxies – Zaritsky, Kennicutt & Huchra 1994 and references therein). However, not all dwarf galaxies comply with the metallicity – luminosity relationship. For example, many blue compact dwarf galaxies (BCDGs), which derive a significant

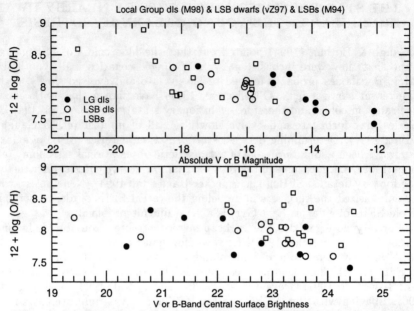

Figure 4. Comparison of HII region oxygen abundance versus absolute magnitude and central surface brightness for Local Group dwarf irregular galaxies from the compilation of Mateo (1998), LSB dwarf irregular galaxies from van Zee et al. (1997), and LSB galaxies from McGaugh (1994).

fraction of their total luminosity from their high surface brightness star forming regions, appear overluminous for their abundance (e.g., Roennback & Bergvall 1995). If the underlying correlation is between mass and metallicity, then this is expected since the BCDGs have lower M/L ratios. Skillman et al. (1997) have shown that fading also contributes to the scatter in the O/H,L relationship.

On the other hand, the fundamental relationship may not be between mass and metallicity, but between surface density and metallicity. Mould, Kristian, & DaCosta (1983) first discovered that dwarf elliptical galaxies also show a strong correlation between metallicity and luminosity, and Aaronson (1986) showed that the metallicity vs. luminosity relationships for the two classes of galaxies are roughly identical in both slope and zero-point (although see Mateo (1998) for a more recent review). Since low luminosity, high surface brightness, and relatively metal rich dwarf ellipticals have been observed, Bothun & Mould (1988) suggested that surface density may be the fundamental parameter determining metallicity (see also Edmunds & Phillips 1989).

Caldwell et al. (1998) have used the HST to obtain metallicities in two dEs in the M81 group and found that they both comform to the metallicity-luminosity relationship as defined by the Local Group dEs, but that the very low central surface brightness object F8D1 lies well off the metallicity – central surface brightness relationship. Thus, they propose that luminosity and not surface brightness is the key parameter determining metallicity.

This argument is based on a sample of one, and it pertains to the gas poor family of dEs. How does a similar test for gas rich galaxies result? I drew upon data from three different studies in the literature to compose Figure 4. First, I used the compilation of Mateo (1998) to compare Local Group dIs. Then I added the LSB dIs of van Zee et al. (1997b,c), and finally the LSBs studied by McGaugh (1994). While the known correlation in O/H,L is clearly shown, the correlation is much weaker for O/H,μ. It would appear that these studies support luminosity as the important parameter.

In the past, LSB galaxies have been characterized as metal poor, with chemical abundances less than one-third of the solar value (McGaugh 1994). A reexamination of all available data indicates that both LSB and HSB galaxies show correlations between metallicity and luminosity, and thus, it is better to characterize LSB galaxies in this way as opposed to all metal poor.

Is this relation universal? Garnett et al. (1997) compared the abundances in different spiral galaxies at similar disk optical surface brightness, and found a clear correlation between the abundance at a given surface brightness and the total luminosity of the galaxy. This suggests that massive spirals are more efficient at enriching their ISM *at fixed surface brightness* than low mass spirals. (Because of the luminosity – gas mass fraction correlation, this can be cast in terms of simple models tying abundance to gas mass fraction, e.g., McGaugh & de Blok 1997). Note that the Garnett et al. sample contains a limited range in overall galaxy surface brightness. Adding LSB galaxies to this plot will rovide a strong test (e.g., Pickering 1998), and it will be interesting to see if surface brightness acts as a second parameter in determining present day abundance. This diagram is telling us something fundamental about galaxy evolution. Perhaps we are closer to defining a fundamental set of parameters which govern disk galaxy evolution (e.g., Mollá, Ferini, & Díaz 1996, and references therein).

Acknowledgments. I would like to thank Liese van Zee for recalculating surface brightness parameters for her LSB dI galaxies so that I could include them in Figure 4. I would also like to thank Chris Taylor and Stacy McGaugh for comments on an earlier draft. Partial support from a NASA LTSARP grant No. NAGW-3189, is gratefully acknowledged.

References

Aaronson, M. 1986, in Star Forming Dwarf Galaxies and Related Objects, eds. D. Kunth, T.X. Thuan, and J.T.T. Van, Editions Frontieres, 125
Aparicio, A., Gallart, C., Chiosi, C., & Bertelli, G. 1996, ApJL, 469, 97
Bertelli, G., Bressan, A., Chiosi, C., Fagotto, F. & Nasi, E. 1994, A&AS 106, 275
Bertelli, G. Mateo, M., Chiosi, C., & Bressan, A. 1992, ApJ, 388, 400
Bothun, G. D., & Mould, J. R. 1988, ApJ, 324, 123
Butcher, H. 1977, ApJ, 216, 372
Caldwell, N., Armandroff, T. E., Da Costa, G. S., & Seitzer, P. 1998, AJ, 115, 535
Dohm-Palmer, R. C., Skillman, E. D., Saha, A., Tolstoy, E., Mateo, M., Gallagher, J., Hoessel, J., & Dufour, R. J. 1997, AJ, 114, 2527
Dohm-Palmer, R.C., Skillman, E.D., Gallagher, J., Tolstoy, E., Mateo, M., Dufour, R.J., Saha, A., Hoessel, J., & Chiosi, C. 1998, AJ, 116, 1227

Edmunds, M. G., & Phillips, S. 1989, MNRAS, 241, 9p
Gallagher, J. S., Tolstoy, E., Dohm-Palmer, R. C., Skillman, E. D., Cole, A., Hoessel, J., Saha, A., & Mateo, M. 1998, AJ, 115, 1869
Gallart, C., Aparicio, A., & Víchez, J. M. 1996, AJ, 112, 1950
Garnett, D. R., Shields, G. A., Skillman, E. D., Sagan, S. P., & Dufour, R. J. 1997, ApJ, 489, 63
Greggio, L., Marconi, G., Tosi, M., & Focardi, P. 1993, AJ, 105, 894
Hodge, P.W. 1989, ARA&A, 27, 139
Impey, C., & Bothun, G. 1997, ARA&A, 35, 267
Impey, C. D., Sprayberry, D., Irwin, M. J., & Bothun, G. D. 1996, ApJS, 105, 209
Israel, F. P. 1997, A&A, 328, 471
Kennicutt, R. C., Jr. 1989, ApJ, 344, 171
Mateo, M. 1998, ARA&A, 36, in press
McGaugh, S. S. 1994, ApJ, 426, 135
McGaugh, S. S., & Bothun, G. D. 1994, AJ, 107, 530
McGaugh, S. S., & de Blok, W. J. G. 1997, ApJ, 481, 689
McGaugh, S. S., & de Blok, W. J. G. 1998, ApJ, 499, 41
Mollá, M., Ferini, F., & Díaz, A. I. 1996, ApJ, 466, 668
Mould, J. R., Kristian, J., and Da Costa, G. S. 1983, ApJ, 270, 471
O'Neil, K., Bothun, G. D., Schombert, J., Cornell, M. E., & Impey, C. D. 1997, AJ, 114, 2448
Pickering, T. E. 1998, PhD. Thesis, University of Arizona
Roennback, J., & Bergvall, N. 1995, A&A, 302, 353
Skillman, E. D., Bomans, D. J., & Kobulnicky, H. A. 1997, ApJ, 474, 205
Skillman, E. D., Kennicutt, R. C. & Hodge, P. W. 1989, ApJ, 347, 875
Smecker-Hane, T. A., Stetson, P. B., Hesser, J. E., & Lehnert, M.D. 1994, AJ, 108, 507
Spaans, M., & Carollo, C. M. 1998, ApJ, 502, 640
Spaans, M., & Norman, C. A. 1997, ApJ, 483, 87
Taylor, C. L., Brinks, E., Pogge, R. W., & Skillman, E. D. 1994, AJ, 107, 971
Taylor, C. L., Kobulnicky, H. A., & Skillman, E. D. 1998, AJ, in press
Tolstoy, E., Gallagher, J. S., Hoessel, J., Saha, A., Skillman, E. D., Dohm-Palmer, R. C., & Mateo, M. 1998, AJ, 116, 1244
Tolstoy, E., & Saha, A. 1996, ApJ, 462, 672
Toomre, A., 1964, ApJ, 139, 1217
Tosi, M. Greggio, L., Marconi, G., & Focardi, P. 1991, AJ, 102, 951
van der Hulst, J. M., Skillman, E. D., Smith, T. R., Bothun, G. D., McGaugh, S. S., & de Blok, W. J. G. 1993, AJ, 106, 548
van Zee, L., Haynes, M. P., Salzer, J. J., & Broeils, A. H. 1997a, AJ, 113, 1618
van Zee, L., Haynes, M. P., & Salzer, J. J. 1997b, AJ, 114, 2479
van Zee, L., Haynes, M. P., & Salzer, J. J. 1997c, AJ, 114, 2497
van Zee, L., Skillman, E. D., & Salzer, J. J. 1998, AJ, 116, 1186
Zaritsky, D., Kennicutt, R. C., & Huchra, J. P. 1994, ApJ, 420, 87

Low Surface Brightness Galaxies Beyond Z = 0.5: Existence, Detection and Properties

G. Bothun
Dept. of Physics, University of Oregon, Eugene OR 97403
nuts@moo2.uoregon.edu

Abstract.
I argue that low surface brightness disks evolve sufficiently slowly that, at any redshift, there should be a high space density of these objects. Cosmological dimming, in combination with the tenacity of Freeman's Law, will make their direct detection very difficult. Given their abundance at z =0, it seems reasonable to suggest that such hidden galaxies will be important absorbers along the line of sight to distant QSOs.

1. Introduction

This talk is a complete exercise in speculation (see also Bothun et al 1997). Why? Well a) because its easier than trying to be definitive and b) no data yet exists on the topic to even build a position, let alone defend it. Let's begin with my concept of what a low surface brightness (LSB) galaxy is. A prime example is shown in Figure 1 which is one of the more extreme LSB objects discovered in O'Neil (1997a).

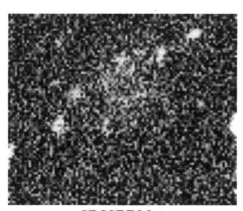

SDSSPDM-1

The reason to introduce the topic with this object is two fold: a) this is exactly the kind of object which will be a challenge for the Sloan Digital Sky Survey (SDSS) to detect (PDM stands for Please Detect Me) and b) how can objects like this be detected at all at moderate redshift due to $(1+z)^4$ surface brightness dilution? The severity of this term will likely be borne out when surface photometry is done on some of the disk galaxies in the Hubble Deep Field. So in some sense were done already. Cosmological dimming will simply exacerbate the already severe selection effects present at $z=0$ in the detection of LSB galaxies and hence this topic can be confined to a single page. However, a single page of speculation is not very interesting so let's press on.

2. Surface Brightness Evolution

Clearly the distribution of disk galaxy surface brightness at any redshift depends upon the rate of surface brightness evolution. Galaxy evolution is usually defined in terms of luminosity, size or color evolution but these are a sub-set of surface brightness evolution. The observed surface brightness is the convolution of the average spacing of stars, the stellar luminosity function and the amount of dust that is present. Hence, LSB galaxies could be so because of:

• Low mass density (an argument favored by de Blok and McGaugh 1996) which leads to large stellar separation

• A high M/L stellar population (e.g. low luminosity per unit baryon mass)

• High obscuration (although Witt et al 1992 clearly show that forward scattering by dust is sufficient so that the outgoing optical luminosity can not be suppressed by more than about a factor of 2).

• Something else.

A useful thought experiment is to consider the long term surface brightness evolution of galaxies. They are born as baryonic gas bags inside of some dark potential and thus start off as LSB as few stars have formed. Eventually, we can expect the disk to run out of gas to make stars (although Tully in this conference has stated that disk galaxies won't do this but continue to operate in some low recycling level) and then it will begin to fade significantly. A portion of this fading process might have already occurred and this could account for the population of red LSB disks recently discovered by O'Neil et al 1997b (see also O'Neil 1998 - this conference).

3. Freeman's Law is Not Easy to Break

As is well known, Freeman's Law establishes that the typical disk galaxy you find in a survey has a central surface brightness about 1 mag brighter than the sky background. In the blue the typical value for the sky background (at a dark site) is 22.5 - 23.0 mag arcsec^{-2}. Freeman's Law would then suggest the typical disk which is selected would have $\mu_B(0) \sim 21.5$ mag arcsec^{-2}. This was the essence of Disney's (1976) original argument. This conference exists largely because Freeman's Law has been broken and even astronomers outside this conference now actually believe that LSB galaxies exist. In fairness to Freeman, his data sample of the time (Freeman 1970) did contain IC 1613, a genuine LSB galaxy by all accounts (see McGaugh and Bothun 1994). Scrutiny of Freeman's result was motivated by the puzzling implication it had for galaxy formation. Namely, it would mean that the physics of disk galaxy formation is rather finely tuned to produced a preferred mass density disk over a larger range of scale lengths.

Now we know that the range of central surface brightness in disk galaxies at a given scale length is large (de Jong 1998 - this conference) . To first order, this implies large variance in the amount of dissipation, baryonic drag and angular momentum transfer that must occur during the initial collapse phase. Although its possible to model this, the input physics is unclear and the relevance of objects such as that shown in Figure 1 is to demonstrate that extremely diffuse flattened galaxies did, apparently, form. I claim no understanding of how such diffuse systems either exist or survived a Hubble time of potential galaxy harassment. Given the extreme difficulty of discovering such diffuse systems, and here I mean objects with $\mu_B(0) = 25.0$ mag arcsec^{-2} and fainter, implies that if you found a few, there must be many. Below is an example of another member of the SDSSPDM class.

SDSSPDM-2

As a testimony to how even more difficult the detection of this population would be at modest redshift, consider an interesting result contained in O'Neil et al 1999. Those authors performed a surface photometry analysis of 210 galaxies in 3 deep WFPC2 frames. The data was all taken through the 814W filter. Most of the detected galaxies are disks with scale lengths of 0.2 to 1.5 arcseconds. The 814W sky brightness was measured to be 23.0 mag arcsec^{-2}(which is equivalent to a Cousins I-band sky brightness of 21.8 mag arcsec^{-2}). The median central surface brightness obtained for the fitted disks was 814W(0) = 21.9 mag arcsec^{-2} ⇒ 1 magnitude brighter than the sky background. Thus, even with a much different detector and observing system, a random sample of galaxies returns Freeman's result. This directly confirms the tenacity of visibility function induced selection effects.

4. Once an LSB Always an LSB?

If star formation in some disk galaxies is episodic in nature, would it be possible for LSB disks to migrate out of this category as "bursts" of star formation serve to elevate the surface brightness. One immediate problem with what seems to be a reasonable scenario is the lack of any color vs. surface brightness relation for disk galaxies. As documented by O'Neil et al (1997b) even at the very lowest levels of central surface brightness, both very red and very blue disks can be found. Another problem may arises if we take the surface mass density argument to its extreme. If $(\delta\rho/\rho)_{gas} \propto (\delta\rho/\rho)_{stars}$ (an unproven assumption) then low values of $\mu_B(0)$ mean low gas densities. Under these conditions, the Jeans length can be 1-2 disk scale lengths (because the low gas density requires a large scale length for the gas mass to exceed the Jeans mass).

The physics of star formation that would occur in this regime is probably very different than that which governs the fragmentation of Giant Molecular Clouds in the ISM of HSB disks. For those disk galaxies detected to date with $\mu_B(0) \sim 25.0$ mag arcsec^{-2}, this begs the question, how could stars ever form in the first place in such a low density environment? Current attempts to explain this as a result of star formation in small clumps of molecular gas (e.g. Spaans 1998 - this conference) don't seem destined to account for all the stars that presently exist in these systems. On a global scale, LSB disks have been shown to be significantly deficient in molecular material and dust (e.g. Schombert et al 1990; de Blok et al 1996) yet they seem to contain as many stars as those galaxies where we know that star formation is occurring in GMCs. This basic point seems under-appreciated. It certainly obscures whether or not star formation would follow a Schmidt Law in LSB disks.

O'Neil et al (1998) have made some toy models to simulate the effects of weak tidal encounters and the initiation of star formation in localized regions of a LSB disk. The starburst strength is 15% of total galaxy gas mass that is available. The key feature of these models is that the low gas density means limited formation of high mass stars as there is a low probability of sufficient gas mass per model cell. Thus, in this scheme there is a natural suppression of high mass star formation and one does wonder if this is the relevant physics of star formation in a low density environment. This suppression of massive star formation would lead to a low rate of chemical evolution of the disk and an under-production of dust mass over a Hubble time. The color and luminosity evolution of such a disk would also not be very dramatic as a result of induced star formation. All of these predicted properties seem to be borne out by the observations to date of LSB disks. This produces the expectation that LSBs are docile over time and never are the hosts of monster starbursts that might drive the evolution of higher surface brightness disk galaxies.

This also means that once $\mu_B(0)$ is low, it will remain low. This directly implies that the space density of LSB disks should be relatively large at all redshifts. This has obvious implications with respect to the kinds of galaxies that produce the observed QSO absorption lines. The two views adopted at this conference were 1) All QSO absorption lines originate in the halos of big galaxies (e.g. Lanzetta 1998) or 2) LSB galaxies make up a substantial portion of the absorbers because the overall space density of galaxies is higher than we think it is (e.g. Linder 1998). Resolution of these two views requires more data. However, it is instructive to remember what Mateo and Armandroff have told us at this conference: halos around big galaxies contain several LSB objects!

To get more substantial evolution in $\mu_B(0)$ requires a model in which there is a threshold $(\delta\rho/\rho)_{gas}$ that is maintained by disk galaxies, independent of their value of $\mu_B(0)$. This allows galaxies to have a larger reservoir of gas to be processed into stars thus increasing their luminosity/color evolution. While the existence of this threshold is actually supported by real observations (Skillman et al 1987, van der Hulst et al 1993, Pickering et al 1997), no disk with $\mu_B(0)$ fainter than 23.0 mag arcsec^{-2} has been mapped at 21-cm. If the threshold argument is correct, this should become self-evident in the Parkes Multi-Beam Survey.

5. Summary

1. Surface brightness selection effects are already severe at z =0. The discovery of z > 0.2 LSB disks will be a challenge because Freeman's Law is robust.

2. The rate of surface brightness evolution of disk galaxies is not well known but it seems reasonable to expect that some disk galaxies have made the bulk of their stars and are beginning to fade. The red LSBs detected by O'Neil et al (1997b) may be the first z=0 detection of this population.

3. Preliminary modeling of the response of an LSB disk to induced star formation indicates that the formation of massive stars is suppressed due to the low initial gas density. This implies a low rate of chemical and surface brightness evolution for these disks and a preservation of their space density over a large range in redshifts. Potentially, these disks are a major contributor to QSO absorption line systems.

4. All disk galaxies are destined to become LSB high M/L systems. I propose that we all come back to Wales in 10 billion years and count the number of LSB galaxies struggling for detection against the noisy background sky.

Acknowledgments. I thank the conference organizers for inviting me and their persistence that I actually show up. As always, my collaborators Chris Impey, Jim Schombert, Stacy McGaugh, Karen O'Neil, Erwin de Blok and Thijs van der Hulst did all the work. Thanks also to the Welsh beer makers.

References

Bothun, G.D., Impey, C., & McGaugh, S. 1997 PASP 109, 745
de Blok, W., and McGaugh, S. 1996, ApJ469, L89
de Blok, W., McGaugh, S. & van der Hulst, J. 1996, MNRAS, 283, 18
de Jong, R. 1998 - this conference
Disney, M. 1976 Nature 263,573
Freeman, K. 1970 ApJ169,811
Lanzetta, K. 1998 - this conference
Linder, S. 1998 - this conference
McGaugh, S. & Bothun, G. 1994, AJ, 107, 530
O'Neil, K. et al 1999 ApJsubmitted
O'Neil, K. 1998 - this conference
O'Neil, K., Bothun, G., Schombert, J. 1998 AJ, in press
O'Neil, K., Bothun, G., & Cornell, M. 1997a AJ, 113, 1212
O'Neil, K. et al 1997b AJ, 114, 2448
Pickering, T., et al 1997 AJ114,1858
Schombert, J. et al 1990 AJ100,1523
Skillman, E., et al 1987 A&A 185,61
Spaans, M. 1998 - this conference
van der Hulse T., et al 1993, AJ106, 548
Witt, A., Thronson, H., and Capuano, J. 1992, ApJ393, 611

Environmental Effects on the Faint End of the Luminosity Function

S. Phillipps, J.B. Jones

Department of Physics, University of Bristol, Bristol, UK

R.M. Smith

Department of Physics and Astronomy, University of Wales, Cardiff, UK

W.J. Couch, S.P. Driver

School of Physics, University of New South Wales, Sydney, Australia

Abstract. Recent studies have demonstrated that many galaxy clusters have luminosity functions (LFs) which are steep at the faint end. However, it is equally clear that not all clusters have identical LFs. In this paper we explore whether the variation in LF shape correlates with other cluster or environmental properties.

1. Introduction

Much recent work has been devoted to measuring the galaxy luminosity function (LF) within rich clusters, particularly with regard to the faint end which has become accessible to detailed study through various technical and observational improvements (see the paper by Smith et al. in these proceedings). These studies suggest that the LF becomes steep (Schechter (1976) slope $\alpha \leq -1.5$) in many clusters, faintwards of about $M_B = -17.5$ or $M_R \simeq -19$ (for $H_0 = 50$ km s^{-1} Mpc^{-1}), where (generally low surface brightness) dwarfs begin to dominate (e.g. Smith, Driver & Phillipps 1997; Trentham 1997a,b). Using deep CCD imaging from the Anglo-Australian Telescope, we have now extended this work (see Driver, Couch & Phillipps 1998), in order to examine the luminosity distribution in and across a variety of Abell and ACO clusters. In particular, we were interested in any possible dependence of the dwarf population (specifically the ratio of the number of dwarfs to the number of giants) on cluster type or on position within the cluster.

2. Dwarfs in Rich Clusters

A number of papers (e.g. Driver et al. 1994; Smith et al. 1997; Wilson et al. 1997) have demonstrated remarkably similar dwarf populations in a number of morphologically similar, dense rich clusters like (and including) Coma. This similarity appears not only in the faint end slope of the LF, around $\alpha = -1.8$, but also in the point at which the steep slope cuts in, $M_R \simeq -19$ (i.e. about $M^* + 3.5$). The latter implies equal ratios of dwarf to giant galaxy numbers in the different clusters.

However, there clearly do exist differences between some clusters. For example, several of the clusters in the Driver et al. (1998) sample do not show a conspicuous turn up at the faint end (see also Lopez-Cruz et al. 1997 for further examples). Either these clusters contain completely different types of dwarf galaxy population or, as we suggest, the turn up occurs at fainter magnitudes. For a composite giant plus dwarf LF, this is equivalent to a smaller number of dwarfs relative to giants.

To simplify the discussion, we will define the dwarf to giant ratio DGR as the number of galaxies with $-16.5 \geq M_R \geq -19.5$ compared to those with $-19.5 \geq M_R$),

i.e. $DGR = \frac{N(-16.5 > M_R > -19.5)}{N(-19.5 > M_R > -23.5)}$.

The DGR does not have any obvious dependence on cluster richness (Driver et al. 1998; see also Turner et al. 1993), but we can also check for variations with morphological characteristics of the clusters. For giant galaxies, it is well known that a cluster's structural and population characteristics are well correlated. For example, dense regular clusters are of early Bautz-Morgan type (dominated by cD galaxies) and have the highest fractions of giant ellipticals (Dressler 1980). In a similar way, we find that the DGR (i.e. the fraction of dwarfs) is *smallest* in these early Bautz-Morgan type clusters (Driver et al. 1998). Next consider the galaxy density. We can characterise the clusters by their central (giant) galaxy number densities, for instance the number of galaxies brighter than $M_R = -19.5$ within the central 1 Mpc^2 area. An alternative would be to use Dressler's (1980) measure of the average number of near neighbours. We then find (solid squares in Figure 1) that the clusters with the least prominent dwarf populations (low DGRs ~ 1) are just those with the highest projected galaxy densities (e.g. the Bautz-Morgan Type I-II cluster A3888). Previously, Turner et al. (1993) had noted that the rich but low density cluster A3574, which is very spiral rich (Willmer et al. 1991), had a very high ratio of low surface brightness (LSB) dwarfs to giants. This is now backed up by the observations of clusters like A204 which are dwarf rich (DGR ~ 3), have low central densities and late B-M types (A204 is B-M III).

To extend the range of environments studied, we can add in further LF results from the literature (Figure 2). A problem here, of course, is the lack of homogeneity due to different observed wavebands, different object detection techniques and so forth. Nevertheless, we can explore the general trends. Several points are shown for surveys of Coma (hexagons). These surveys (Thompson & Gregory 1993, Lobo et al. 1997, Secker & Harris 1996 and Trentham 1998) cover different areas and hence different mean projected densities (see also the next section). All these lie close to the relation defined by our original data, with the larger area surveys having higher DGRs. Points (filled triangles) representing the rich B-M type I X-ray selected clusters studied by Lopez-Cruz et al. (1997) fall at somewhat lower DGR than most of our clusters at similar densities. However we should note that these clusters were selected (from a larger unpublished sample) *only* if they had LFs well fitted by a single Schechter function. This obviously precludes clusters with steep LF turn-ups at intermediate magnitudes

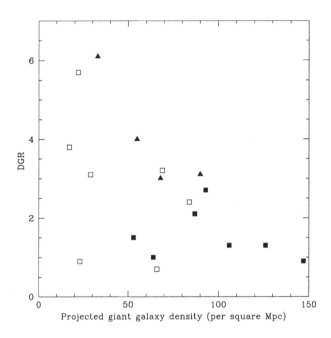

Figure 1. Variation of the dwarf-to-giant ratio (DGR), as defined in the text, with projected density of cluster giants (per square Mpc). Solid boxes represent the central 1 square Mpc regions of the clusters, the open boxes the outer regions (data from Driver at al 1998). The triangles show the variation over a wider range of radii for Abell 2554 (data from Smith et al 1997). Note that typical error bars (due to the combination of Poisson errors and background subtraction errors) are 10% in density and 20% in DGR for the denser regions, rising to 30% in density and 50% in DGR at the lowest densities (and hence object numbers). The outlier at low density and low DGR (the outskirts of A22) has a very large error in DGR ($\sim 100\%$).

and hence rules out high DGRs. The one comparison cluster they do show *with* a turn up (A1569 at DGR $\simeq 4.2$) clearly supports our overall trend.

Ferguson & Sandage (1991 = FS), on the other hand, deduced a trend in the opposite direction, from a study of fairly poor groups and clusters, with the early type dwarf-to-giant ratio *increasing* for denser clusters. However, this is not necessarily as contradictory to the present result as it might initially appear. For instance, FS select their dwarfs morphologically, not by luminosity (morphologically classified dwarfs and giants significantly overlap in luminosity) and they also concentrate solely on early type dwarfs. If, as we might expect, low density regions have significant numbers of late type dwarf irregulars (e.g. Thuan et al. 1991), then the FS definition of DGR may give a lower value than ours for these regions. Furthermore FS calculate their projected densities from *all* detected galaxies, down to very faint dwarfs. Regions with high DGR will therefore be forced to much higher densities than we would calculate for giants only. These two effects may go much of the way to reconciling our respective results. This is illustrated by the open triangles in Figure 2, which are an attempt to place the FS points on our system; magnitudes have been adjusted approximately for the different wavebands, DGRs have been estimated from the LFs and the cluster central densities (from Ferguson & Sandage 1990) have been scaled down by the fraction of their overall galaxy counts which are giants (by our luminosity definition). Given the uncertainties in the translation, most of the FS points then lie close to those of our overall distribution. Finally, a field LF with a steep faint end tail ($\alpha \simeq -1.5$; e.g. Marzke, Huchra & Geller 1994, Zucca et al. 1997, Morgan, Smith & Phillipps 1998)) would also give a point (filled pentagon) at DGR $\simeq 4$, again consistent with the trend seen in the clusters.

Nevertheless, there are exceptions. The FS points of lowest density (the Leo and Dorado groups) also have low DGR (and lie close to *our* main 'outlier', the point for the outer region of A22). The Local Group (shown by the star) would also be in this regime, at low density and DGR = 2, as would the 'conventional' field with $\alpha \simeq -1.1$ (Efstathiou, Ellis & Peterson 1988; Loveday et al. 1992) and hence DGR $\simeq 1.5$ (open pentagon). This may suggest that at very low density the trend is reversed (i.e. is in the direction seen by FS), or that the cosmic (and/or statistical) scatter becomes large. More data in the very low density regime is probably required before we can make a definitive statement on a possible reversal of the slope of the DGR versus density relation. In particular, the scatter in the derived faint end of the field LF between different surveys (see, e.g., the recent discussion in Metcalfe et al. 1998) precludes using this to tie down the low density end of the plot.

2.1. Population Gradients

It was suggested by the results on A2554 (Smith et al. 1997), that the dwarf population was more spatially extended than that of the giants, i.e. the dwarf to giant ratio increased outwards. This type of population gradient has now been confirmed by the results in Driver et al. (1998) illustrated in Figure 1, where we contrast the inner 1 Mpc^2 areas (solid symbols) with the outer regions of the same clusters (open symbols). The triangles show in slightly more detail the run of DGR with radius (and hence density) across A2554. A similar effect can be seen for Coma in Figure 2 and can explain the discrepancy between the

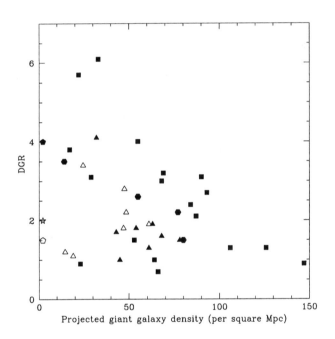

Figure 2. As Figure 1, but including data from other observers. Squares are our data repeated from Figure 1, hexagons are for various Coma surveys detailed in the text, filled triangles are from Lopez-Cruz's sample and open triangles are for Ferguson and Sandage's poor clusters and groups. The open pentagon at low density represents a conventional 'flat' field LF, the filled pentagon a possible steep ($\alpha \simeq -1.5$) field LF and the Local Group is represented by the star at DGR = 2.

LFs derived for the core and as against larger areas. It is found, too, in Virgo (Phillipps et al. 1998a; Jones et al., these proceedings), where the dwarf LSBG population has almost constant number density across the central areas while the giant density drops by a factor ~ 3.

3. A Dwarf Population Density Relation

The obvious synthesis of the above results is a relationship between the *local* galaxy density and the fraction of dwarfs (i.e. the relative amplitude of the dwarf LF). The inner, densest parts of rich clusters have the smallest fraction of dwarfs, while loose clusters and the outer parts of regular clusters, where the density is low, have high dwarf fractions. It is particularly interesting to note the clear overlap region in Figure 1, where regions of low density on the outskirts of dense clusters (open squares) have similar DGRs to the regions of the same density at the centres of looser clusters (solid squares).

The proposed relation of course mimics the well known morphology - density relation (Dressler 1980), wherein the central parts of rich clusters have the highest early type galaxy fraction, this fraction then declining with decreasing local galaxy density. Putting the two relations together, it would also imply that dwarfs preferentially occur in the same environments as spirals. This would be in agreement with the weaker clustering of low luminosity systems in general (e.g. Loveday et al. 1995), as well as for spirals compared to ellipticals (Geller & Davies 1976). Thuan et al. (1991) have previously discussed the similar spatial distributions of dwarfs (in particular dwarf irregulars) and larger late type systems.

4. The Origin of the Relation

As with the corresponding morphology - density relation for giant galaxies, the cause of our population - density relation could be either 'nature' or 'nurture', i.e. initial conditions or evolution. Some clues may be provided by the most recent semi-analytic models of galaxy formation, which have been able to account successfully for the excess of (giant) early type galaxies in dense environments (e.g. Baugh, Cole & Frenk 1996), basically through different merging histories for different types of galaxy. Does this also work for the dwarfs?

The steep faint end slope of the LF appears to be a generic result of hierarchical clustering models (e.g. White & Frenk 1991; Frenk et al. 1996; Kauffmann, Nusser & Steinmetz 1997 = KNS), so is naturally accounted for in the current generation of models. The general hierarchical formation picture envisages (mainly baryonic) galaxies forming at the cores of dark matter halos. The halos themselves merge according to the general Press-Schechter (1974) prescription to generate the present day halo mass function. However the galaxies can retain their individual identities within the growing dark halos, because of their much longer merging time scales. The accretion of small halos by a large one then results in the main galaxy (or cluster of galaxies, for very large mass halos) acquiring a number of smaller satellites (or the cluster gaining additional, less tightly bound, members).

KNS have presented a detailed study of the distribution of the luminosities of galaxies expected to be associated with a single halo of given mass. We can thus easily compare the theoretically expected numbers of dwarf galaxies per unit giant galaxy luminosity with our empirical results (Phillipps et al. 1998b).

The KNS models mimic a "Milky Way system" (halo mass $5 \times 10^{12} M_\odot$), a sizeable group (halo mass $5 \times 10^{13} M_\odot$) and a cluster mass halo ($10^{15} M_\odot$). Their results imply that the Milky Way and small group halos have similar numbers of dwarf galaxies per unit giant galaxy light, whereas the dense cluster environment has a much smaller number of dwarfs for a given total giant galaxy luminosity. Thus the predictions of the hierarchical models (which depend, of course, on the merger history of the galaxies) are in qualitative agreement with our empirical results if we identify loose clusters and the outskirts of rich clusters with a population of (infalling?) groups (cf. Abraham et al. 1996), whereas the central dense regions of the clusters originate from already massive dark halos. If we renormalise from unit galaxy light to an effective giant galaxy LF amplitude (see Phillipps et al. 1998b) then the actual expected ratios ($\sim$ 1 to a few) are also consistent with our observational results.

By inputting realistic star formation laws etc., KNS could further identify the galaxies in the most massive halos with old elliptical galaxies, and those in low mass halos with galaxies with continued star formation. This would imply the likelihood that our dwarfs in low density regions may still be star forming, or at least have had star formation in the relatively recent past (cf. Phillipps & Driver 1995 and references therein). Note, too, that these galaxy formation models would also indicate that the usual (giant) morphology - density relation and our (dwarf) population - density relation *do* arise in basically the same way. Finally, we can see that if these semi-analytic models are reasonably believable, then we need not necessarily expect the field to be even richer in dwarfs than loose clusters; the dwarf to giant ratio seems to level off at the densities reached in fairly large groups.

5. Summary

To summarise, then, we suggest that the current data on the relative numbers of dwarf galaxies in different clusters and groups can be understood in terms of a general dwarf population versus local galaxy density relation, similar to the well known morphology - density relation for giants. Low density environments are the preferred habitat of low luminosity galaxies; in dense regions they occur in similar numbers to giants, but at low densities dwarfs dominate numerically by a large factor. This fits in with the general idea that low luminosity galaxies are less clustered than high luminosity ones (particularly giant ellipticals). Plausible theoretical justifications for the population - density relation can be found within the context of current semi-analytic models of hierarchical structure formation.

References

Abraham R.G., et al., 1996, ApJS, 471, 694
Baugh C.M., Cole S., Frenk C.S., 1996, MNRAS, 283, 1361

Dressler A., 1980, ApJ, 236, 351
Driver S.P., Phillipps S., Davies J.I., Morgan I., Disney M.J., 1994, MNRAS, 268, 393
Driver S.P., Couch W.J., Phillipps S., 1998, MNRAS, in press
Efstathiou G., Ellis R.S., Peterson B.A., 1988, MNRAS, 232, 431
Ferguson H.C., Sandage A., 1990, AJ, 100, 1
Ferguson H.C., Sandage A., 1991, AJ, 96, 1520
Frenk C.S., Evrard A.E., White S.D.M., Summers F.J., 1996, ApJ, 472, 460
Geller M.J., Davis M., 1976, ApJ, 208, 13
Jones J.B., Phillipps S., Schwartzenberg J.M., Parker Q.A., 1998, The Low Surface Brightness Universe, p.xxx
Kauffmann G., Nusser A., Steinmetz M., 1997, MNRAS, 286, 795
Lobo C., et al., 1997, A&A, 317, 385
Lopez-Cruz O., Yee H.K.C., Brown J.P., Jones C., Forman W., 1997, ApJL, 475, L97
Loveday J., Maddox S.J., Efstathiou G., Peterson B.A., 1995, ApJ, 442, 457
Loveday J., Peterson B.A., Efstathiou G., Maddox S.J., 1992, ApJ, 390, 338
Marzke R., Huchra J.P., Geller M.J., 1994, ApJ, 428, 43
Metcalfe N., Ratcliffe A., Shanks T., Fong R., 1998, MNRAS, 294, 147
Morgan I., Smith R.M., Phillipps S., 1998, MNRAS, 295, 99
Phillipps S., Driver S.P., 1995, MNRAS, 274, 832
Phillipps S., Driver S.P., Couch W.J., Smith R.M., 1998b, ApJ, 498, L119
Phillipps S., Parker Q.A., Schwartzenberg J.M., Jones J.B., 1998a, ApJ, 493, L59
Press W.H., Schechter P.L., 1974, ApJ, 187, 425
Schechter P., 1976, ApJ, 203, 297
Secker J., Harris W.E., 1996, ApJ, 469, 623
Smith R.M., Driver S.P., Phillipps S., 1997, MNRAS, 287, 415
Smith R.M., Phillipps S., Driver S.P., Couch R.M., 1998, The Low Surface Brightness Universe, p.xxx
Thompson L.A., Gregory S.A., 1993, AJ, 106, 2197
Thuan T.X., Alimi J.M., Gott J.R., Schneider S.E., 1991, ApJ, 370, 25
Trentham N., 1997a, MNRAS, 286, 133
Trentham N., 1997b, MNRAS, 290, 334
Trentham N., 1998, MNRAS, 293, 71
Turner J.A., Phillipps S., Davies J.I., Disney M.J., 1993, MNRAS, 261, 39
White S.D.M., Frenk C.S., 1991, ApJ, 379, 52
Willmer C., Focardi P., Chan R., Pellegrini P., da Costa L., 1991, AJ, 101, 57
Wilson G, Smail I., Ellis R.S., Couch W.J., 1997, MNRAS, 284, 915
Zucca E., et al., 1997, in Wide-Field Spectroscopy, eds. Kontizas E. et al., Dordrecht; Reidel, p.247

Morphology and Stellar Populations in the Gas-Rich, Giant LSBGs

P. Knezek[1]

Center for Astrophysical Sciences, Bloomberg Center, The Johns Hopkins University, Baltimore, MD 21218-2695

Abstract. An unexpected characteristic of low surface brightness galaxies (LSBGs) is that a significant number are massive and possess substantial amounts of atomic gas. We present preliminary results of an ongoing program to obtain BVRIJHK imaging, along with some nuclear spectroscopy, of a well-defined sample of LSBGs which are gas-rich and of similar size to giant, high surface brightness spiral galaxies (HSBGs). These LSBGs span the entire range of Hubble disk morphologies. While their disks are bluer, on average, than comparable HSBGs, the optical morphology of massive LSBGs indicates that many of these systems have undergone previous star formation episodes. They typically have long disk scale lengths, and range from $M_B = -16$ to -22 ($H_0 = 75$ km s^{-1} Mpc^{-1}). About half of the LSBGs with bulges show evidence of nuclear activity, and $\sim$30% appear to be barred. These massive, gas-rich LSBGs apparently have varied, and often complex, evolutionary histories.

1. Introduction

One of the discoveries in the work on low surface brightness galaxies (LSBGs) was that a significant number are massive and possess significant quantities of atomic gas. Many disk galaxies with sizes approaching that of Malin 1 ($M_{HI} \sim 2\times10^{11}$ $M_\odot$; D $\sim$ 147 kpc, $H_0 = 75$ km s^{-1} Mpc^{-1}; Impey and Bothun 1989) have been identified through the use of the Palomar Sky Surveys (Schombert et al. 1992) and UK Schmidt plates (Impey et al. 1996), with follow-up studies in HI. While many of these galaxies have large amounts of atomic gas and unusually blue disk colors, they have only weak regions of Hα emission, indicating little ongoing massive star formation, and their low surface brightnesses suggest extremely low stellar surface densities. The blue disk colors have ruled out the hypothesis that all LSBGs are faded galaxies, but the difficulty in disentangling the difference between low metallicity and young stellar populations based on broadband optical colors has hindered understanding their stellar evolutionary history. It has been suggested (McGaugh 1992) that some LSBGs may, in fact, be undergoing their first episodes of star formation since there is little evidence of a difference in the distribution of light in $UBVRI$ images. The LSBGs in his sample are largely *dwarf* systems, however, and the situation is not necessarily comparable for more massive LSBGs. In fact, as can be seen below, the optical morphology of massive LSBGs in B and R is consistent with the idea that many of these systems have undergone previous episodes of star formation.

We are conducting an extensive survey of both broadband optical ($BVRI$) and near-infrared (JHK) imaging of a sample of LSBGs which are gas-rich ($M_{HI} \geq 5\times10^9 M_\odot$) and of similar size to giant spiral galaxies ($D_{25} \geq 25$kpc). We have also obtained spectroscopy of the nuclear regions of a small sample of HI-rich giant LSBGs which indicate that some LSBGs, particularly those with bulges, *do* have low level active galactic nuclei (Knezek & Schombert 1999).

2. The Sample

The primary sample consists of galaxies identified from the *Uppsala General Catalog of Galaxies* (Nilson 1973) based on the first Palomar Observatory Sky Survey, and supplemented by galaxies from the second Palomar Sky Survey (Schombert and Bothun 1988; Schombert et al. 1992). These galaxies are selected to be gas-rich ($M_{HI} \geq 5\times10^9 M_\odot$) and of similar size to giant spiral galaxies ($D_{25} \geq 25$kpc), assuming $H_o = 75$ km s^{-1} Mpc^{-1}. No morphological criteria is applied other than the galaxies must be identified as disks. Atomic hydrogen data is from Schneider et al. (1990, 1992). Only galaxies with galactic latitudes $> \pm 30°$ are included. Galaxies are then separated into "high" and "low" surface brightness samples based on their mean blue surface brightnesses, which are initially estimated from published blue magnitues and sizes from either Nilson (1973) or Schombert et al. (1992), then refined using our own data. Those galaxies with $\mu_B < 24.5$ mag arcsec^{-2} are designated giant high surface brightness galaxies (HSBGs), and those with $\mu_B > 24.5$ mag arcsec^{-2} are designated giant low surface brightness galaxies (LSBGs). This is a mean blue surface brightness measured within $\mu_B = 26.0$ mag arcsec^{-2} from our optical data.

3. Observations

The broadband optical and near-infrared (NIR) imaging presented here are part of a larger, ongoing, project to image the entire sample of 175 LSBGs selected according to the criteria above, as well as a sample of corresponding HSBGs to use for comparison. Imaging has primarily been accomplished through the use of Michigan-Dartmouth-MIT Observatory (MDM; 1993-1995), San Pedro Martir Observatory (1996-present), and Las Campanas Observatory (1995-1997). A few observations of galaxies too large to be imaged at those facilities were obtained using the Kitt Peak National Observatory 0.9m in 1993. $BVRI$ imaging has been obtained of $\sim$ half the sample of 175 LSBGS, and JHK imaging has been obtained for $\sim$ one-third of the sample. These data have been corrected for galactic extinction using the interpolation program provided by Burstein and Heiles. No intrinsic extinction correction is applied. Spectroscopic observations of a sample of HI-rich galaxies were obtained using the MDM 2.4m in September, 1992 and January, 1993 (Knezek & Schombert 1999).

4. Discussion

As can be seen in Figure 1, the most obvious trait of these gas-rich LSBGs is that their optical morphology spans the *entire* Hubble Sequence for disk galaxies.

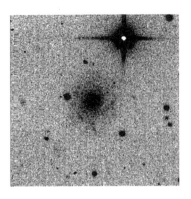 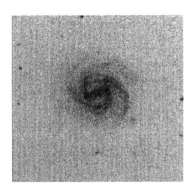
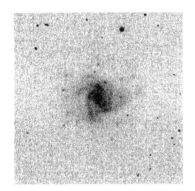 

Figure 1. B images of UGC 12740, UGC 2712, UGC 9022, and UGC 10313.

Not all LSBGs are late-type disk galaxies. Furthermore, it is not possible to determine whether a disk galaxy is a dwarf or a giant simply by its optical morphology. Many of these gas-rich, massive LSBGs were originally classified as dwarfs by Nilson (1973). Their optical morphologies are often indistinguishable from the true "dwarf spirals" studied by Pildis et al. (1997) and Matthews & Gallagher (1997). Yet the average disk scale length for a LSBG from our sample is ∼7 kpc, versus less than 1 kpc for the "dwarf spirals". Also, based on their optical morphology, ∼30% of the LSBGs are barred systems, a comparable number to that found for HSBGs.

We find that the gas-rich disk galaxies have very blue disks on average, whether they are high or low surface brightness systems. LSBGs with a prominent bulge typically have redder *disks*, with $< B - R > \sim 1.0$. The bulges themselves have $< B - R > \sim 1.4 - 1.6$, comparable to the bulge colors of HSBGs. Many of the bluest systems, with $< B - R > \sim 0.75$, can be fit by "pure disks", i.e. require a single exponential component to characterize their radial light distributions. These systems are too blue to be explained by old, metal poor stellar populations. Comparing to models by Worthey (1994), we find that even if the metallicity of a galaxy is only 1% solar, stars with ages of 8×10^9 years produce a system that is too red. Yet most LSBGs have only a few HII regions, and metallicities that are closer to 30% solar (McGaugh 1994). Furthermore, preliminary results based on the addition of the NIR data suggest that the colors of these systems are inconsistent with a Salpeter IMF and a constant star formation rate. Apparently, despite their low stellar surface densities, many LSBGs have had a complex star formation history.

Studies of the molecular emission of these LSBGs (Knezek 1993) indicate that only the redder LSBGs, and those with bulges, possess measureable molecular hydrogen. Furthermore, it is only these same redder, gas-rich LSBGs with bulges that have evidence of nuclear activity (Knezek & Schombert 1999). Of those which do exhibit nuclear activity, over half have line ratios indicative of active galactic nuclei (AGNs) rather than star formation. Those with AGNs show no evidence of a broad line region. Finally, we find that there is a cut-off in the relationship between disk central surface brightness and disk scale length, in the sense that *no* galaxies with bright disk central surface brightnesses have long disk scale lengths. These correlations may provide a clues to the formation and evolution of these LSBGs, and the underlying physics of star formation in disk galaxies in general.

5. Conclusions

Results of our ongoing project indicate that:

• LSBGs have optical morphologies which span the entire range of Hubble disk types. It is *not* possible to determine whether a disk galaxy is a dwarf, in luminosity and mass, simply by it's optical morphology. Furthermore, some bulge-dominated LSBGs have likely been missed simply because sample selections generally use mean surface brightnesses or late-type morphologies as a criteria.

• A significant fraction of the LSBGs (~30%) appear to possess optical bars.

• Gas rich, massive disk galaxies have very blue disk colors on average. For those galaxies with bulges, the bulge colors are normal. Preliminary results based on the additon of optical and near-infrared data suggests that the colors are inconsistent with a Salpeter IMF with a constant star formation rate. We have compared our data to models of the evolution of single burst stellar populations with varying ages and metallicities, assuming a standard IMF (Worthey 1994). If we assume metallicities of 30% solar and an age typical for globular clusters, 12×10^9 years, the resultant color is $B - R = 1.43$. If we assume the most extreme case, where the metallicity is 1% solar and the age is only 8×10^9 years, the modelled color is $B - R = 1.04$. While this is consistent with our average disk color for the sample as a whole, it is still too red for the pure disk systems, which have colors of $B - R = 0.749 \pm 0.166$.

- Only the redder, early-type LSBGs with bulges show evidence for molecular gas and AGN activity.

Acknowledgments. I am grateful to S. Lawrence and I. Cruz-Gonzalez for their continuing help with the observations of this ongoing project.

References

Impey, C. & Bothun, G. 1989, ApJ, 341, 89
Impey, C. et al. 1996, ApJS, 105, 2091
Knezek, P. M. 1993, Ph.D. thesis, Univ. of Massachusetts
Knezek, P. M. & Schombert, J. 1999, in prep
Matthews, L. S., & Gallagher III, J. S. 1997, AJ, 114, 1899
McGaugh, S. S. 1994, ApJ, 426, 135
McGaugh, S. S. 1992, Ph.D. thesis, Univ. of Michigan
Nilson, P. N. 1973, *Uppsala General Catalog of Galaxies, Uppsala Astron. Obs. Ann.*, 6
Pildis, R. et al. 1997, ApJ, 481, 157
Schneider, S. E. et al. 1992, ApJS, 81, 5
Schneider, S. E. et al. 1990, ApJS, 72, 245
Schombert, J., & Bothun, G. 1988, AJ, 95, 1389
Schombert, J. et al. 1992, AJ, 103, 1107
Worthey, G. 1994, ApJS, 95, 107

The Extreme Outer Regions of Disk Galaxies: Star Formation and Metal Abundances

Annette Ferguson

Institute of Astronomy, Univ. of Cambridge, Cambridge UK CB3 0HA

Rosemary Wyse

Dept. of Physics & Astronomy, JHU, Baltimore, MD, USA 21218

Jay Gallagher

Dept. of Astronomy, University of Wisconsin, Madison, WI, USA 53706

Abstract. The extreme outer regions of disk galaxies, lying at or beyond the classical optical radius defined by R_{25}, present an opportunity to study star formation and chemical evolution under unique physical conditions, possibly reminscent of those which existed during the early stages of disk evolution. We present here some of the first results from a large study to measure star formation rates and metallicities in the extreme outer limits of a sample of nearby spiral galaxies. Despite their low gas column densities, massive star formation is often observed in these outer parts, but at an azimuthally-averaged rate much lower than that seen in the inner disk. Gas-phase O/H abundances of roughly 10% solar characterize the gas at 1.5–2 R_{25}. The implications of our results for star formation 'laws' and models of disk evolution are discussed.

1. Introduction

Distinguishing between competing models of disk galaxy formation and evolution requires observational constraints on the radial variations of the present star formation rate, the star formation history, and the gas-phase chemical abundance. Unfortunately, most observational studies to date have focused only on the bright, easily-observed inner regions of galactic disks, lying at or within the classical optical radius, R_{25} (defined by the B-band 25th magnitude isophote). It is well known, however, that disk galaxies have HI disks which extend to typically $\gtrsim$ 1.5–2 R_{25}, and in some rare cases to $\gtrsim$ 3 R_{25}. Knowledge of the star formation rates and metallicities in these optically-faint, extreme outer reaches of disks is of particular importance for a variety of reasons. First of all, the predictions of various star formation laws and chemical evolution models often diverge most strongly in the outer parts of galaxies (eg. Prantzos & Aubert 1995, Tosi 1996), hence observational constraints are needed as far out in the disk as possible. Furthermore, the outer regions of disk galaxies provide a unique opportunity to study star formation and chemical evolution under the extreme physical conditions of low gas surface density (yet high gas fraction), low metallicity and long dynamical times; similar conditions are also inferred for many high-redshift damped Lyman-α systems (eg. Pettini et al 1997), and low surface brightness galaxies (Pickering et al 1997).

We have carried out a large observational project to study the extreme outer disks of a sample of nearby spiral galaxies. Results are presented here for three galaxies (NGC 628, NGC 1058 and NGC 6946) studied so far which exhibit recent massive star formation at particularly large radii (and which perhaps not surprisingly have unusually large HI-to-optical sizes). A full discussion of these results is provided in Ferguson et al (1998a, 1998b).

2. Observations

Deep wide-field Hα images were obtained to map the distribution of recent star formation, using the KPNO 0.9 m and the Lowell 1.8 m telescopes (deep BVR images were also obtained to study the extent and morphology of the underlying stellar disks, but these data will not be discussed here). In Figure 1, we show a continuum-subtracted Hα images of one of the galaxies, NGC 6946. Radial Hα surface brightness profiles were constructed via elliptical aperture photometry on the Hα images.

Long-slit spectroscopy in the range 3700–7000Å was carried out for a small sample ($\sim$ 10) of our newly-discovered outer disk HII regions, in order to obtain metallicities. Oxygen and nitrogen abundances were derived via well-established 'semi-empirical' methods, based on the inter-relationship between metallicity and the intensities of the strong lines, [OII] λ3727 and [OIII] $\lambda\lambda$4959,5007, via the parameter R_{23}. We adopted the particular calibrations proposed by McGaugh (1991) and Thurston et al (1996) to derive O/H and N/O respectively.

3. Star Formation Beyond the Optical Edge of Disk Galaxies

3.1. Morphology

Our deep images reveal the discovery of massive star formation out to the extent of our imagery ($\gtrsim$2 R_{25}) in all three galaxies. The inner and outer disk HII regions appear strikingly different, in that star formation in the outer disk occurs in smaller, fainter and more isolated HII regions. The brightest outer disk HII regions detected here have diameters of 150–500 pc and Hα luminosities of only 1–80 $\times$ 10^{37} erg s^{-1} (for reference, the Orion nebula has $L_{H\alpha} \sim 10^{37}$ erg s^{-1}). These luminosities imply enclosed ionizing populations of 0.2–20 equivalent O5V stars. Establishing whether the populations of inner and outer disk HII regions are actually intrinsically different, perhaps reflecting different modes of star formation, or if they only appear that way due to poor statistical sampling of the HII region luminosity function in the outer disk (ie. fewer HII regions, hence fewer luminous ones) is an important issue that we will address in the future.

The outer disk star formation appears remarkably organized, with the HII regions delineating narrow spiral arms. Our deep broad-band images reveal the existence of faint (B $\sim$ 26–28 mag/$\square''$) stellar arms in all three galaxies, associated with these HII arms, and inspection of published HI maps also reveals

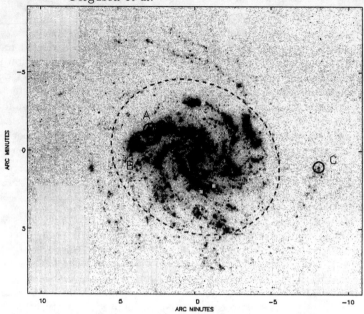

Figure 1. An Hα continuum-subtracted image of NGC 6946. The marked HII regions indicate those for which metallicities have been obtained. R$_{25}$ is marked in each case by the large dashed circle.

similar structures in the underlying neutral gas (eg. Shostak & van der Kruit 1984; Dickey et al 1990; Kamphuis 1993). The relationship between the inner and outer spiral structure remains unclear, as indeed is the dynamics underlying the outer arms.

3.2. Star Formation Rates

The radial variation of the massive star formation rate per unit area across the disk is traced by the azimuthally–averaged Hα surface brightness ($\Sigma_{H\alpha}$) distribution. Since most models for star formation invoke dependences on some form of the gas density, it is informative to plot $\Sigma_{H\alpha}$ against various components of the interstellar gas (Figure 2). Different components of the gas are seen to correlate in very different ways with the star formation rate. While a roughly linear correlation is seen between $\Sigma_{H\alpha}$ and Σ_{CO} (except for NGC 1058, where the relation is somewhat steeper), a very complicated non-linear behaviour is seen between $\Sigma_{H\alpha}$ and Σ_{HI}. Abrupt steepenings are seen at low total gas column densities in all cases, where they occur at azimuthally-averaged total gas surface densities of 5–10 M$_\odot$/pc^2 or 6–10 $\times$ 10^{20} cm^{-2}. Inspection of Figure 2 indicates that the location of the steepenings is approximately coincident with the radius where the disk undergoes the transition from being dominated by (warm) molecular to atomic gas, suggesting that a high covering factor of the molecular phase might be a requisite for significant star formation (eg. Elmegreen & Parravano 1994).

The existence of abrupt declines in $\Sigma_{H\alpha}$ at low gas surface densities make it impossible to describe the observations with a single-component Schmidt law,

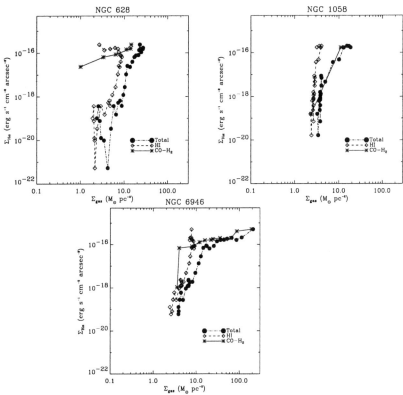

Figure 2. The variation of the deprojected, azimuthally-averaged Hα surface brightness, $\Sigma_{H\alpha}$, versus various components of gas surface density, Σ_{gas} (both quantities have been determined as a function of galactocentric radius in a series of elliptical annuli). The total gas surface density is the sum of the Σ_{CO} and Σ_{HI}, corrected for heavy elements.

with a dependence on total gas surface density alone. These steepenings are due almost entirely to sharp declines in the *covering factor* of star formation however, and not to changes in the rate at which stars form locally (see Ferguson et al 1998b). In other words, whenever star formation occurs, it occurs with roughly the same local intensity in both the inner and outer disks.

4. What Drives Star Formation at Low Gas Surface Densities?

The instability driving star formation may well be gravitational; in this case the Toomre-Q criterion (Toomre 1964) yields a critical gas surface density above which one expects local instability to axisymmetric perturbations. For an infinitely thin, one component isothermal gas disk, the critical gas surface density above which self-gravity overcomes shear and pressure is given by

$$\Sigma_{crit} = \frac{\alpha \sigma \kappa}{\pi G} \qquad (1)$$

where α is a constant of order unity, σ is the velocity dispersion of the gas and κ is the epicyclic frequency[1]. Following Kennicutt (1989), we adopt $\alpha = 0.67$.

We have used published gas data (rotation curves and velocity dispersions) to calculate the radial variation of Σ_{crit} for each galaxy. Both NGC628 and NGC1058 have low inclinations, which means that the amplitudes and shapes of the rotation curves are somewhat uncertain (but, on the other hand, the radial variation of the velocity dispersions are known with good accuracy). For these two galaxies, we have normalised the rotation curves using the Shostak (1978) relation between M_B and the maximum rotation velocity, and assumed that the curves remain flat in the outer regions. Two estimates of the critical gas density were made, one assuming a constant gas velocity dispersion (6 km s^{-1} for consistency with Kennicutt 1989) and the other including the radial variation determined directly from the HI observations.

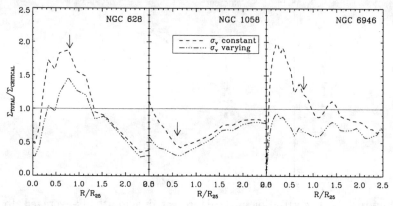

Figure 3. The radial variation of the ratio of the azimuthally-averaged (observed) gas surface density to the critical surface density for gravitational instability, calculated using both a constant velocity dispersion (dashed line) and a radially-varying one (dashed-dotted line). The solid horizontal line indicates the value of the ratio above which instability is expected.

The observed gas surface densities typically lie within a factor of two of the estimated critical densities at all radii, with the agreement often being even better when a radially-varying velocity dispersion is used (eg. NGC 6946, see Figure 3). This general agreement is very encouraging, in view of that fact that many of the input data – eg. rotation curves, gas surface densities, value of α etc – used to estimate Σ_{crit} have significant uncertainties. It therefore appears that the outer disks of these galaxies lie close enough to the Q-stability limit so that processes such as swing amplification can operate and trigger star formation locally. Realistic uncertainties in the quantities used in this derivation are not likely to change the value of the stability parameter (Σ/Σ_{crit}) by more than a factor of two at any given radius, and hence should not affect our principal conclusion that the gas disks are marginally unstable over a large radial zone.

---

[1] The epicyclic frequency is defined as $\kappa = \sqrt{2}\, \frac{V}{R} \sqrt{1 + \frac{R}{V}\frac{dV}{dR}}$.

On the other hand, the abrupt decreases in star formation rate at low gas surface density appear uncorrelated with changes in the gravitational stability of the disk (see Figure 3); that is, the locations where $\Sigma_{H\alpha}$ plummets (indicated by the small arrows in Figure 3) are no more or no less stable than any other location in the disk. Further, while the stability parameter changes by only a factor of a few at most across each disk, the star formation rate per unit area typically changes by a factor of 10^2–10^3! The sizes of the star formation regions are also difficult to reconcile with the predictions of the gravitational instability model, the theoretically most unstable length being much larger (ie. a few kiloparsecs). Thus, while local gravitational instability can probably account for the existence of star formation at large radii in these three galaxies, it fails to explain either the rates of star formation or the scales of star formation across the disks.

What alternative explanations are there for the reduced star formation rates seen in the outer disk? It may be that star formation occurs only when the local gas column exceeds some fixed threshold value (eg. Skillman 1987), and perhaps reaching that threshold becomes more difficult at larger radii. Such a local threshold could be related to a critical column density of dust necessary to shield molecular gas from UV radiation. Indeed, Elmegreen & Parravano (1994) argue that the low thermal pressures make it increasingly difficult to sustain a cool molecular phase beyond ~ 3 disk scale-lengths. Comparison of our Hα images to HI maps shows star formation down to local HI columns of a few times 10^{20} cm^{-2}, but the HI maps are for the most part low resolution (few kiloparsecs) and better data are needed to test this idea. Yet another explanation for the reduced rates could be an intrinsic correlation between azimuthally-averaged star formation rate and gas *volume* density, combined with a vertical flaring of the gas disk, such that the transformation between gas surface density and volume density varies with galactocentric radius (see also Madore et al 1974). It is well known that gaseous disks exhibit an increase of scaleheight at large radius (eg. Merrifield 1992) whereas the young stars are confined to a thin plane. The combination of an increase in gas scale height with radius with the slow decline of gas surface density could conspire to produce a rapid decline in areal star formation rate between the inner and the outer disks, as is observed. We will investigate the success of these models in more detail in future work.

5. Chemical Abundances at Large Radii

Figure 4 presents the derived O/H abundances for the galaxies as a function of the deprojected galactocentric radius, normalised to R_{25}. The outermost abundances in all cases are $\sim$ 10-15% solar, measured at radii in the range 1.5–2 R_{25}. Although these represent some of the lowest abundances ever measured in spiral disks, they indicate that the outermost gas is far from being pristine.

We have compared the O/H gradients derived from fitting the entire set of datapoints in each galaxy and from fitting only those points lying within the optical disk (see Figure 4). Clearly, the outer disk abundances play a crucial role in defining the abundance gradient across the disk, leading to significantly steeper gradients for NGC 1058 and NGC 6946. Within the limits of the current dataset, it appears that the radial abundance gradients can be adequately described by

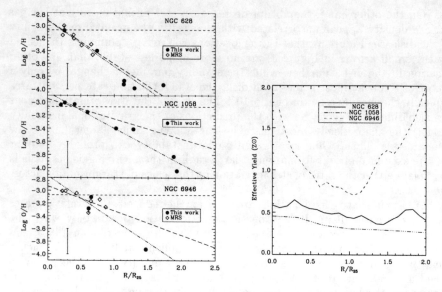

Figure 4. (Left) Radial variation of O/H, expressed in terms of the optical radius. The solar value is indicated by the horizontal dashed line. The dashed-dotted lines indicate fits to all the abundance measurements, whereas the long-dashed lines indicate fits to only the inner disk points. (Right) The radial variation of the effective yield, as derived for the closed box model.

single log-linear relationships. This result would appear to imply a continuity in the star formation (ie. metal production) process from inner to extreme outer disk, whereas our direct observations of the present star formation rate indicate that this is not the case. Gas flows and/or significant pre-enrichment of the disk gas may be required to reconcile these observations.

To assess the importance that gas flows might have had in the evolution of outer galactic disks, we have compared our data with the predictions of the simple 'closed box' model (no inflow or outflow from each radial zone). The closed box model can be represented by a simple relation $Z = -p \, ln \, \mu$ where p is the yield of the element in question and μ is the gas fraction, defined as baryonic mass in gas to the total baryonic mass (stars + gas). We have used our deep B-band surface photometry and published HI and CO maps to calculate the gas fractions. Figure 4 (right) shows the radial variation of the effective yield required to reconcile our observations with the closed box model. Interestingly, both NGC 628 and NGC 1058 are consistent with relatively constant effective yields across their disks, with values similar to that found in the solar neighbourhood (0.5 $Z_\odot$; Wyse & Gilmore 1995). In these cases, it would therefore appear that the role of gas flows in the evolution of the disk is similar to that for the solar neighbourhood. On the other hand, NGC 6946 exhibits gas fractions that are too high for the observed metallicity at both small and large radius.

6. Conclusions

The first results from our ongoing study of extreme outer galactic disks indicate that, in at least some galaxies, these regions are the sites of ongoing massive star formation and have metallicities which are low, although far from pristine. Observations of star formation in these low gas surface density parts are of particular importance for testing current ideas about the processes which govern large scale formation in galaxies. Our analysis suggests that considerations of gravitational instability alone cannot explain why the rate of star formation drops so abruptly beyond the optical edges of disk galaxies.

The low metallicities in these regions are similar to those measured in some high-redshift damped Lyman-α systems (eg. Pettini et al 1997) and suggest that outer disks are relatively unevolved at the present epoch. Perhaps star formation has only recently began in these parts, or perhaps the extreme outer disk has been forming stars for a significant period of time, but at a such low rate that little evolution has had the chance to take place. Distinguishing between these two alternatives requires establishing the mean age of the bulk of the outer disk stars, and we are currently investigating this issue via deep HST photometry of resolved stars in the outer parts of galaxies.

References

Dickey, J. M., Hanson, M. & Helou, G. 1990, ApJ, 352, 522

Elmegreen, B. G. & Parravano, A. 1994, ApJ, 435, 21

Ferguson, A. M. N., Gallagher, J. S. & Wyse, R. F. G. 1998a, AJ, 116, 673

Ferguson, A. M. N, et al D. A. 1998b, ApJ, 506, L19

Kamphuis, J. 1993, Ph. D. Thesis, Groningen University

Kennicutt, R. C. 1989, ApJ, 344, 689

Madore, B. F., van den Bergh, S. & Rogstad, D. H. 1974, ApJ, 191, 317

McGaugh, S. S. 1991, ApJ, 380, 140

Merrifield, M. R. 1992, AJ, 103, 1552

Pettini, M., Smith, L. J., King, D. L. & Hunstead, R. W. 1997, ApJ, 486, 665

Pickering, T., Impey, C., van Gorkom, J. & Bothun, G. 1997, AJ, 114, 1858

Prantzos, N. & Aubert, O. 1995, A&A, 302, 69

Shostak, G. S. 1978, A&A, 68, 321

Shostak, G. S. & van der Kruit, P. C. 1984, A&A, 132, 20

Skillman, E. D. 1987 in Star Formation in Galaxies, ed. C. J. Lonsdale Persson, NASA Conference Publication, 2466, p. 263

Thurston, T. R., Edmunds, M. G. & Henry, R. B. C. 1996, MNRAS, 283, 990

Toomre, A. 1964, ApJ, 139, 1217

Tosi, M. 1996 in From Stars to Galaxies, eds. C. Leitherer, U. Fritze-von Alvensleben & J. Huchra (Astronomical Society of the Pacific, San Francisco), p. 299

Wyse, R. F. G. & Gilmore, G. 1995, AJ, 110, 2771

The Low Surface Brightness Galaxy HIPASS1126-72

Virginia Kilborn[1], Erwin de Blok[1], Lister Staveley-Smith[2], Rachel Webster[1]

1. School of Physics, University of Melbourne, Parkville, Vic 3052, Australia
2. ATNF, PO Box 76, Epping, NSW 2121, Australia

Abstract. The low surface brightness galaxy HIPASS1126-72 was detected in the HI Parkes All Sky Survey (HIPASS). The galaxy was previously listed in the Southern Galaxy Catalogue under the name SGC1124.8-7221. This galaxy represents a class of galaxies that we will readily detect in the HIPASS survey, which have low surface brightness in the optical, but are easily detectable in neutral hydrogen.

1. Introduction

The HIPASS survey is being conducted at the 64m Parkes Radio-telescope in Australia, mounted with a 13 beam "multibeam" detector (Webster et al 1998, Barnes et al, 1998). The beam size for the survey is ~ 14', and the velocity resolution is 18 km s^{-1}. HIPASS1126-72 was detected by eye during routine searches for galaxies in the HIPASS data cubes. The HIPASS detection represents the first HI observation of this galaxy, and it was selected for higher resolution observations on the ATCA since its velocity profile was only just resolved in the HIPASS data. The velocity profile from HIPASS can be seen in Figure 1(a).

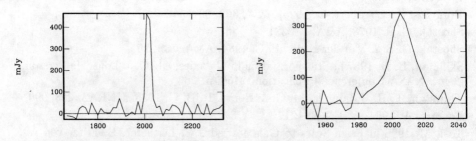

Figure 1. HI Spectrum from (a) the HIPASS survey, and (b) the Compact Array.

2. Observations

To obtain higher resolution of HIPASS1126-72, about 7 hours of data was taken in May 1998 at the Australia Telescope Compact Array (ATCA), with a 750m array. The resultant beam size was $\sim 60'' \times 40''$, and the velocity resolution was 3.3 km s^{-1}. The data reduction package MIRIAD was used to reduce the ATCA data. The HI distribution overlaid on the optical image (Fig. 2), shows a depression in the HI distribution at the optical center of the galaxy, which is also the brightest star forming region. This suggests the HI has been depleted in the formation of stars in this region. The ATCA spectrum can be seen in Figure 1(b).

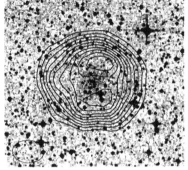

Figure 2. HI Spectrum overlaid on the DSS image (image size: $6.4' \times 6'$) - the contours are 7% contours, starting with the peak flux.

HIPASS1126-72 was observed in B and R-bands at the Siding Spring 40-inch telescope in January, 1998. The resulting images can be seen in Figure 3. We can see that the galaxy is more luminous in the B-band, which is a characteristic common to low surface brightness galaxies (de Blok, 1997). Ellipse fitting to the B band image gives an inclination of $\sim 25°$, and the central surface brightness in B was found to be $\sim$23.8 mag arcsec^{-2}. This galaxy lies close to the galactic plane, and foreground stars partially obscure one side of the galaxy; these are especially prominent in the R-band image, and have obstructed us from doing photometry in that band. The galactic extinction at these galactic coordinates (l=296,b=-10) is 1.42 magnitudes in the B-band (Schlegel et al, 1998).

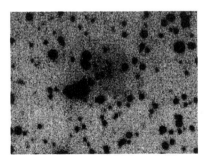 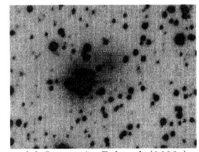

Figure 3. MSSSO 40' Observations: (a) Image in B-band (3600s); (b) Image in R-band (1800s).

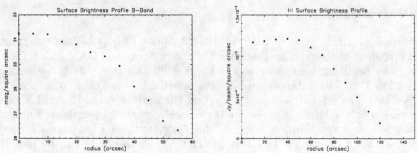

Figure 4. (a) Surface Brightness Profile B-band and (b) HI Surface Brightness Profile.

3. Characteristics of HIPASS1126-72

Position (J2000): RA: 11:26:28 Dec: -72:37:15 (ATCA)
Central Surface Brightness (B-band): 23.8 mag arcsec^{-2}
Heliocentric Velocity: 2012 km s^{-1} (ATCA)
Velocity Width (20%, 50%): 27 km s^{-1}, 16 km s^{-1} (ATCA)
HI Mass: $3.4 \times 10^9 M_\odot$ (HIPASS)

The 20% velocity width is less than twice the velocity resolution of the HIPASS survey. Thus the survey will detect galaxies unresolved in velocity, making interference rejection harder. Figure 4. shows the HI emission extends to a much greater radius (120") than the optical light (40-50"). The HI surface brightness is almost constant to the optical radius of the galaxy, then it decreases.

4. Conclusions

HIPASS1126-72 has one of the narrowest integrated profile widths of any known galaxy. The low velocity width appears to be the result of its low inclination, low gravitational potential and low surface density of star-formation. In general, the HIPASS survey is very sensitive to such low surface brightness, HI-rich galaxies. Optical and radio follow-up observations of such galaxies will do much to elucidate their nature.

References

Barnes D. G. et al 1998, ADASS VII, ASP conference series 145, p89
de Blok W. J. G. 1997, PhD thesis, University of Groningen
Schlegel D. J., Finkbeiner D. P. & Davis M. 1998, ApJ 500, p525
Webster R. L. et al 1998, this issue

The Structure of the Superthin Spiral Galaxy UGC7321

L. D. Matthews
NRAO, Charlottesville, VA, USA

J. S. Gallagher
University of Wisconsin, Madison, WI, USA

W. van Driel
Unité Scientifique Nançay, Observatoire de Paris, France

Abstract. UGC 7321 is an edge-on low surface brightness (LSB) spiral galaxy with a number of extraordinary properties. Its vertical scale height ($\sim$70 pc) is one of the smallest ever measured for a galaxy disk. Its disk also exhibits strong vertical and radial color gradients. UGC 7321 appears to be an extremely unevolved galaxy in both a dynamical and in a star-formation sense.

1. Motivation

Among the latest-type spiral galaxies there exist objects that when viewed edge-on, exhibit extraordinarily thin disks, and often, very low optical surface brightnesses. These and other features suggest these are some of the least evolved nearby galaxies in terms of their dynamical properties and star-formation histories. Studies of the vertical structures, color gradients, and global properties of these "superthin" spirals can shed new insight into the structure, stability, and evolution of galaxies.

We have obtained new multiwavelength observations of the nearby, LSB, superthin spiral UGC 7321 (Fig. 1). From these data we have measured the global properties of this galaxy (see Fig. 1) and have undertaken an analysis of its vertical structure and disk color gradients. Details concerning the analysis and interpretation of our new UGC 7321 observations can be found in Matthews (1998).

2. Summary of Key Results

The superthin galaxy UGC 7321 exhibits strong radial color gradients: $B-R \sim 2.0$ near the galaxy center, while along the major axis, $B - R$ reaches $\sim$0.20 at the easternmost edge of the disk, and $\sim$0.4 at the western edge. This is one of the strongest color gradients ever observed in a galaxy disk. UGC 7321 also exhibits vertical color gradients of several tenths of a magnitude in $B - R$.

UGC 7321 has one of the smallest disk scale heights ever measured for a spiral galaxy: $\bar{h}_z = 70$ pc at the disk center (r=0). The ratio of the disk scale length to the disk scale height at r=0 is ~20. UGC 7321 thus appears to have undergone little dynamical heating and has likely remained unperturbed over its lifetime.

The disk of UGC 7321 is not purely isothermal over most of its radial extent (cf. van der Kruit & Searle 1981). At r=0, the vertical light profile can be fit with a single exponential function (with scale height $\bar{h}_{z,c}$=70 pc; Fig. 2). For $|r| \leq 0\rlap{.}'5$, the vertical light profile becomes less peaked than an exponential and can be adequately characterized by the sum of two sech functions of differing scale heights ($\bar{h}_{z,2}$=60 pc and $\bar{h}_{z,3}$=109 pc). We interpret the exponential profile at the disk center as the superposition of a third disk component with a very small scale height ($h_{z,1}$ <46 pc). We also find marginal evidence for a fourth, faint disk component with $h_{z,4}$ >111 pc.

We attribute the smallest scale height disk component of UGC 7321 to a "nuclear disk" with a very small radial extent. The $h_{z,2}$ and $h_{z,3}$ disk components may be analogous to the "young disk" and the "thin disk" of the Milky Way, respectively. Counterparts to these components are visible in our optical images and color maps. The nature of the fourth disk component is uncertain.

UGC 7321 is underluminous for its observed rotational velocity compared with the prediction of the Tully-Fisher relation. This indicates the galaxy is likely to be dark matter-dominated even within its inner disk. This assertion is strengthened by simple analytical arguments that indicate that a significant fraction of the total mass of UGC 7321 must lie in a dark matter halo in order to prevent "firehose" instabilities. *Taken together, the global, kinematic, and structural properties of UGC 7321 suggest it is an extremely unevolved galaxy.*

Acknowledgments. LDM acknowledges the financial support of the WFPC2 IDT and a Peter B. Kahn Fellowship from the State University of New York at Stony Brook.

References

Matthews, L. D. 1998, Ph.D. Thesis, State University of New York at Stony Brook
Tully, R. B., Shaya, E. J., & Pierce, M. J. 1992, ApJS, 80, 479
van der Kruit, P. C. & Searle, L. 1981, A&A, 95, 105

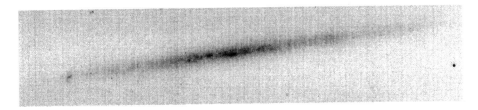

Figure 1. *R*-band CCD image of UGC 7321 obtained with the WIYN telescope. Field size is $5'.0 \times 0'.8$. We measure absolute magnitude: $M_B = -14.64$ (for $A_B = 0.04$), linear diameter: $A_{25} = 8.14$ kpc, H I mass: $\mathcal{M}_{HI} = 2.8 \times 10^8 M_\odot$, radial velocity: $V_{HI} = 406.7$ km/s, and H I line width: $W_{20} = 233$ km/s. We assume a distance of 5 Mpc (Tully et al. 1992).

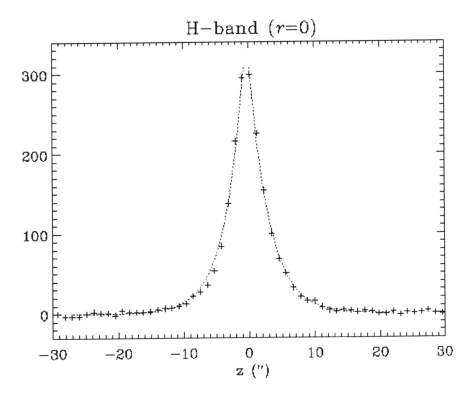

Figure 2. *H*-band vertical brightness profile of UGC 7321 along the galaxy minor axis ($r = 0$). The best-fitting exponential model with $h_z \sim 70$ pc is overplotted as a dotted line. Axes are distance from the galaxy plane, in arcseconds, versus detector counts in arbitary units.

HST WFPC-2 Imaging of Four Nearby LSB Galaxies

Karen O'Neil

Arecibo Observatory, PO Box 995, Arecibo, PR 00613
koneil@naic.edu

Abstract. Between May, 1996 and January, 1997 HST WFPC-2 images of four nearby LSB galaxies were taken through both the 8140Å and 3000Å filters. The first galaxy imaged, UGC 12695, is a nearby (z~0.021) LSB disk galaxy. UGC 12695 has an unusual morphology, consisting of a Y-shaped nucleus surrounded by a faint spiral arm. Additionally, numerous HI regions are spread throughout the galaxy. One of the surprising discoveries with the WFPC2 imaging was that a number of what were previously believed to be structural peculiarities in the galaxy are actually background galaxies. Once the effects of these galaxies are removed, the resultant U−I color of UGC 12695 is only −0.2, making it possibly the bluest galaxy in the local universe. When combined with the metallicity studies of McGaugh (1992), these colors indicate UGC 12695 to be a highly unevolved galaxy.

The other three galaxies imaged – V1L4, V2L8, and V7L3 – are dwarf elliptical galaxies located in the Virgo cluster. The intent of their images was to determine the galaxies' small scale structure and place limits on the density and type of giant branch stars within each galaxy. Placed at the distance of the Virgo cluster, luminosity fluctuations indicate the galaxies to contain only from 4 − 13 stars per pixel, coinciding with a K/M giant ratio ranging from 6 to ∞ (no M giant stars). Additionally, we found no evidence for stellar clumping in these galaxies although an extremely red, extremely small bulge was found at the core of V2L8.

1. UGC 12695: A Remarkable Unevolved Galaxy at Low Redshift

Utilizing the 8140Å (F814W) and 300Å (F300W) filters, short exposure Hubble Space Telescope Wide Field Planetary Camera-2 (WFPC2) images were taken of UGC 12695 a nearby ($z \sim 0.021$) low surface brightness disk galaxy. UGC 12695 has an unusual morphology, consisting of a Y-shaped nucleus surrounded by a faint spiral arm with a number of bright H II regions interspersed throughout the galaxy. Surface photometry indicates the majority of recent star formation in this galaxy occurred in these very localized regions, most of which have a radius of ≤ 2".

Some of the structural peculiarities of this galaxy arise because a number of background galaxies, previously thought to be morphological components of this galaxy (i.e. McGaugh, Schombert & Bothun 1995; McGaugh 1994; Klein, et al 1992), are showing through both the outer nucleus and spiral arms of UGC

Table 1. Global Properties of UGC 12695

RA	Dec	V (km/s)	d (Mpc)	M_{HI}	M_{dyn}	M_{HI}/L_B
23:36:02.0	12:52:32	6182	4590	9.62	9.98	1.28

f^B_g	h (kpc)	M_B	$\mu_B(0)$
0.62	8.4	-18.92	23.8

Table 2. WFPC2 Color of UGC 12695

	inner 2"		inner 10"			inner 30"	
	814	U−I	814	300−814	U−I	814	U−I
a:	22.22	-0.54	2.16	19.43	-0.44	2.26	18.07	-2.75	-0.09
b:	22.22	-0.54	2.16	19.51	-0.98	1.72	18.15	-2.88	-0.18
c:	22.22	-0.54	2.16	19.51	-0.97	1.73	18.15	-2.88	-0.18

Row a lists the colors with both the background galaxies and all H-α regions included.
Row b lists the results of masking the background galaxies.
Row c lists the results of masking both the background galaxies and the H-α regions.
All colors are within ± 0.05.

12695. When these galaxies are masked out, the resultant U−I color of UGC 12695 is −0.2 ± 0.1, making it perhaps the bluest galaxy every measured in this color system and confirming its nature as a very unevolved galaxy at low redshift.

McGaugh (1992) performed a metallicity study on a number of UGC 12695's H-α regions and found the average metallicity of the studied regions to be ∼$0.4Z_\odot$. When these are combined with the extremely blue colors of the same regions this indicates they are likely young starburst areas which have not yet enriched much of the gas.

In all, UGC 12695 proved to be a remarkable galaxy. It has an exceptionally high gas mass fraction, very low metallicity, very diffuse morphology, perhaps the bluest colors known for a galaxy, and an extremely transparent nature. Combined these indicate UGC 12695 to be a highly unevolved galaxy. Since UGC 12695 is at fairly low redshift, its properties indicate some potentials may well have late collapse and formation timescales. Additionally, and unlike most galaxies, UGC 12695 has the advantage of being relatively isolated, with its only nearest neighbor at a projected distance of 200 kpc. It is therefore likely UGC 12695 has been affected minimally (if at all) by other galaxies, and as such it provides an ideal study for low density galaxy evolution. Understanding this evolutionary process may shed considerable light into galaxy formation scenarios as a whole. More information on the WFPC2 study of UGC 12695 can be found in O'Neil, et al (1998).

Table 3. Luminosity Fluctuations (in Electrons) From the Inner regions of V1L4, V2L8, V7L3

Galaxy	μ_I	Galaxy	Sky	σ	N	M_I(1 star)
V1L4	21.57	182.2 ± 20.3	146.1 ± 17.7	0.279	13	-2.03
V2L8	22.88	178.0 ± 19.5	168.2 ± 19.0	0.450	5	-1.66
V7L3	22.67	168.2 ± 19.3	155.0 ± 18.2	0.493	4	-2.01

The central surface brightnesses are given in mag arcsec^{-2}, while the fluctuations are in electrons.

2. Three Low Surface Brightness Dwarf Elliptical Galaxies in the Virgo Cluster

WFPC2 images through the F814W and F300W filters were taken of V1L4, V2L8, & V7L3, three LSB dwarf elliptical galaxies in the Virgo cluster. The intent of these observations was to determine the small scale structure in these enigmatic galaxies and to attempt to learn something about the nature of their giant branch through the detection of luminosity fluctuations. These luminosity fluctuations, found in the inner constant surface brightness regions, were unambiguously detected. At the nominal distance of the Virgo cluster, the measured luminosity fluctuations in the F814W band yields a density of 4-13 red giants/pixel. In the most extreme case, V7L3, this is equivilent to a surface density of giant stars of $\sim$ 1 per 10 pc^2. Additonally, we find no evidence for small scale clumping of stars in these systems at this much improved spatial resolution scale. In the case of V2L8, however, we have discovered what is likely the smallest bulge measured to date, having an effective radius of only 50 pc. This bulge is quite red (as red as giant ellipticals) and thus may well be substantially more metal-rich than the rest of the galaxy. Moreover, while we are able to derive differences in the mean spectral type of the giant branch causing the observed fluctuation we found 1) K/M giant ratios ranging from 6 to ∞ (no M giant stars) and 2) no correlation between mean spectral type and surface brightness in our small sample. The latter point is important because it indicates that conditions of low I-band surface brightness in these dEs are caused both by low giant star luminosity per pixel (*e. g.* K0 giants) and by large physical separation between individual giants (e.g. V7L3). Such large physical separation is a manifestation of the extremely low density of these systems and when imaged at high spatial resolution of the WFPC2 ($\sim$ 6 pc per pixel), the galaxies are easy to look right through without evening knowing they are present in the middle of the WFPC2 frame. More details on these observations can be found in O'Neil, Bothun, & Impey (1998).

Acknowledgments. Thanks to Greg Bothun, Chris Impey, and Stacy McGaugh for help with this work, which was done while at the University of Oregon.

References

Klein, U., Giovanardi, C., Altschuler, D., & Wunderlich, E., 1992, A&A, 255, 49

McGaugh, S., Schombert, J., & Bothun, G., 1995, AJ, 109, 2019
McGaugh, S. 1994, ApJ, 426, 135
McGaugh, S. 1992, Ph.D. thesis, University of Michigan, Ann Arbor
O'Neil, K., Bothun, G., & Impey, C. 1998, AJ, submitted
O'Neil, K., Bothun, G., Impey, C., & McGaugh, S. 1998, AJ, 116, 657

Kinematics of Giant Low Surface Brightness Galaxies

T. E. Pickering

Kapteyn Institute, Postbus 800, 9700 AV Groningen, The Netherlands

Abstract. High sensitivity H I observations now exist for six giant low surface brightness (LSB) disk galaxies including the two prototypes, Malin 1 (Bothun *et al.* 1987; Impey & Bothun 1989) and F568-6 (also known as Malin 2; Bothun *et al.* 1990). Their H I surface brightnesses are generally low, but proportionally not as low as their optical surface brightnesses. Their total H I masses and radial extents are quite large, however, with $M_{\rm HI} \sim 10^{10} M_\odot h_{75}^{-2}$ and with detectable H I out to 100 kpc h_{75}^{-1} or more in a couple of cases. The rotation curves of these systems rise slowly and are consistent with negligible disk contribution, similar to many previously observed dwarf galaxies. However, the peak rotation velocities of these galaxies are high (>200 km s^{-1}) and infer high dynamical masses. These galaxies provide the some of the first examples of galaxies that are both massive and dark matter dominated.

1. Introduction

The serendipitous discovery of Malin 1, whose remarkable properties are described by Bothun *et al.* (1987) and Impey & Bothun (1989), demonstrated the existence of very large, very luminous, but very diffuse, galaxies. As an example of an L^* galaxy that is extremely hard to detect, Malin 1 helped confirm the idea, first clearly articulated by Disney (1976), that the brightness of the night sky can and does bias our understanding of galaxies and of the range of galaxy properties. It also helped to spur subsequent deliberate searches for low surface brightness (LSB) galaxies such as those of Schombert & Bothun (1988) and Impey *et al.* (1996). While these surveys didn't find any objects with properties as extreme as Malin 1 (indeed, none of these surveys was deep enough to detect Malin 1 itself), they did turn up several giant LSB galaxies whose properties approach those of Malin 1. F568-6 (Bothun *et al.* 1990), 1226+0105 (Sprayberry *et al.* 1993), UGC 6614 (van der Hulst *et al.* 1993), and two other galaxies from Sprayberry *et al.* (1995a) are the closest "cousins" to Malin 1 and have physically large disks with scale lengths greater than 10 kpc h_{75}^{-1}, total luminosities of L^* or greater, central surface brightnesses in B fainter than 23 mag arcsec^{-2}, and total H I masses of greater than $10^{10} M_\odot h_{75}^{-2}$.

These giant LSB galaxies probe a previously unexplored region of parameter space for galactic disks. Knezek (1993) and Sprayberry *et al.* (1995b) have tabulated optical and single-dish H I properties, while previous H I synthesis studies of LSBs by McGaugh (1992), van der Hulst *et al.* (1993), de Blok *et al.* (1996), and de Blok & McGaugh (1997) concentrated largely on smaller, less luminous LSB galaxies. Rotation curve studies so far have shown that the shape of a rotation curve is closely related to the rotation velocity of the galaxy. Slow rotators (i.e., the dwarfs mentioned above) have slowly rising rotation curves. Galaxies with rotation speeds in excess of 100 km s^{-1} tend to have flat rotation curves while the fastest rotators (> 200 km s^{-1}) have slightly declining rotation curves (Casertano & van Gorkom 1991; Broeils 1992). These results for the faster rotators are entirely based on HSB galaxies, however.

In the inner parts of HSB galaxies the rotation curves are often dominated by the luminous mass, i.e., the rotation curve can be fit by scaling the luminous component by a constant mass-to-light ratio (the "maximal disk" hypothesis). Only at large radii does the dark matter component become important. However, within the massive HSB galaxies a slight trend was found with surface brightness: less bright, less compact galaxies tend to be more dark matter dominated (Casertano & van Gorkom 1991). At a given total mass and mass-to-light ratio, a disk with a larger scale length will reach it's peak rotation velocity at a larger radius and will have a lower peak rotation velocity than the disk with the smaller scale length. If the galaxies have comparable halo properties, the halo should contribute proportionally more to the kinematics in the large scale length case. With their much larger and thus more diffuse disks giant LSB galaxies provide an important contrast to massive HSB galaxies and an important probe into the structure of galactic halos.

2. The Data

To date we have obtained VLA H I synthesis observations for five giant LSB galaxies. The data for the first four including Malin 1 and F568-6 have been published in Pickering *et al.* (1997) and the fifth, UGC 2936, has been published as part of my thesis work, Pickering (1998). The H I velocity fields overlayed on H I intensity maps for these five galaxies are shown in Figure 1. Australia Telescope Compact Array H I data for another giant, gas-rich LSB galaxy, NGC 289, has been published by Walsh *et al.* (1997).

Figure 2 shows the radially-averaged H I surface density profiles, derived rotation curves, and critical density profiles calculated from the rotation curves for the galaxies from Pickering *et al.* (1997) and Pickering (1998). The critical density profiles are calculated using the method of Kennicutt (1989) assuming a gas velocity dispersion of 6 km s^{-1}.

3. Stability of Giant LSB Disks

In two of the cases shown in Figure 2 the H I surface density lies below the critical density at all radii which is consistent with the low rate of star formation in those disks. However, in the other three cases significant portions of the disks have H I surface densities that exceed the critical density. In the case of UGC 2936 there is a fairly significant amount of star formation occurring. Figure 3 shows the optical rotation curve based on Hα for UGC 2936 plotted with the H I position-velocity map and rotation curve. The last measured points in the optical rotation curve occur at radii of about 14 kpc h_{75}^{-1} which is in excellent agreement with the radii at which the H I surface density shown in Figure 2 falls below the critical

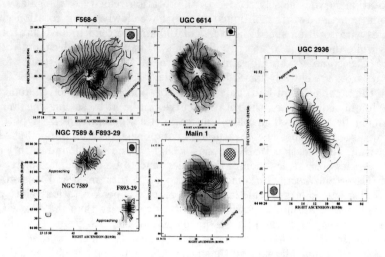

Figure 1. H I isovelocity contours overlayed on H I intensity greyscale for 5 giant LSB galaxies: Malin 1, F568-6, UGC 6614, NGC 7589, and UGC 2936. The velocity contour spacing is 20 km s^{-1} in all cases.

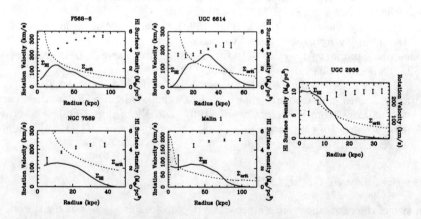

Figure 2. Radially-averaged H I surface density profiles (*solid lines*), rotation curves (*data points*), and critical density profiles (*dotted lines*) for Malin 1, F568-6, UGC 6614, NGC 7589, and UGC 2936. The critical density profiles are calculated assuming a gas velocity dispersion of 6 km s^{-1}.

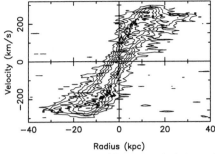

Figure 3. Optical rotation curve (*filled circles*) for UGC 2936 plotted with the H I rotation curve (*filled squares*) and major axis position-velocity map (*contours*). The position-velocity map is corrected for $i = 61°$ with contour levels of $9.0 \times 10^{-3} \times (1\ 2\ 3\ 4\ 5\ 6\ 7)$ Jy/beam.

density. In the case of Malin 1 the large uncertainty in the inclination, and thus the absolute scaling of the rotation curve, leads to a large uncertainty in the determination of the critical density as well as in the line-of-sight corrections to the surface density. The inclination is much more well-determined in the case of UGC 6614, though, so it's more clearly a case where the critical density criterion predicts widespread star formation where none is observed to occur. This is an indication that the conditions in a giant LSB disk are quite different than in the HSB galaxies used to calibrate the critical density criterion. One might expect this to be the case since the dark halo will be more kinematically dominant as a disk becomes more diffuse resulting in a lower restoring force in the disk. The lower restoring force results in a thicker and more stable disk at a given velocity dispersion which might explain the results for UGC 6614.

4. Mass-to-Light Ratios in Giant LSB Galaxies

If these rotation curves are modeled by a disk component alone, the amplitude of the rotation curves in these galaxies requires R-band M/L's (assuming $h_{75} = 1$) of up to 15–20 in the cases of F568-6 and UGC 6614. The shapes of the curves, however, are not well-produced by the disk-only fits. Quillen & Pickering (1997) used the strengths of the optical and H I spiral arms along with the strengths of the velocity perturbations due to the spiral arms to place much better constraints on the disk M/L's in UGC 6614 and F568-6. They find that in both cases the R-band M/L's are in the range of about 1 to 5 which clearly rule out the maximal disk-only results. If the bulge component is included and assumed to have the same M/L as the disk, the M/L constraints are even tighter: < 1.5 for F568-6 and < 2.5 for UGC 6614. Thus there's no indication that giant LSB galaxies have disk mass-to-light ratios that differ significantly from more normal HSB galaxies.

The global M/L's for giant LSB galaxies derived by measuring the mass enclosed by the last measured points in the rotation curves are similarly not far out of line with what is found for HSB galaxies. This is rather remarkable since their last measured points are at radii approaching 100 kpc h_{75}^{-1} in a couple of

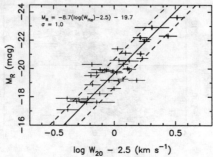

Figure 4. Absolute R magnitude, M_R, versus H I linewidth, W_{20} for the giant LSB galaxies (*filled circles*) and other LSB galaxies from the samples of Zwaan et al. (1995) and Sprayberry et al. (1995b) (*filled squares*).

cases and greater than 35 kpc h_{75}^{-1} in all cases. The five galaxies from Pickering et al. (1997) and Pickering (1998) all have R-band global M/L's in the range 11–15. A typical HSB galaxy such as NGC 3198 (Begeman 1989), for example, has a total R-band M/L of 10 at a radius of 22 kpc h_{75}^{-1}. Zwaan et al. (1995) and Sprayberry et al. (1995b) showed that LSB galaxies follow the same Tully-Fisher relation as comparable HSB galaxies which also implies that their global mass-to-light ratios must be at least roughly similar. Figure 4 shows the results for the giant LSBs plotted with results for other LSB galaxies. The giant LSB galaxies are completely consistent within the uncertainties with the Tully-Fisher relation derived from the other LSB galaxies.

5. Shapes of Giant LSB Rotation Curves and NFW Halos

The high resolution numerical simulations of galactic halos by Navarro et al. (1996) showed that dark matter halos over several orders of magnitude in mass have density profiles that can all be well-described by a single two parameter density profile. This so-called NFW halo profile has a halo circular velocity profile of the form

$$V_{\rm NFW}(r) = V_{200}\sqrt{\frac{1}{x}\frac{\ln(1+cx) - cx/(1+cx)}{\ln(1+c) - c/(1+c)}} \quad (1)$$

where the two free parameters are V_{200}, the circular velocity of the halo at the virial radius, and c, the concentration of the halo. Navarro et al. (1996) compared some observed rotation curve shapes with those predicted by a NFW halo profiles within a Standard Cold Dark Matter (SCDM) universe and found that while HSB galaxies can be made consistent by tweaking M/L, LSB dwarfs cannot be made consistent with SCDM.

Navarro et al. (1997) developed this further and showed that the NFW halo profile is a generic result of hierarchical clustering and that different cosmologies result in different relations between V_{200} and c. Using this, one can turn the comparison around and see what cosmologies are consistent with the values of

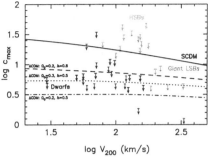

Figure 5. Results of maximal NFW fits to samples of LSB and HSB rotation curves taken from the literature plotted with the predicted V_{200}—c relations for different cosmologies. The predictions were calculated using the method described by Navarro et al. (1997)

V_{200} and c one measures from galactic rotation curves. A quick and dirty way to estimate this and place upper limits on c is to fit galactic rotation curves using only Equation 1, basically a maximal NFW halo fit. The results of some maximal halo fits to the giant LSB galaxies as well as other samples of LSB and HSB rotation curves taken from the literature are shown in Figure 5. As in Navarro et al. (1996), the HSB galaxies could be made consistent with SCDM. They could also be made consistent with everything else and hence don't have much predictive power when fitting with a halo only. This is because much of the mass concentration in HSB galaxies is due to the baryonic component rather than the dark halo component. In the LSB cases the baryonic component is much less concentrated and less kinematically important so they offer better constraints on the halo properties. Almost all of those cases fall below and are not consistent with SCDM. Most of them are consistent with lower density models that include a cosmological constant (ΛCDM). However, a few have very low best-fit concentrations when fitted with an NFW halo only. Those are cases where the rotation curve is solid-body-like and linearly rising and thus has a shape that is not well-described by Equation 1.

6. Conclusions

Giant LSB galaxies are some of the first known examples of massive galaxies whose kinematics are dark halo dominated at almost all radii. Their rotation curve shapes cannot be reproduced by scaling their stellar components by constant mass-to-light ratios. Even the amplitudes of their rotation curves would require higher mass-to-light ratios than are allowed by other constraints such as spiral arm strengths. The shapes of their rotations curves cannot be made consistent with the predictions of SCDM, but are consistent with the predictions of low density ΛCDM models.

While at a given radius a giant LSB galaxy's halo dominates its kinematics to a greater extent than in an HSB, the global mass-to-light ratios are about the same in both cases. The giant LSB galaxies also follow, or are at least consistent with, the same Tully-Fisher relation as other LSB and HSB galaxies.

References

Begeman, K. G. 1989, A&A, 223, 47
Broeils, A. H. 1992, Ph.D. thesis, University of Groningen
de Blok, W. J. G., McGaugh, S. S., & van der Hulst, J. M. 1996, MNRAS, 283, 18
de Blok, W. J. G., & McGaugh, S. S. 1997, MNRAS, 290, 533
Bothun, G. D., Impey, C. D., Malin, D. F., & Mould, J. R. 1987, AJ, 94, 23
Bothun, G. D., Schombert, J. M., Impey, C. D., & Schneider, S. E. 1990, ApJ, 360, 427
Casertano, S., & van Gorkom, J. H. 1991, AJ, 101, 1231
Disney, M. J. 1976, Nature, 263, 573
Impey, C. D., & Bothun, G. D. 1989, ApJ, 341, 89
Impey, C. D., Sprayberry, D., Irwin, M. J., & Bothun, G. D. 1996, ApJS, 105, 209
Kennicutt, R. C. 1989, ApJ, 344, 685
Knezek, P. M. 1993, Ph.D. thesis, University of Massachusetts
McGaugh, S. S. 1992, Ph.D. thesis, University of Michigan
Navarro, J. F., Frenk, C. S., & White, S. D. M. 1996, ApJ, 462, 563
Navarro, J. F., Frenk, C. S., & White, S. D. M. 1997, ApJ, 490, 493
Pickering, T. E., Impey, C. D., van Gorkom, J. H., & Bothun, G. D. 1996, AJ, 114, 1858
Pickering, T. E. 1998, Ph.D. thesis, University of Arizona
Quillen, A., & Pickering, T. E. 1997, AJ, 113, 2075
Schombert, J. S., & Bothun, G. D. 1988, AJ, 95, 1389
Sprayberry, D., Impey, C. D., Irwin, M. J., McMahon, R. G., & Bothun, G. D. 1993, ApJ, 417, 114
Sprayberry. D., Impey, C. D., Bothun, G. D., & Irwin, M. J. 1995a, AJ, 109, 558
Sprayberry, D., Bernstein, G. M., Impey, C. D., & Bothun, G. D. 1995b, ApJ, 438, 72
van der Hulst, J. M., Skillman, E. D., Smith, T. R., Bothun, G. D., McGaugh, S. S., & de Blok, W. J. G. 1993, AJ, 106, 548
Walsh, W., Staveley-Smith, L., & Oosterloo, T. 1997, AJ, 113, 1591
Zwaan, M. A., van der Hulst, J. M., de Blok, W. J. G., & McGaugh, S. S. 1995, MNRAS, 273, L35

What Causes the HI Holes in Gas-Rich LSB Dwarfs?

Katherine L. Rhode

Astronomy Department, Yale University, New Haven, CT 06520

John J. Salzer

Astronomy Department, Wesleyan University, Middletown, CT 06459

David J. Westpfahl

Department of Physics, New Mexico Institute of Mining & Technology, Socorro, NM 87801

Abstract. We have carried out a deep, multi-color imaging study of Holmberg II (Ho II) and several other nearby LSB dwarf galaxies for which detailed HI maps exist. The formation of the HI holes in these galaxies has been attributed to multiple supernovae (SNe) occurring within wind-blown shells around young, massive star clusters. To search for evidence of the clusters, we have compared optical images with the HI maps and measured magnitudes and colors of objects in and around the HI holes.

Although the SN scenario requires that detectable star clusters should often be present in the centers of the HI holes, our observations have in most cases failed to reveal these clusters at the expected magnitudes. In fact, many of the HI holes are located in regions of very low optical surface brightness, which show no evidence of recent star formation.

1. Introduction

Several nearby LSB dwarf galaxies have been mapped at high resolution in HI using the Very Large Array (Puche et al. 1992, Puche & Westpfahl 1994). The maps show intricate structures in the galaxies' interstellar gas; analysis of the structures reveals many expanding HI holes. In the galaxy Ho II, for example, the holes are roughly spherical and surrounded by shells of higher density. Measured expansion velocities are typically $4-10$ km s^{-1}, and diameters are $100-1600$ pc.

Puche et al. (1992; hereafter P92) proposed that the HI holes in Ho II and other dwarf galaxies were created by multiple, sequential SNe, occurring within wind-blown cavities in the interstellar medium (ISM). Using the expansion velocities and sizes of the holes, and the volume density of the surrounding medium, P92 calculated that many of the holes require the energy of $\sim 10-200$ SNe to create them, and that the holes typically have ages of 10^7-10^8 years.

The SN scenario proposed by P92 provides us with a direct observational test. Multiple SNe are thought to occur in massive clusters or OB associations. The hole ages, if accurate, imply that the late-B, A, and F dwarf population should still be present in these clusters. For example, in a cluster which produces 50 SNe, a Salpeter IMF (Salpeter 1955) predicts ~300 upper main sequence stars that should still be present after 10^8 years. At the distance of Ho II, the brightness of such a cluster would be $B\sim22$, making it stand out against the low surface brightness level of the underlying older population of stars.

To search for the star clusters which would provide evidence for the SN scenario, we carried out a deep, multi-color optical imaging study of Ho II and the other dwarfs for which detailed HI maps exist. Our goal was to measure accurate magnitudes and colors of objects located in or near the HI holes. The published hole ages and energies can be used to calculate magnitudes and colors of the clusters that should be present if the SN scenario is correct. These predicted cluster properties can be directly compared to our observational data, to help refute or confirm the scenario described in P92.

2. Observations and Photometry

BVR and Hα images of Ho II and ~10 other LSB dwarf galaxies were obtained in Feb 1994 and/or Apr 1995 with the 0.9-m telescope at Kitt Peak National Observatory. To create the deepest possible images with which to do photometry, the BVR data from separate observing runs were merged. The HI maps from P92 and Puche & Westpfahl (1994) were placed on the same pixel scale and aligned with our optical images to permit direct comparison.

Results from analysis of the data for Ho II are given here. Work on the other galaxies is still in progress. For Ho II, the information in P92 was used to mark the sizes and positions of the holes on the HI map and optical images. Photometry of our deepest BVR frames was carried out at the locations of the hole and inter-hole regions. Photometry was carried out on 44 of the 51 hole regions identified in P92 (some HI hole locations coincided with foreground stars in the optical images, and had to be excluded from the sample). Twenty of these 44 holes had one or more faint objects located in or near the hole; photometry was executed on those objects individually, since they may be the star clusters which produced sequential SNe.

3. Modelling the Putative Clusters

To compare our observations with predictions arising from the SN scenario, we used the published hole properties to derive the observable properties of the clusters which should exist if the SN hypothesis is correct. Given the kinetic energy requirements for each hole, the number of SNe needed to create it was calculated. The energy imparted to the ISM by one supernova was taken to be $\sim10^{51}$ ergs (McCray & Kafatos 1987). For each hole, a Salpeter IMF (Salpeter 1955) was used to calculate the mass distribution for a model star cluster. This model cluster was scaled so that the number of stars with mass ≥ 7 M$\odot$ equalled the number of SNe required to create the hole. Stars that would leave the main sequence over a time period equal to the age of the hole were removed from the distribution. Finally, composite magnitudes and colors were calculated for the model cluster, for comparison with observations. A similar process was repeated using a Miller-Scalo IMF (Miller & Scalo 1979), and assuming stars with mass

≥ 8 M⊙ become SNe, to provide a reasonable range of predicted magnitudes and colors to compare with observations.

4. Results and Conclusion

Results for Ho II are summarized here; details may be found in Rhode et al. (1998):

- *Many of the well-defined holes in the HI map of Ho II are located in extremely LSB regions in the optical image, which show no evidence of recent star formation.* Fourteen of the 44 holes in our sample lie beyond the Holmberg radius ($\mu_B = 26.6$ mag arcsec^{-2}), where star formation appears unlikely.

- *There is no hint of a point source in 24 of the 44 holes in the sample.* The 4σ-limit on the brightness of a point source in our data is $B=23$. Our models suggest that the remnant clusters which caused many of the holes should have $B \sim 21-22$; we would detect the clusters if they were present.

- We measured magnitudes and colors for 29 sources in the 20/44 holes which had objects located within the hole or just outside it. *Ten of the sources have $B-V > 1.0$, thereby excluding them as likely candidates for the putative clusters.*

- A total of 12 holes, in the main optical body of the galaxy, contain objects that could be the putative clusters; i.e., the objects have magnitudes and colors that *may be* consistent with a massive young cluster. *We cannot rule out the SN hypothesis for those 12 HI holes.* Note, however, that such holes appear in regions of relatively high galaxian background, which makes it difficult to determine whether the objects that appear to coincide with them are actually associated with a given hole.

The observational evidence strongly suggests that at least some of the HI holes in Ho II did not originate in precisely the manner envisioned by P92. A number of alternative explanations for the origin of the holes have been put forth, including the recent suggestion that the holes are remnants of Gamma-Ray Burst events (Efremov et al. 1998; Loeb & Perna 1998). These alternative explanations, and how they relate to our findings, are discussed thoroughly in Rhode et al. (1998).

References

Efremov, Y.N., Elmegreen, B.G., & Hodge, P.W. 1998, ApJ, 501, L163
Loeb, A., & Perna, R. 1998, ApJ, 503, L35
McCray, R., & Kafatos, M. 1987, ApJ, 317, 190
Miller, G.E., & Scalo, J.M. 1979, ApJS, 41, 513
Puche, D., & Westpfahl, D. 1994, in *Proceedings of the ESO/OHP Workshop on Dwarf Galaxies*, eds. G. Meylan & P. Prugniel (ESO: Garching), 273
Puche, D., Westpfahl, D., Brinks, E., & Roy, J.-R. 1992, AJ, 103, 1841
Rhode, K.L., Salzer, J.J., Westpfahl, D.J., & Radice, L.A. 1998, AJ, submitted
Salpeter, E.E. 1955, ApJ, 121, 161

VLA HI Imaging of the Low Surface Brightness Dwarf Galaxy DDO 47

F. Walter

Radioastronomisches Institut, Bonn, Germany

E. Brinks

Departamento de Astronomía, Guanajuato, México

Abstract. We present high resolution VLA-observations of the nearby Low Surface Brightness Dwarf Galaxy DDO 47. This object shows many hole–like structures in its neutral interstellar medium. The majority of the detected H I–shells are found to be expanding. Their origin is therefore believed to be due to stellar winds of the most massive stars and their subsequent supernova (SN type II) explosions within regions of recent star formation (SF). Current SF in DDO 47 is predominantly present on the rims of the H I–shells suggesting propagating SF. At a projected distance of 20 kpc (adopting a distance to DDO 47 of 4 Mpc), a companion galaxy was detected at almost the same systemic velocity (DDO 47 B). A search for an optical identification suggests that CGCG 087-033 is the optical counterpart of the companion. A preliminary dynamical analysis based on DDO 47's rotation curve yields that it is dark matter dominated (about 80% of its dynamical mass is in some non–visible form). A simple mass model suggests that DDO 47 is one of the 'thickest' dwarf galaxies studied so far.

1. Introduction

Recent H I–studies of nearby dwarf galaxies show that holes and shell–like structures dominate the appearance of their interstellar medium (ISM). These holes are thought to be created as a result of the formation and rapid evolution of massive stars within a group or association, via the interaction of their collective winds and supernova explosions with the surrounding ISM (see Tenorio–Tagle and Bodenheimer 1988). This picture is not without its critics (see Rhode *et al.* 1997; also this volume). But whatever energetic events cause these structures, they leave a much more dramatic impression on a dwarf galaxy than on a more massive spiral galaxy (Walter & Brinks 1998). This is attributed to dwarf galaxies having a lower overall gravitational potential.

Despite the fact that dwarf galaxies are ideal laboratories for ISM studies, only few papers dealing with detailed H I observations have been published thusfar. Examples are the Large and the Small Magellanic Clouds (see Kim et al. 1997) and Holmberg II (Puche et al. 1992). We therefore started a programme to observe a sample of nearby dwarf galaxies. The first paper, on IC 2574 has been submitted (Walter & Brinks 1998). First results on a second object, DDO 47, are presented here. DDO 47 (UGC 3974) was observed in the 21-cm line of neutral hydrogen with the NRAO Very Large Array (VLA) in its D, DnC, C and B–configurations. In total, 16 hours were spent on source. The velocity and spatial resolution are 2.5 km s^{-1} and 7″ (resulting in a linear resolution of 120 pc at an adopted distance of 4 Mpc). In the course of the data reduction we discovered a companion galaxy at nearly the same systemic velocity (see Fig 1).

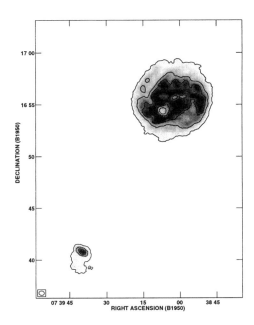

Figure 1.   H I surface brightness map of DDO 47 (north) and its companion (south). The projected distance is about 20 kpc.

2. The H I–Hα connection of DDO 47

In total, we detected 19 H I–holes in DDO 47. An overlay of the positions and sizes of the holes with the H I surface brightness map is given in Fig. 2 (left). Most of the holes were found to be expanding. This lends support to the picture that they are wind– and/or supernova driven. The theory of propagating star formation predicts that secondary sites of star formation form close to the rims of expanding H I holes. The idea behind this is that the holes, while expanding, sweep up ambient matter, to such a point that the mass density on the rim gets high enough for star formation to commence. To test this scenario in the case of DDO 47, the Hα–regions detected by Strobel et al. (1991) were overlaid on the H I surface density map (Fig. 2, right). Note that virtually all Hα emission is situated outside the holes which at least qualitatively fits this description.

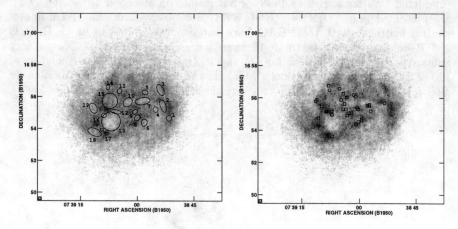

Figure 2. The distribution of the neutral hydrogen in DDO 47 overlaid with the position and sizes of the H I holes detected by us (left) and the H II regions from Strobel et al. (right).

3. Dynamics of DDO 47 and the thickness of its H I disk

After a preliminary analysis of the observed rotational velocities and rotation curve fitting (see Fig. 3), the dynamical mass of DDO 47 was determined to be about $5 \times 10^9\,M_\odot$. Assuming a gaseous mass of $3 \times 10^8\,M_\odot$ (H I, corrected for the contribution of primordial helium) and a stellar mass of $2 \times 10^8\,M_\odot$ (assuming a solar mass to light ratio for the stars), yields that most of the mass of DDO 47 is present in the form of Dark Matter. Fig. 3 shows the rotation curve of DDO 47 (boxes, with appropriate error bars). The triangles present the contribution from neutral hydrogen, the open square shows the contribution from the stellar population out to a radius of $4'$. From this simple analysis it is already clear that most of the mass must be present in the form of Dark Matter. Using a simple model which relates the velocity dispersion of the gas to its scale height, we derive the thickness of the H I–layer of DDO 47 to be about 700 pc. DDO 47 is therefore one of the thickest dwarf galaxies studied so far.

4. Summary

- DDO 47 is a gas–rich dwarf irregular galaxy at distance of 4 Mpc.

- A companion has been detected at almost the same systemic velocity. Its H I mass is about 7% of that of DDO 47. The companion has been previously catalogued in the optical as CGCG 087–033. The geometry of the pair suggests that the relative orbits are within the plane of the sky. The projected distance (20 kpc) is therefore likely a good approximation of the true seperation.

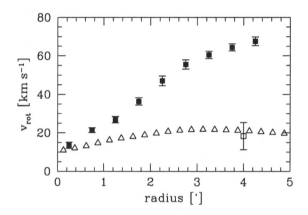

Figure 3. The measured rotation curve of DDO 47 (filled squares). For comparison, the contribtion from neutral hydrogen (triangles) and the stellar population (open square) is presented as well.

- To explain the derived rotation curve, DDO 47 must have a high Dark Matter content (about 80% of its dynamical mass).
- In total, 19 H I holes were detected, most of which are expanding. There is a striking correlation between the positions of the holes and current star forming regions (as traced by Hα–observations) in the sense that current star formation is restricted to the rims of the shells.
- The thickness of the H I disk is about 700 pc. DDO 47 is therefore one of the thickest dwarf galaxies studied so far.

References

Kim, S., Staveley–Smith, L., Sault, R.J., Kesteven, M.J., McConnell, D., & Freeman, K.C. 1997, PASA, 14, 119
Puche, D., Westpfahl, D., Brinks, E., & Roy, J.-R. 1992, AJ, 103, 1841
Rhode, K.L., Salzer, J.J., & Westpfahl, D.J. 1997, BAAS, 191, 81.09
Strobel, N.V., Hodge, P., Kennicutt, R.C. Jr. 1991, ApJ, 383, 148
Tenorio–Tagle, G., & Bodenheimer, P. 1988, ARA&A, 26, 145
Walter, F. & Brinks, E. 1998, AJ, submitted

Searching for LSB - IV

Call yourselves astronomers, it's up there - on the ceiling.

The fate of LSB galaxies in clusters and the origin of the diffuse intra-cluster light

Ben Moore

Department of Physics, University of Durham, South Road, Durham, DH1 3LE, UK

George Lake, Joachim Stadel & Thomas Quinn

Department of Astronomy, University of Washington, Seattle, WA 98195, USA

Abstract.
We follow the evolution of disk galaxies within a cluster that forms hierarchically in a standard cold dark matter N-body simulation. At a redshift $z = 0.5$ we select several dark matter halos that have quiet merger histories and are about to enter the newly forming cluster environment. The halos are replaced with equilibrium high resolution model spirals that are constructed to represent luminous examples of low surface brightness (LSB) and high surface brightness (HSB) galaxies. Whilst the models have the same total luminosity, $\sim L_*$, they have very different internal mass profiles, core radii and disk scale lengths, however they all lie at the same place on the Tully-Fisher relation. Due to their "soft" central potentials, LSB galaxies evolve dramatically under the influence of rapid encounters with substructure and strong tidal shocks from the global cluster potential – galaxy harassment. As much as 90% of the LSB disk stars are tidally stripped and congregate in large diffuse tails that trace the orbital path of the galaxy and form the diffuse intra-cluster light. The bound stellar remnants closely resemble the dwarf spheroidals (dE's) that populate nearby clusters, with large scale lengths and low central surface brightness.

1. Introduction

Clusters of galaxies provide a unique environment wherein the galaxy population has been observed to rapidly evolve over the past few billion years (Butcher & Oemler 1978, Dressler *et al.* 1998). At a redshift $z \gtrsim 0.4$, clusters are dominated by spiral galaxies that are predominantly faint irregular or Sc-Sd types. Some of these spirals have disturbed morphologies; many have high rates of star-formation (Dressler *et al.* 1994a). Conversely, nearby clusters are almost completely dominated by spheroidal (dSph), lenticulars (S0) and elliptical galaxies (Bingelli *et al.* 1987, 1988, Thompson & Gregory 1993). Observations suggest that the elliptical galaxy population was already in place at much higher redshifts, at which time the S0 population in clusters is deficient compared to nearby clusters (Couch et al. 1998, Dressler et al. 1998). This evolution of the morphology-density relation appears to be driven by an increase in the S0 fraction with time and a corresponding decrease in the luminous spiral population.

Low surface brightness (LSB) galaxies appear to avoid regions of high galaxy densities (Bothun et al. 1993, Mo et al. 1994). This is somewhat puzzling since recent work by Mihos et al. (1997) demonstrated that LSB disk galaxies are actually *more* stable to close tidal encounters than HSB disk galaxies. In fact, LSB galaxies have lower disk mass surface densities and higher mass-to-light ratios, therefore their disks are less susceptible to internal global instabilities, such as bar formation. However, in a galaxy cluster, encounters occur frequently and very rapidly, on a shorter timescale than investigated by Mihos et al. and the magnitude of the tidal shocks are potentially very large.

Several physical mechanisms have been proposed that can strongly affect the morphological evolution of disks: ram-pressure stripping (Gunn & Gott 1978), galaxy merging (Icke 1985, Lavery & Henry 1988, 1994) and galaxy harassment (Moore et al. 1996a, 1998). The importance of these mechanisms varies with environment: mergers are frequent in groups but rare in clusters (Ghigna et al. 1998), ram pressure removal of gas is inevitable in rich clusters but will not alter disk morphology (Abadi & Moore, in preparation). The morphological transformation in the dwarf galaxy populations ($M_b > -16$) in clusters since $z = 0.4$ can be explained by rapid gravitational encounters between galaxies and accreting substructure - galaxy harassment. The impulsive and resonant heating from rapid fly-by interactions causes a transformation from disks to spheroidals.

The numerical simulations of Moore et al. focussed on the evolution of fainter Sc-Sd spirals in static cluster-like potentials and their transition into dSph's. In this work we shall examine the role of gravitational interactions in driving the evolution of luminous spirals in dense environments. We will use more realistic simulations that follow the formation and growth of a large cluster that is selected from a cosmological simulation of a closed CDM universe. The parameter space for the cluster model is fairly well constrained once we have adopted hierarchical structure formation. The structure and substructure of virialised clusters is nearly independent of the shape and normalisation of the power spectrum. Clusters that collapse in low Omega universes form earlier, thus their galaxies have undergone more interactions. The cluster that we follow virialises at $z \sim 0.3$, leaving about 4 Gyrs for the cluster galaxies to evolve.

The parameter space for the model spirals is much larger. Mihos et al. examined the effects of a single encounter at a fixed number of disk scale lengths, whilst varying the disk surface brightness and keeping other properties fixed. The key parameter that determines whether or not dark matter halos survive within a cluster N-body simulation is the core radius of the substructure, which is typically dictated by the softening length (Moore, Katz & Lake 1996b). We suspect that the "softness" of the dark matter potentials may also be the key factor that governs whether or not a given disk galaxy will survive within a real cluster.

2. The model galaxies

We use the technique developed by Hernquist (1989) to construct equilibrium spiral galaxies with disk, bulge and halo components, that represent "standard" HSB and LSB disk galaxies. In each model the disk mass is $4.0 \times 10^{10} M_\odot$ and the rotation curves both peak at 200 km s^{-1}. They are a little less massive than "L_*" galaxies, the characteristic luminosity of the break in the galaxy luminosity function and would have absolute magnitudes ~ -17.8 for a mass to light ratio of 5. The "HSB" spiral has an exponential disk scale length, $r_d = 3.0$ kpc and a bulge with a mass of one third of the total disk mass. The "LSB" disk scale length is 10 kpc and has no bulge. The scale height, r_z, of each disk is $0.1r_d$ and they are constructed with a Toomre Q parameter of 1.5. Each galaxy has a dark halo modeled by truncated isothermal spheres with core radii set equal to the disk scale length. This scaling ensures that each galaxy lies at the same point on the Tully-Fisher relation, yet the galaxies will have different internal mass distributions (Zwaan et al. 1995).

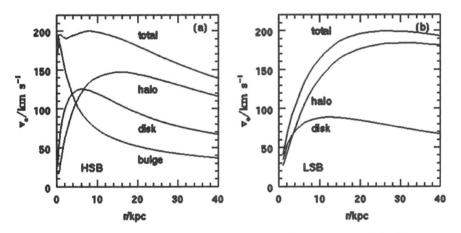

Figure 1. The curves show the contributions from stars and dark matter to the total rotational velocity of the disk within (a) the HSB galaxy and (b) the LSB galaxy.

Figure 1 shows the contribution to the rotation velocity of the disks from each component. Note that the bulge component of the HSB galaxy has ensured that the rotation curve is close to flat over the inner 5 disk scale lengths, whereas the rotation curve of the LSB galaxy rises slowly over this region. These rotation curves are typical of that measured for LSB galaxies (de Blok & McGaugh 1996) and HSB galaxies (Persic & Salucci 1997).

Each disk is modeled using 20,000 star particles of mass $2 \times 10^6 M_\odot$ and 40,000 dark matter halo particles of mass $2 \times 10^7 M_\odot$ in the LSB and $6 \times 10^6 M_\odot$ in the HSB galaxy. The force softening is $0.1r_d$ for the star particles and $0.5r_d$ for the halo particles. Their disks are stable and they remain in equilibrium when simulated in isolation. Discreteness in the halo particles causes the disk scale height to increase with time as quantified in Section 4 for the LSB galaxy.

3. The response to impulsive encounters

For a given orbit through a cluster, the visible response of a disk galaxy to a tidal encounter depends primarily upon its internal dynamical timescale. Galaxies with cuspy central mass distributions, such as ellipticals, have short orbital timescales at their centres and they will respond adiabatically to tidal perturbations. Sa-Sb spirals have flat rotation curves, therefore a tidal encounter will cause an impulsive disturbance to a distance $\sim v_c b/V$, where b is the impact parameter, V is the encounter velocity and v_c is the galaxy's rotation speed. LSB galaxies and Sc-Sd galaxies have slowly rising rotation curves, indicating that the central regions are close to a uniform density. The central dynamical timescales are constant throughout the inner disk and an encounter that is impulsive at the core radius will be impulsive throughout the galaxy.

The strength of an encounter is $\propto M_p^2/V^2$, where M_p is the perturbing mass. The typical galaxy-galaxy encounter within a virialised cluster occurs at a relative velocity $\sim \sqrt{2}\sigma_{1d}$. Substituting typical parameters for an Sa–Sb spiral orbiting within a cluster, we find that such encounters will not perturb the disk within $\sim 3r_d$. However, tidal shocks from the mean cluster field also provides a significant heating source for those galaxies on eccentric orbits (Byrd & Valtonen 1990, Valluri 1993). Ghigna et al. (1998) studied the orbits of several hundred dark halos within a cluster that formed hierarchically in a cold dark matter universe. The median ratio of apocenter to pericenter was 6:1, with a distribution skewed towards radial orbits. More than 20% of the halos were on orbits more radial than 10:1. A galaxy on this orbit would move past pericenter at several thousand km s^{-1} and would be heated across the entire disk.

We illustrate the effect of a single impulsive encounter on each of our model disks in Figure 2 and Figure 3. At time t=0 we send a perturbing halo of mass $2 \times 10^{12} M_\odot$ perpendicular to the plane of the disk at an impact parameter of 60 kpc and velocity of 1500 km s^{-1}. This encounter would be typical of that occurring in a rich cluster with a tidally truncated L_* elliptical galaxy near the cluster core. Any one galaxy in the cluster will suffer several encounters stronger than this since the cluster formed. Although we simulate a perpendicular orbit here, we do not expect the encounter geometry to make a significant difference since the difference between direct and retrograde encounters will be small i.e. $V \gg v_c$.

At t=0.1 Gyrs after the encounter, the perturber has moved 150 kpc away, yet the visible response to the encounter is hardly apparent. After 0.2 Gyrs, we can begin to see the response to the tidal shock as material is torn from the disk into extended tidal arms. Even at this epoch their is a clear difference to the response of the perturbation by each galaxy. After 0.4 Gyrs, the LSB galaxy is dramatically altered over the entire disk and a substantial fraction of material has been removed past the tidal radius. Remarkably, the central disk of the HSB galaxy remains intact and only the outermost stars have been strongly perturbed.

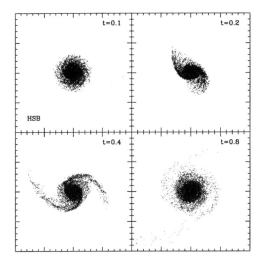

Figure 3. Snapshots of the distribution of disk stars from a HSB galaxy after a single high-speed encounter with a massive galaxy. Each frame is 120 kpc on a side and encounter takes place perpendicular to the disk at the box edge (60 kpc).

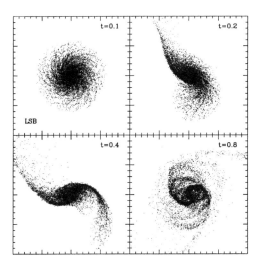

Figure 4. Snapshots of the distribution of disk stars from an LSB galaxy after a single high-speed encounter with a massive galaxy. Each frame is 120 kpc on a side and the encounter takes place perpendicular to the disk at the box edge (60 kpc).

4. Simulating disk evolution within a hierarchical universe

Previous simulations of tidal shocks and galaxy harassment focussed upon the evolution of disk galaxies in static clusters with substructure represented by softened potentials with masses drawn from a Schechter function (Moore et al. 1996a & 1998). Here we use a more realistic approach of treating the perturbations by following the growth of a cluster within a hierarchical cosmological model. The cluster was extracted from a large CDM simulation of a closed universe within a 50 Mpc box and was chosen to be virialised by the present epoch. (We assume a Hubble constant of 100 km s^{-1} Mpc^{-1} .) Within the turn-around region there are $\sim 10^5$ CDM particles of mass $10^{10} M_\odot$ and their softening length is 20 kpc. At a redshift $z = 0$ the cluster has a one dimensional velocity dispersion of 700 km s^{-1} and a virial radius of $2h^{-1}$ Mpc . The tidal field from the mass distribution beyond the cluster's turn-around radius is simulated with massive particles to speed the computation.

Between a redshift z=2 to z=0.5 we follow the merger histories of several candidate dark matter halos from the cosmological simulation that end up within the cluster at later times. We select three halos with circular velocities ~ 200 km s^{-1} that have suffered very little merging over this period and would therefore be most likely to host disk galaxies. We extract these halos from the simulation at z=0.5 and replace the entire halo with the pre-built high resolution model galaxies. We rescale the disk and halo scale lengths by $(1+z)^{-1}$ according to the prescription of Mao et al. (1998) to represent the galaxies entering the cluster at higher redshifts. On a 32 node parallel computer, each run takes several hours; three runs were performed in which the halos were replaced with LSB disks and a further three runs using HSB disks.

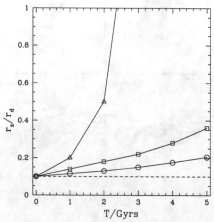

Figure 4. The vertical scale height, r_z, of the disk in units of the initial disk scale length, r_d, measure at r_d and plotted against time. The circles show the HSB galaxy placed in a void to test the numerical heating. The squares and triangles show one of the HSB and LSB galaxies that enters the cluster respectively.

At z=0.5, the cluster is only just starting to form from a series of mergers of several individual group and galaxy sized halos. The cluster quickly virialises, although several dark matter clumps survive the collapse and remain intact orbiting within the clusters virial radius. Between $z = 0.4 - 0.3$ the model galaxy receives a series of large tidal shocks from the halos that are assembling the cluster. Once the galaxy enters the virialised cluster, it continues to suffer encounters with infalling and orbiting substructure. By a redshift z=0.1, many stars have been stripped from the disk and now orbit through the cluster - closely following the rosette orbit of the parent galaxy. Of the three LSB galaxy runs, between 60% and 90% of the stars were harassed from the disk, whereas the stellar mass loss in the HSB runs was between 10%-30%.

5. Summary

The response of a disk galaxy to tidal shocks is governed primarily by the concentration of the mass distribution that encompasses the visible disk. LSB galaxies have slowly rising rotation curves and dynamical timescales that are constant within their central regions. LSB galaxies cannot survive the chaos of cluster formation; gravitational tidal shocks from the merging substructure literally tear these systems apart, leaving their stars orbiting freely within the cluster and providing the origin of the intra-cluster light.

Recent observations of individual planetary nebulae within clusters, but outside of galaxies, lends support to this scenario. Estimates of the total diffuse light within clusters, using CCD photometry (Bernstein et al. 1995, Tyson & Fischer 1995) or the statistics of intra-cluster stars (Theuns & Warren 1997, Feldmeier et al. 1998, Mendez et al. 1998, Ferguson et al. 1998), ranges from 10% to 45% of the light attached to galaxies. Presumably, these stars must have originated within galactic systems. The integrated light within LSB galaxies may be equivalent to the light within "normal" spirals (Bothun, Impey & McGaugh 1997, and references within). This is consistent with the entire diffuse light in clusters originating from harassed LSB galaxies.

High surface brightness disk galaxies and galaxies with luminous bulges have steep mass profiles that give rise to flat rotation curves over their visible extent. The orbital time within a couple of disk scale lengths is short enough for the disk to respond adiabatically to rapid encounters. Tidal shocks cannot remove a large amount of material from these galaxies, nor transform them between morphological types, but will heat the disks and drive instabilities that can funnel gas into the central regions (Lake et al. 1998). A few Gyrs after entering a cluster, their disks are thickened and no spiral features remain. If ram-pressure is efficient at removing gas from disks, we speculate that these galaxies will rapidly evolve into S0's. Since the harassment process and ram-pressure stripping are both more effective near the cluster centers, we expect that a combination of these effects may drive the morphology–density relation within clusters.

References

Bernstein, G.M., Nichol R.C., Tyson J.A., Ulmer M.P. & Wittman D. 1995, AJ, 110, 1507.

de Blok W.J.G. & McGaugh S.S. 1996, ApJ, 469, L89.

Bothun G.D., Schombert J.M., Impey C.D., Sprayberry D. & McGaugh S.S. 1993, AJ, 106, 530.

Butcher, H. and Oemler, A. 1978, ApJ, 219, 18.

Byrd, G. and Valtonen, M. 1990, ApJ, 350, 89.

Couch, W.J., Barger, A.J., Smail, I., Ellis R.S. & Sharples R.M. 1998, in press.

Dressler, A, Oemler, A., Butcher, H. and Gunn, J.E. 1994a, ApJ, 430, 107.

Dressler A., Oemler A., Couch W.J., Smail I., Ellis R.S., Barger A., Butcher H., Poggianti B.M., Sharples R.M. 1998, ApJ, in press.

Dubinski J., 1998, ApJ, in press.

Feldmeier J, Ciardullo R. & Jacoby G. 1998, ApJ, in press.

Ferguson H.C., Tanvir N.R. & von Hippel T. 1998, accepted to Nature (astro-ph/9801228).

Ghigna, S., Moore, B., Governato, F., Lake, G., Quinn, T. & Stadel, J. 1998, MNRAS, in press.

Gunn J.E. & Gott J.R. 1972, ApJ, 176, 1.

Hernquist, L. 1993, ApJS, 86, 389.

Icke, V. 1985, Astr. Ap. 144, 115-23.

Lake, G., Katz, N. and Moore, B. 1998, ApJ, 495, 152.

Lavery R.J. & Henry J.P. 1988, ApJ, 330, 596.

Lavery R.J. & Henry J.P. 1994, ApJ, 426, 524.

Mao S., Mo. H.J & White S.D.M. 1998, MNRASin press.

Mendez, R.H., Guerrero M.A., Freeman K.C., Arnaboldi M., Kudritzki R.P., Hopp U., Capacciolo M. & Ford H. 1997, ApJ, 491, 23.

Mihos J.C., McGaugh S.S. & de Blok W.J.G. 1997, ApJ, 477, L79.

Mo H.J., McGaugh S.S. & Bothun G.D. 1994, MNRAS, 267, 129.

Moore, B., Katz N., Lake G., Dressler, A. and Oemler, A. 1996a, Nature 379, 613.

Moore, B., Katz, N. and Lake, G. 1996b, ApJ, 457, 455.

Moore, B., Lake, G. & Katz, N. 1998, ApJ, 495, 139.

Persic M. & Salucci P., 1997, Dark and visible matter in galaxies ASP Conference series, 117 ed. M. Persic P. Salucci.

Theuns T. & Warren S.J. 1997, MNRAS, 284, L11.

Thompson, L.A. and Gregory, S.A. 1993, AJ, 106, 2197.

Tyson J.A. & Fischer P. 1995, ApJ, 446, L55.

Valluri, M. and Jog, C. J. 1991, ApJ, 374, 103.

Zwaan M.A., van der Hulst J.M., de Blok W.J.G. & McGaugh S.S. 1995, MNRAS, 273, L35.

The Structure of the Multi-Phase ISM in Low Surface Brightness Galaxies

M. Spaans[1]

Harvard-Smithsonian Center for Astrophysics, 60 Garden Street, Cambridge, MA 02138

Abstract. The multi-phase structure of the interstellar medium in low surface brightness galaxies is investigated and compared to observations. It is found that the ambient pressure and metallicity very strongly influence the abundances of molecular hydrogen and carbon monoxide. The emissivity of the latter is computed and found to agree naturally with the upper limits measured for low surface brightness galaxies. The implications for star formation efficiency and galaxy evolution are discussed.

1. Introduction

Low surface brightness (LSB) disk galaxies represent a class of galactic systems which have experienced very slow evolution since their formation epoch. Their low surface brightnesses (< 1 mag/arcsec^2 below the canonical Freeman (1970) value of $\mu_0^B = 21.65 \pm 0.3$ mag/arcsec^2) indicate that, over the age of the Universe, their mean stellar birthrate per unit area has been significantly lower than that of typical high surface brightness (HSB) disks. Their current rate of star formation is similarly low — while some HII regions do exist in LSBs, the global star formation rate in LSBs is lower by an order of magnitude than comparably sized HSBs (McGaugh 1992). The lack of significant star formation is reflected in the low metallicities of LSBs, which are typically $< 1/3$ solar (McGaugh 1992). Not coincidentally, LSBs are also very gas-rich systems — McGaugh & de Blok (1997) found that the gas mass fraction of galaxy disks correlates strongly with surface brightness, such that in LSBs, as much as 50% of the disk mass is in the form of gas, compared to $\sim 10\%$ at high surface brightnesses.

The suppressed rate of star formation in LSB disks must ultimately be connected to the differing physical conditions of the ISM between LSB and HSB disk galaxies. As star formation is presumed to take place in molecular clouds, the molecular content of LSBs is of particular interest. In typical HSB spirals, the mass of molecular gas is comparable to that in neutral HI (e.g., Young & Knezek 1989). The situation in LSBs may be quite different — while several CO surveys of LSBs have been made (e.g., Schombert et al. 1990 (S90)), CO emission has not been detected in any LSB disk galaxy. If CO emission traces molecular gas content in the same way as in normal HSB galaxies, then the

---
[1]Hubble Fellow

upper limits on molecular gas in LSBs are $M_{H_2}/M_{HI} < 0.1$. These upper limits have led to the speculation that the low disk surface densities in LSBs preclude molecular cloud formation and, in turn, inhibit star formation (e.g., S90; van der Hulst et al. 1993; Bothun et al. 1997). Alternatively, the lack of CO detection may simply reflect the fact that the CO/H_2 conversion factor is not a universal constant, so that perhaps large quantities of H_2 exist despite the lack of detected CO emission.

Unfortunately, an observational answer to the question of the molecular content of LSBs is inexorably tied to the CO/H_2 conversion factor and its dependency on environment. For example, Wilson (1995) and Israel (1997) recently showed that the CO/H_2 conversion factor is a strong function of metallicity; this dependency raises the upper limits on the derived molecular content of LSBs. Nonetheless, even accounting for metallicity effects, previous CO surveys should have detected CO in LSBs if they had M_{H_2}/M_{HI} ratios similar to HSBs. Unfortunately other dependencies also play a role, such as the local gas density and temperature (e.g., Maloney & Black 1988, Scoville & Sanders 1987), which in turn are affected by the ionizing radiation field and density structure ("clumpiness") of the ISM. In LSBs all these factors may well be significantly different than expected for HSBs.

To explore the ISM properties of LSB galaxies in a manner independent of the CO/H_2 conversion factor, we take a complementary, theoretical route towards understanding the molecular content of LSB galaxies. We construct models of an inhomogeneous ISM under varying physical conditions, spanning a range of disk galaxy types. The models employ a Monte Carlo approach to radiative transfer (see Spaans 1996), and explicitly solve for the CO emissivity and M_{H_2}/M_{HI} ratio in galactic disks. We investigate models on a grid of metallicity, surface brightness, and ISM density structure, tracking the changing physical conditions between LSB and HSB disk galaxies.

2. Model Description

The code developed by Spaans (1996) and its extensions as discussed in Mihos, Spaans & McGaugh (1998) is used to derive the physical and chemical structure of the ambient ISM in LSBs. The interested reader is referred to these papers for a detailed description of the code's structure. The main features can be summarized as follows.

1) For a given metallicity, geometry, global pressure structure and distribution of illuminating (ultraviolet) sources, the thermal and chemical balance of the medium is computed in three dimensions. The continuum (dust attenuation) and line transfer is modeled through a Monte Carlo method. The self-shielding of H_2 and CO and the shielding of CO by H_2 absorption lines is explicitly included. The heating processes include photo-electric emission by dust grains, cosmic ray heating, collisional de-excitation of ultraviolet pumped H_2, and H_2 dissociation heating. The cooling processes include fine-structure emission of C^+, C and O, rotational line emission of CO, and vibrational ($v = 1 - 0$) H_2 emission.

2) The solutions to the thermal balance equations allow, for a given hydrodynamic pressure and metallicity, multiple solutions (Norman & Spaans 1997).

These constitute the possible multi-phase structure of the ISM as first suggested by Field, Goldsmith, & Habing (1969). If multiple solutions exist, then one finds from a stability analysis that there is a $\sim 10^4$ K diffuse medium and a ~ 50 K dense component. It is the density structure derived from these solutions which couples strongly with the chemical balance of interstellar gas, and therefore with the amount of molecular gas which is supported by the stellar radiation field and the ambient pressure of the galaxy. The cold component has a typical density of $\sim 50-300$ cm^{-3} and is representative of diffuse and translucent clouds in the Milky Way.

2.1. Model Parameters and Their Implementation

To investigate the molecular content of the ISM the following model parameters are considered: average gas density, the average interstellar radiation field (ISRF) in the LSB, metallicity, surface density, and ISM density structure. These parameters are not all independent. To capture the essential dependencies of the ISM structure on ambient physical conditions the following scaling relations are adopted. The HI volume density $n_{\rm HI}$ correlates with HI surface density $\Sigma_{\rm HI}$ according to

$$n_{\rm HI} = \Sigma_{\rm HI}/H, \qquad (1)$$

where $H = 300$ pc is the scale height of the galaxy model. Using data from de Blok et al. (1996), one can derive a rough correlation between local surface brightness μ^B and local HI density:

$$\log \Sigma_{\rm HI} \approx -0.12 * \mu^B + 3.6. \qquad (2)$$

This relationship shows that HI surface density and stellar surface brightness do not drop off in lockstep; instead, the HI surface density falls off more slowly. As surface brightness decreases, the gas mass fraction of the disk increases, such that very low surface brightness disks ($\mu_0^B > 23$) have fully half their baryonic mass in the form of gas.

If one relates the surface brightness and radius r as

$$\mu^B = \mu_0^B + 1.086 * (r/h), \qquad (3)$$

with μ_0^B the central surface brightness in B mags per square arcsecond one finds

$$\log \Sigma_{\rm HI} \approx -0.12 * \mu_0^B - 0.13 * (r/h) + 3.6. \qquad (4)$$

Again, the relationship implies that, as a function of radius, the HI surface density drops off more slowly than the stellar surface brightness, reproducing the extended gaseous disks observed in disk galaxies. In this parametrization, the gas surface density is exponential, but with a scale length 3.3 times larger than that for the stars. See Mihos et al. (1998) for further details.

To parameterize the strength of the ISRF in our models, we assume that the ISRF is dominated by the contribution from the stellar populations in galaxies. Under this assumption, the ISRF scales with surface brightness:

$$I_{\rm UV} = I_{\rm UV}(MW) * 10^{0.4*(\mu_0^B({\rm MW})-\mu_0^B)}, \qquad (2)$$

where I_{UV}(MW) is the strength of the ISRF in the Milky Way given by Draine (1978), and μ_0^B(MW) is the central surface brightness of the Milky Way disk (assumed to be 21 mag arcsec^{-2}). Furthermore, we assume that the spectral shape is *independent* of surface brightness, or, equivalently, that the stellar populations which give rise to the ISRF do not drastically change as a function of surface brightness.

Finally, we need to characterize the inhomogeneity of the dense phase, if it is supported, in the models. This inhomogeneity can be parameterized by *choosing* a certain volume fraction F of the gas in high density clumps with a fixed density contrast R. The size of the clumps is not varied and assumed equal to 2 pc, typical for translucent clouds in the Milky Way. By investigating a range of density contrasts, and therefore clump extinction, this somewhat arbitrary length does not strongly influence the results. We calculate one model, "H", which is completely homogeneous and lacks any density structure. Two more models are explored which have modest amounts of structure ("I1,I2", with $R = 2, 4$ and $F = 0.5, 0.25$; see Mihos et al. 1998). Finally, the clumpy ISM models ("C1,C2", $R = 20, 60$ and $F = 0.25, 0.1$; see Mihos et al. 1998) are chosen to represent our own Galaxy at high ISM pressure.

With these parameterizations, we are left with four variables describing the model galaxies: disk surface brightness, metallicity, pressure, and clumpiness. We create a grid of models spanning a range of plausible values: central surface brightness $\mu_0^B = 21 \to 24$, metallicity $Z/Z_\odot = 1 \to 0.1$, and ISM types H (homogeneous, $P \sim 10^3$ K cm^{-3}), I1 and I2 (intermediate, $P \sim 2 \times 10^3$ K cm^{-3}), and C1 and C2 (clumpy, $P \sim 10^4$ K cm^{-3}). These models thus capture the properties of both high surface brightness spirals as well as low surface brightness disks. For each model we calculate the H$_2$ gas mass fraction as a function of radius, as well as the CO emissivity and mass averaged gas temperature.

3. Results

3.1. Molecular Gas Fractions

Figure 1 shows Σ_{H_2}/Σ_{HI} as a function of radius for several characteristic models. Several trends are immediately obvious:

- At fixed metallicity and ISM structure, lower surface brightness models have *higher* molecular fractions (Figure 1a). Because the ISRF scales with surface brightness, the molecules in the low surface brightness models are less apt to be dissociated by the background ISRF.

- At fixed surface brightness and ISM density structure, models with lower metallicity have lower molecular hydrogen gas content (Figure 1b). This result is due to the fact that dust grains act as formation sites for molecules; lower metallicities mean fewer dust grains to drive molecule formation.

- At fixed surface brightness and metallicity, clumpier ISM models have higher molecular gas fractions (Figure 1c). In clumpy models, a larger mass fraction of the gas is found in denser cores, and are shielded from the background ISRF. Molecules in diffuse ISM models lack this shielding, and are more easily dissociated by the UV background.

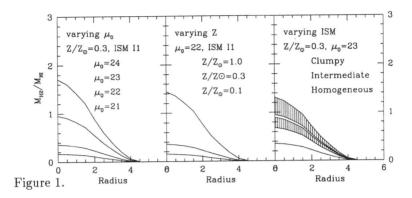

Figure 1.

How well do these models describe actual disk galaxies? One point of constraint is provided by the Milky Way model. The high surface brightness, solar metallicity, and clumpy ISM model shows a mean H_2/HI mass ratio ~ 1 averaged across the inner scale length of the disk, similar to that inferred for Milky Way-like Sb galaxies (Young & Knezek 1989). This is not surprising, since the ISM models were scaled to the ISRF and structure of the Milky Way's ISM, but nonetheless it is reassuring that we recover the correct physical description for the given model inputs.

Assigning a model to LSB galaxies is not as straightforward. Certainly LSB disks are lower in metallicity (McGaugh 1992) and mean ISRF than the Milky Way. The density structure of the ISM in LSBs is not well determined, precisely due to the fact that CO measurements have not yielded any detections. Because of the lowered mass surface density of LSB disks (de Blok & McGaugh 1996), it is likely that the ISM pressures are too low to support the amount of multi-phase structure found in the Milky Way. Models H (homogeneous) and I1 and I2 (intermediate) are therefore likely candidates to describe the density structure of LSB galaxies.

The models suggest that typical LSB galaxies have molecular contents which are only factors of 2–3 below that of normal HSB spirals. The mass averaged gas temperatures in the molecular phase are by no means very cold, in contrast with their multi-phase counterparts (Mihos et al. 1998). Typical temperatures are around 30–50 K, similar to Spitzer-type HI clouds in our own Milky Way. Such high temperatures argue against efficient star formation, but self-consistent rates of the order of ~ 0.05 $M_\odot$ yr^{-1} appear feasible in these low metallicity environments (Norman & Spaans 1997). In conclusion, The lack of detected CO emission in LSBs does not preclude the presence of modest amounts of molecular gas.

3.2. CO Intensity and the CO/H_2 Conversion Factor

To calculate the CO intensity of the models, the root mean square velocity of the interstellar clouds, the dispersion along the z-axis, is taken equal to 10 km s^{-1}, a typical value in the Milky Way and other galaxies. The turbulent velocity width of individual clouds is assumed equal to 3 km s^{-1}. The integrated

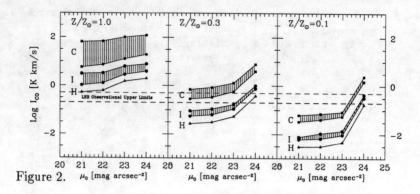

Figure 2.

CO intensities are presented for face-on galaxies, integrated over the inner scale length. Figure 2 shows the variation in I(CO), the CO 1-0 intensity in K km s^{-1}, as a function of metallicity, surface brightness, and ISM structure. As with the H_2/HI mass ratio, several trends are immediately apparent: lower metallicity, higher surface brightnesses, and a more diffuse ISM all act to lower the CO intensity in the models. All these trends are as expected. Lower metallicities mean fewer carbon and oxygen atoms are available to form the CO molecule; higher surface brightnesses result in a stronger ISRF which destroys the CO molecule; and a diffuse ISM is less effective at shielding the CO molecules against radiative dissociation.

Also plotted on Figure 2 are the observational upper limits to the CO intensity of LSB galaxies determined by S90 and de Blok & van der Hulst (1998). If LSBs have solar metallicity, these observations should have detected CO emission. But the subsolar metallicities of LSBs result in lowered CO intensities, making detection difficult. At $Z/Z_\odot \sim 0.3$, the CO emission is only a factor of $\sim 2-5$ below the observational limits, suggesting that deeper CO mapping may in fact reveal the molecular ISM of moderately metal poor LSBs. However, reducing the metallicity by another factor of three reduces the CO emission to levels 30 times fainter than the current observational limits; detecting these LSBs in CO will be very hard indeed.

Perhaps most germane to the observational status of molecular gas in LSB disk galaxies the the conversion factor $X = n(H_2)/I(CO)$ (in units of 10^{21} cm^{-2} (K km s^{-1})$^{-1}$). Figure 3 shows this value calculated for the grid of ISM models. Far from being a constant value, X shows significant and systematic variation between the different models. At solar metallicities, $X \sim 0.1 - 1$, spanning the "standard" value of X derived from Milky Way observations ($\sim 0.2 - 0.5$; see, e.g., Scoville & Sanders 1987). Because the CO intensity scales non-linearly with density, X has a strong dependence on the density structure of the ISM. Our models calculate the properties of the ISM over the inner disk scale length (3 kpc), averaging over both cloud and inter-cloud regions. As the ISM becomes more clumpy, X decreases as the CO intensity rises faster than the H_2 mass fraction. The value of X determined in the Milky Way may therefore be quite different from that applicable to galaxies with a more homogeneous ISM.

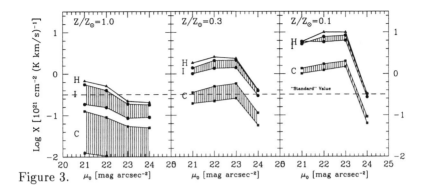

Figure 3.

Aside from the dependence on ISM density structure, there is also a clear correlation between X and metallicity: as metallicity drops, the value of X increases. Such a trend has also been seen in observational data (e.g., Wilson 1995; Israel 1997), and in models of low metallicity clouds (Maloney & Black 1988). The strength of this trend is still quite uncertain; Israel (1997) finds a strong dependence ($\partial \log X/\partial \log Z = -2.7 \pm 0.3$), whereas Wilson (1995) derives a weaker relationship, $\partial \log X/\partial \log Z = -0.67 \pm 0.1$. In our models, the relationship is dependent on the ISM phase structure, but falls in the range $\partial \log X/\partial \log Z = -1$ to -2.

Given the strong dependence on metallicity and ISM density structure, it is clear that use of the standard Milky Way value of X is suspect in LSB galaxies. Instead, we can now turn the problem around and ask, given our theoretical calculation of X, what are the inferred constraints on the molecular gas fraction of LSBs from the CO studies of S90 and dBvdH. If our models are correct, the upper limits on LSB molecular gas are an order of magnitude higher than inferred by these studies due to their use of the standard (low) value of X. Unfortunately, these upper limits then become rather weak: $M_{H_2}/M_{HI} < 1$. While ruling out molecular gas as a dominant component of "dark matter" in LSB disks, these limits still allow for significant molecular gas fractions. More stringent limits on the molecular content of LSBs must await deeper CO observations.

4. Discussion

Our models indicate that even very low surface brightness galaxies may not be completely void of molecular gas – instead, 10–20% of the ISM may be in molecular form. The physical conditions in this gas may be very different from the conditions in the molecular ISM of the Milky Way. If the ISM pressure is extremely low, as might be expected due to the low surface mass density of LSB disks, the molecular phase of the ISM will be diffuse and generally warmer than found in Galactic GMCs. These models also shed light on the lowered efficiency of star formation in LSB disks. Compared to HSBs, LSB galaxies have a lower fraction of molecular material from which they can produce stars. In addition, whatever molecular gas exists, it is in a more diffuse, probably warmer state than is typical for molecular material in HSBs. These warm temperatures and low densities act to help stabilize any existing molecular clouds against gravitational collapse.

The different evolutionary histories of HSB and LSB galaxies can be traced to differences in their disk surface densities and in the conditions of their ISMs. A plausible evolutionary scenario for HSB galaxies has been outlined by Norman & Spaans (1997). In this scenario, once the proto-HSB gas disk forms, star formation begins at a retarded rate in the primordial molecular hydrogen ISM. This star formation generates supernovae and enriches the ISM, leading to a "phase transition" which leads to a multi-phase ISM that is able to cool efficiently and form stars efficiently, leading to a present-day HSB disk galaxy. In contrast, when a proto-LSB forms, it, too, forms a molecular ISM, but with a smaller molecular mass fraction and at lower surface density. At these low surface densities, it is difficult to trigger star formation or form/maintain a multi-phase ISM. As a result, the LSB evolves little from its primordial conditions, maintaining its low surface brightness and metallicity, and high gas fraction.

Acknowledgments. This work was supported by NASA through HF grant HF-01101.01-97A awarded by the STSCi.

References

Bothun, G.D., Impey, C., & McGaugh, S.S. 1997, PASP, 109, 745
de Blok, W. J. G., & McGaugh, S.S. 1996, ApJ, 469, L89
de Blok, W.J.G., & van der Hulst, J.M. 1998, A&A, submitted
de Blok, W.J.G., McGaugh, S.S., & van der Hulst, J.M. 1996, MNRAS, 283, 18
Draine, B.T. 1978, ApJS, 36, 595
Field, G.B., Goldsmith, D., & Habing, H.H. 1969, ApJ, 155, L149
Freeman, K.C. 1970, ApJ, 160, 811
Israel, F.P. 1997, A&A, 328, 471
Maloney, P., & Black, J.H. 1988, ApJ, 325, 389
McGaugh, S.S. 1992, Ph.D. thesis, University of Michigan
McGaugh, S.S., & de Blok, W.J.G. 1997, ApJ, 481, 689
Mihos, J.C., Spaans, M., & McGaugh, S.S. 1998, ApJ submitted
Norman, C.A., & Spaans, M. 1997, ApJ, 480, 145
Scoville, N.Z., & Sanders, D.B. 1987, in Interstellar Processes, eds. D.J. Hollenbach & H.A. Thronson (Dordrecht: Reidel), 21
Schombert, J.S., Bothun, G.D., Impey, C.D., & Mundy, L.G. 1990, AJ, 100, 1523 (S90)
Spaans, M. 1996, A&A, 307, 271
van der Hulst, J.M., Skillman, E.D., Smith, T.R., Bothun, G.D., McGaugh, S.S. & de Blok, W.J.G. 1993, AJ, 106, 548
Wilson, C.D. 1995, ApJ, 448, L97
Young, J.S., & Knezek, P.M. 1989, ApJ, 347, L55

The Low Surface Brightness Universe, IAU Col. 171
ASP Conference Series, Vol. 170, 1999
J. I. Davies, C. Impey and S. Phillipps, eds.

The star formation histories of Low Surface Brightness galaxies

E. F. Bell, R. G. Bower, R. S. de Jong, B. J. Rauscher

Department of Physics, University of Durham, Science Laboratories, South Road, Durham DH1 3LE, UK

D. Barnaby, D. A. Harper, Jr., M. Hereld, R. F. Loewenstein

Department of Astronomy and Astrophysics, University of Chicago, 5640 S. Ellis Ave., Chicago, IL 60637, USA

Abstract. Near-infrared images of a sample of red, blue and giant low surface brightness disk galaxies (LSBGs) were combined with optical data with the aim of constraining their star formation histories. Most LSBGs have strong colour gradients consistent with mean stellar age gradients. We find that LSBGs have a large range of ages and metallicities, spanning those observed in normal disk galaxies. In particular, red and blue LSBGs have very different star formation histories and represent two independent routes to low B band surface brightness. Blue LSBGs are well described by models with low, roughly constant star formation rates, whereas red LSBGs are better described by a 'faded disk' scenario.

1. Introduction

There has been much recent debate on the star formation histories of low surface brightness disk galaxies (LSBGs; galaxies with B band central surface brightnesses fainter than 22.5 mag arcsec^{-2}). The best studied LSBGs are blue (McGaugh & Bothun 1994; de Blok, van der Hulst & Bothun 1995), indicating a young mean stellar age and/or low metallicity. Their measured H II region metallicities are low, at around or below 1/3 solar abundance (McGaugh 1994; Rönnback & Bergvall 1995; de Blok 1997). Morphologically, the best studied LSBGs appear to have disks, but little spiral structure (McGaugh, Schombert & Bothun 1995). The massive star formation rates (SFRs) in LSBGs are an order of magnitude lower than those of high surface brightness (HSB) galaxies (van der Hulst et al. 1993). H I observations show that LSBGs have high gas mass fractions, sometimes even approaching unity (de Blok, McGaugh & van der Hulst 1996; McGaugh & de Blok 1997). As yet, there have been no CO detections of LSBGs, only upper limits on the CO abundances which indicate that LSBGs have CO/H I ratios significantly lower than those of HSB galaxies (Schombert et al. 1990; de Blok 1997). These observations are consistent with a scenario in which LSBGs are relatively unevolved, low mass surface density, low metallicity systems, with roughly constant or even increasing SFRs (de Blok 1997).

This scenario has been called into question however by the recent discovery of a population of red LSBGs (O'Neil, Bothun & Cornell 1997; O'Neil et al. 1997). The optical colours of these galaxies are similar to those of old stellar populations, but the red colours could be caused by age or metallicity effects. The same age-metallicity degeneracy plagues the analysis of the colours of blue LSBGs. Padoan, Jiminez and Antonuccio-Delogu (1997) question the apparent youth of blue LSBG stellar populations: they find that their optical colours are consistent with those of old, very low metallicity stellar populations.

Another puzzle is posed by the results of Quillen and Pickering (1997) who recently obtained near-infrared (NIR) H band imaging of two LSBG giants (galaxies similar to but less extreme than Malin 1). They concluded that the optical-NIR colours of both the central and outer regions of their galaxies were compatible with those seen in old stellar populations (such as E/S0 galaxies). This is difficult to understand, especially in the context of the relatively high gas fractions and low (but non-negligible) star formation rates observed in giant LSBGs.

Here, we explore the star formation histories (SFHs) of different types of LSBG. Because of the age/metallicity degeneracy inherent in optical broad-band colours, it is impossible to tell, using optical colours alone, what causes the colour differences between these classes of LSBG, and what drives their optical colour gradients (de Blok, van der Hulst & Bothun 1995). NIR images, in conjunction with optical data, offer the first chance to break this degeneracy, and constrain plausible SFHs.

2. Observations and data reduction

Our sample consists of galaxies in both the northern and southern hemisphere. The northern hemisphere sample, imaged in the near-infrared K' passband (1.94–2.29 μm) using the Apache Point Observatory 3.5-m, is taken from de Blok, McGaugh & van der Hulst (1996), O'Neil, Bothun & Cornell (1997) and Sprayberry et al. (1995) and is selected to have estimated $22.5 \leq \mu_{B,0} \leq 23.5$ mag arcsec^{-2} and $R_{25} \geq 16$ arcsec where $\mu_{B,0}$ denotes the B band intrinsic disc central surface brightness, and R_{25} denotes the major axis radius to the 25 B mag arcsec^{-2} isophote. The pixel scale is 0.473 arcsec/pixel, and typical on-source exposure times range between 15 and 25 minutes. Our southern hemisphere sample, imaged in the near-infrared K_{dark} passband (2.27–2.45 μm) using the South Pole 0.6-m, is selected from the ESO-Uppsula Catalogue (Lauberts & Valentijn 1987) to have larger sizes and lower surface brightnesses: $\mu_{B,0} \geq 23.0$ mag arcsec^{-2}, 65 arcsec $\leq R_{eff} \leq$ 150 arcsec and inclination less than 67° (where R_{eff} denotes the half-light radius). The pixel scale is 4.2 arcsec/pixel, and typical on-source exposure times range between 3.5 and 7.5 hours. We also require B and R optical images for our analysis. We use existing data from the source papers where available, and have acquired B and R images for those galaxies without existing optical data. Further details of our sample, data acquisition and reduction will be presented in Bell et al. (1999); details for a subset of five northern hemisphere galaxies are presented in Bell et al. (1998). Our sample is by no means complete, but instead is meant to explore the range of disk LSBG star formation histories.

3. Results

Surface photometry was carried out using the IRAF task *ellipse*. The centroid of the brightest region of the galaxy in R band is taken to be the galaxy centre. The galaxy ellipticity and position angle were determined from the R band outermost isophotes. Due to the low surface brightness of our sample in all passbands, the error in the sky level dominates the uncertainty in the photometry. Estimation of the sky level involves both the mean values in the outermost annuli of the surface photometry and the mean sky level in areas of the image that were free of galaxy emission and contamination from starlight. This sky level, which is an average over large areas, is typically accurate to a few parts in 10^5 for the K' images, and better than ~ 0.4 per cent for the K_{dark} and optical images. Galactic extinction corrections are from Schlegel, Finkbeiner & Davis (1998), and range between 0.07 and 0.20 mag in the B band. K-corrections are applied to our two LSBG giants using non-evolving Sbc spectra from King & Ellis (1985), and are typically 0.2 mag in B, 0.02 mag in R and -0.2 mag in K. Our findings are robust to reasonable uncertainties in the assumed K-correction. Galactic extinction corrected galaxy colours in three radial bins ($0 < r/h < 0.5$, $0.5 < r/h < 1.5$ and $1.5 < r/h < 2.5$ where h is the R band disk scale length) using images degraded to the same angular resolution are shown in Fig. 1.

It is clear that there are colour gradients in our sample of LSBGs. However, in order to interpret the colour gradients, it is necessary to compare the data with the results of a stellar population synthesis code, such as that of Bruzual & Charlot (1998). In Fig. 1, we use single metallicity stellar populations with a Salpeter (1955) IMF and a star formation rate described by an exponential star formation timescale τ. Note that the effects of chemical evolution are not included in this model. The solid lines represent the colours of stellar populations with a fixed metallicity and a range of star-forming timescales from an instantaneous burst to a constant SFR. The dashed lines represent the colours produced with a given star formation timescale and a range of metallicities. The arrow denotes the dust reddening vector given by a screen model using the extinction curve of Rieke & Lebofsky (1985) for a visual extinction of 0.3 mag. There is some uncertainty in the placement of the model grid. Charlot, Worthey & Bressan (1996) discussed the sources of error in stellar population synthesis models, and concluded that the uncertainty in model calibration is about 0.08 mag in $B - R$ colour, and around 0.13 mag in $R - K$ colour, which is comparable to the calibration error bar in Fig. 1.

The observed colour gradients are consistent with a mean stellar age gradient (parameterised by the exponential star forming timescale τ), along with an expected contribution from metallicity effects. Note that we can rule out any strong metallicity gradients in our sample ($\gtrsim 0.5$ dex per disk scale length). The colour gradients may also have a contribution from the effects of differential dust reddening: this is expected to be a small effect however and is discussed further in the next section. The existence of these colour gradients is insensitive to any zero point uncertainties, and are very robust to the maximum flat fielding and sky level uncertainties.

It is clear from Fig. 1 that LSBGs span a wide range of luminosity-weighted stellar ages and metallicities, indicating a wide diversity in their inferred star formation histories. Red LSBGs appear to be older and more metal-rich than their

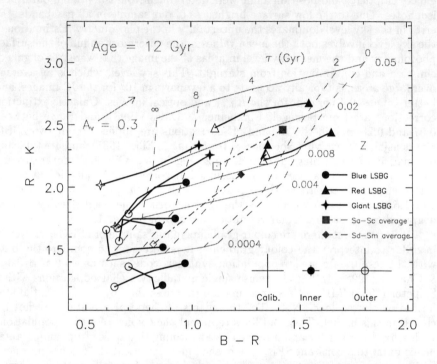

Figure 1. A colour-colour plot comparing the different stellar populations in our LSBG sample. The solid symbols denote central colours, and the open symbols the colours at 2 disk scale lengths. The solid lines are for our sample of LSBGs, where symbol style denotes galaxy type: circles are blue LSBGs, triangles are red LSBGs, and 4-pointed stars are giant LSBGs. Dot-dashed lines connect average colours of Sa–Sc (squares) and Sd–Sm (diamonds) galaxies from the sample of de Jong (1996). Typical colour errors caused by sky level uncertainty are shown by the error bars in the lower right hand corner, where the central colour errors are denoted by the solid circle and those at 2 disk scale lengths are denoted by the open circle. The calibration uncertainties are also shown. The model grid, based on the stellar population synthesis models of Bruzual and Charlot (1998), is discussed in the text.

blue counterparts, indicating an epoch of more vigorous star formation. Blue LSBGs appear to be consistent with models with low, roughly constant star formation rates (implying young luminosity-weighted ages) and low stellar metallicities. Giant LSBGs are metal-rich, but appear to have significant younger stellar populations. Note that the stellar metallicities, and stellar metallicity differences between galaxy types, are consistent with the results of H II region spectroscopy, where available (McGaugh 1994; Rönnback & Bergvall 1995; de Blok 1997). Also, for the blue LSBGs, the optical and NIR colours of our galaxies imply stellar metallicities typically a factor of 20 higher than those proposed by Padoan, Jiminez and Antonuccio-Delogu (1997), thus ruling out their proposed SFH. Our conclusions are unchanged if other combinations of optical-NIR colours, or other stellar population synthesis codes, are used in the analysis.

4. Discussion

4.1. Are the stellar population differences real?

The differences between blue and red LSBGs, and the colour gradients, look as if they may be consistent with the effects of differing amounts of dust reddening in a foreground screen. However, the lack of strong dust features in optical LSBG images (McGaugh & Bothun 1994; de Blok, van der Hulst & Bothun 1995) and generally low ($A_B \sim 1$ mag) Balmer decrements towards LSBG H II regions (McGaugh 1994) tend to argue against large amounts of dust in LSBGs. Furthermore, when more realistic mixtures of dust and stars are used, it is found that the reddening vector steepens considerably, and that larger amounts of dust (typically a few times more) are required to produce the same displacements on a colour-colour plot (de Jong 1996). It is therefore unlikely that the effects of differential dust reddening dominate the colour trends in our sample.

Uncertainties in the high-mass end of the stellar IMF do not alter these conclusions. The above analysis assumes a Salpeter (1955) IMF, however, use of a Scalo (1986) or Miller & Scalo (1979) IMF only significantly changes the high metallicity, large τ end of the grid. While this could change the absolute interpretation of colour gradients in terms of values of metallicity and τ, it is still possible to spot relative metallicity and age trends in and between galaxies, and remain largely unaffected by this uncertainty. In fact, this serves to illustrate the inherent uncertainty in using broad band colours to attempt to unambiguously determine the star formation history of galaxies.

4.2. Do red and blue LSBGs share common ancestors?

Fig. 1 suggests that the red LSBGs have undergone a period of more vigorous star formation (both from the dominance of the older stellar populations, and from the inferred stellar metallicities), whereas blue LSBGs are well described by stellar populations with ongoing low metallicity star formation. However, is it possible that the red and blue LSBGs share the same origin: that is, can red and blue LSBGs transform from one into the other readily? In Fig. 2, we address this question.

Our red LSBGs, using the models of Bruzual & Charlot (1998), have the optical-NIR colours of an old stellar population with roughly solar metallicity. If

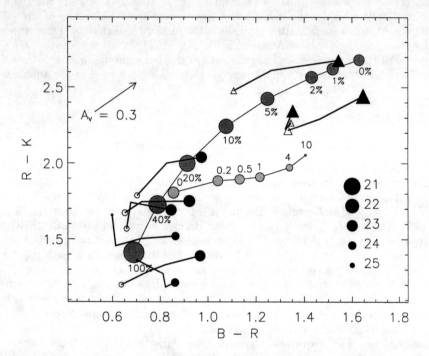

Figure 2. A plot showing the possibilities for transformation between red and blue LSBGs. Central and outer colours for our sample are denoted by solid and empty dark symbols (triangles for red LSBGs and circles for blue LSBGs), with their sizes indicating the B band surface brightnesses at those radii. The upper track (dark grey) shows the colours of a 12 Gyr old solar metallicity stellar population. Added to it is a 1.25 Gyr old single burst population with $Z = 0.004$ with mass fractions ranging from 0 to 100 per cent. The lower track (light grey) shows colours of a $\tau = 16$ Gyr model with an age of 12 Gyr, with the star formation truncated between 0 and 10 Gyr ago. Plot symbol size denotes changes in B band surface brightness in mag arcsec^{-2}.

one adds 20 to 30 per cent, by mass, of a young ~ 1 Gyr stellar population with low metallicity $Z \lesssim 0.004$ (dark grey filled circles) it is possible to reproduce the colours of the blue LSBGs. However, the addition of young stars at all radii (to reproduce the colour of blue LSBGs at all radii) will increase the blue surface brightness by ~ −2 mag, giving central surface brightnesses of ~ 21 mag arcsec^{-2} for these 'transformed' red LSBGs. Thus, it is impossible to transform a red LSBG into a blue LSBG due to surface brightness constraints.

Alternatively, to transform a blue LSBG into a red one, the SFH must be truncated (light grey filled circles). Truncation of star formation will cause the optical colours to redden (note however that the stellar metallicity would still appear lower than those observed in our red LSBGs) at the expense of surface brightness (dimming of ~ 1.5 mag in B, and ~ 0.8 mag in K is expected after 4 Gyr, when compared to a similar galaxy at the time of truncation which is still forming stars with $\tau \sim 16$ Gyr). Therefore, blue LSBGs cannot reproduce *our* sample of red LSBGs because of metallicity constraints, and because the surface brightness would be too faint. Fading from an approximately solar metallicity HSB galaxy to a red LSBG would be possible. Note that an IMF heavily biased towards low-mass stars would also show similar colours, as the optical-NIR colours in essence indicate that red LSBGs simply lack high-mass stars.

5. Conclusions

As part of an ongoing study into the stellar populations of LSBGs, we obtained NIR images of a number of LSBGs spanning a wide range of star formation properties. With the addition of optical B and R images, we found the following.

- Optical-NIR radial colour gradients are common in LSBGs and are consistent with mean stellar age gradients, with the outer regions of LSBGs being younger than the central regions.

- Red LSBGs have the optical-NIR colours of old roughly solar metallicity stellar populations. In contrast, blue LSBGs are still actively forming stars, albeit at a low overall rate (van der Hulst et al. 1993), inconsistent with the SFH proposed by Padoan, Jiminez & Antonuccio-Delogu (1997). LSBG giants have roughly solar metallicity, but have significant contributions from young stellar populations.

- These results suggest that red and blue LSBGs have very different star forming histories, and represent two independent routes to low B band surface brightness: the blue LSBGs are well described by models with a low, roughly constant SFRs, whereas the red LSBGs are more consistent with a 'faded HSB disk' scenario.

Acknowledgments. We would like to thank Erwin de Blok, Stacy McGaugh, Karen O'Neil and David Sprayberry for providing surface photometry and images of galaxies in their sample, and for helpful discussions. In particular, we would like to thank Karen O'Neil for providing the coordinates of her LSBG sample before their publication. We also thank the U. Chicago TAC for regular time allocations to make the K' observations. Some of the observations described here were made during service time at the Isaac Newton Telescope and at the United Kingdom Infrared Telescope.

References

Bell, E. F., Bower, R. G., de Jong, R. S., Hereld, M. & Rauscher, B. J. 1998, submitted to MNRAS

Bell, E. F., Barnaby, D., Rauscher, B. J., Bower, R. G., de Jong, R. S., Harper, Jr., D. A., Hereld, M. & Loewenstein, R. F. 1999, in preparation

Bruzual, A. G. & Charlot, S. 1998, in preparation

Charlot, S., Worthey, G. & Bressan, A. 1996, ApJ, 457, 625

de Blok, W. J. G. 1997 PhD thesis, Groningen University

de Blok, W. J. G., McGaugh, S. S. & van der Hulst, J. M. 1996, MNRAS, 283, 18

de Blok, W. J. G., van der Hulst, J. M. & Bothun, G. D. 1995, MNRAS, 274, 235

de Jong, R. S. 1996, A&A, 313, 377

King, C. R. & Ellis, R. S. 1985, ApJ, 288, 456

Lauberts, A. & Valentijn, E. A. 1989, The Surface Photometry Catalogue of the ESO-Uppsula Catalogue

McGaugh, S. S. 1994, ApJ, 426, 135

McGaugh, S. S. & Bothun, G. D. 1994, AJ, 107, 530

McGaugh, S. S., Schombert, J. M. & Bothun, G. D. 1995, AJ, 109, 2019

McGaugh, S. S. & de Blok, W. J. G. 1997, ApJ, 481, 689

Miller, G. E. & Scalo, J. M. 1979, ApJS, 41, 513

O'Neil, K., Bothun, G. D. & Cornell, M.E. 1997, AJ, 113, 1212

O'Neil, K., Bothun, G. D., Schombert, J. M., Cornell, M. E. & Impey, C. D. 1997, AJ, 114, 2448

Padoan, P., Jiminez, R. & Antonuccio-Delogu, V. 1997, ApJ, 481, L27

Quillen, A. C. & Pickering, T. E. 1997, astro-ph/9705115

Rieke, G. H. & Lebofsky, M. J. 1985, ApJ, 288, 618

Rönnback, J. & Bergvall, N. 1995, A&A, 302, 353

Salpeter, E. E. 1955, ApJ, 121, 61

Scalo, J. M. 1986, Fundam. Cosmic Phys., 11, 1

Schlegel, D. J., Finkbeiner, D. P. & Davis, M. 1998, ApJ, 500, 525

Schombert, J. M., Bothun, G. D., Impey, C. D. & Mundy, L. G. 1990, AJ, 100, 1523

Sprayberry, D., Impey, C. D., Bothun, G. D. & Irwin, M. J. 1995, ApJ, 438, 72

van der Hulst, J. M., Skillman, E. D., Smith, T. R., Bothun, G. D., McGaugh, S. S. & de Blok, W. J. G. 1993, AJ, 106, 548

Gas-rich LSB Galaxies – Progenitors of Blue Compact Dwarfs?

John J. Salzer

Astronomy Department, Wesleyan University, Middletown, CT 06459

Stuart A. Norton

Lick Observatory, Univ. California Santa Cruz, Santa Cruz, CA 95064

Abstract. We analyze deep CCD images of nearby Blue Compact Dwarf (BCD) galaxies in an attempt to understand the nature of the progenitors which are hosting the current burst of star formation. In particular, we ask whether BCDs are hosted by normal or low-surface-brightness dI galaxies. We conclude that BCDs are in fact hosted by gas-rich galaxies which populate the extreme high-central-mass-density end of the dwarf galaxy distribution. Such galaxies are predisposed to having numerous strong bursts of star formation in their central regions. In this picture, BCDs can only occur in the minority of dwarf galaxies, rather than being a common phase experienced by all gas-rich dwarfs.

1. Introduction

If one wished to adopt the simplest scheme for classifying dwarf galaxies, most could be lumped into one of two categories. The first would be the dwarf irregulars (dIs), which can be characterized as having plenty of gas and usually some level of recent or current star formation. The other would be dwarf ellipticals (dEs), which by comparison to dIs have little if any gas, and usually no significant recent star formation (although exceptions to this latter point certainly exist). Interestingly, both types of dwarfs tend to have surface brightness distributions that are well fit by simple exponential profiles. Structural parameters derived from surface photometry of both dIs and dEs show a large range of values: galaxies of both types are observed with both high and low central surface brightnesses (μ_0), and both large and small exponential scale-lengths (α). In fact, the two types overlap completely in the the $\mu_0 - \alpha$ plane.

A group of low-luminosity galaxies which do not readily lend themselves to classification in the above scheme are the blue compact dwarfs (BCDs). These are dwarf galaxies which are currently undergoing an extremely strong burst of star formation, such that the optical appearance of the galaxy is dominated by the energy output of the young stars. In some cases the starburst is so dominant that the presence of an underlying older population of stars is not clearly evident.

Given their extreme nature, it has been difficult to determine with any confidence the type of galaxy that typically hosts BCDs. Since BCDs are usually observed to be gas rich (Thaun & Martin 1983, Salzer et al. 1999a), the most common assumption is that BCDs represent bursts of star formation occurring in dI galaxies. But can any dI initiate a large star-formation episode and appear as a BCD? This is a key question, since it impacts our picture of how dwarf galaxies evolve. Is the BCD phenomenon a stage of galaxy evolution common to all gas-rich dwarfs? We attempt to answer this question in the current study.

2. Relevant Facts about BCDs

Before attempting to address the question posed in the previous section, we review some of the relevant characteristics of BCDs:

• **Optical appearance dominated by light from starburst.** As mentioned above, BCDs are dwarf galaxies whose optical light output is dominated by the energy released by the starburst component. This includes the light from the young O and B stars, plus the nebular emission (both line and continuum) which represents reprocessed UV radiation from the same massive stars. The latter can be a major contributor to the broad-band fluxes measured in BCDs, in extreme cases exceeding the light of the stellar component in the optical. This characteristic of BCDs has made classification of the underlying host galaxy all but impossible in many cases.

• **Very intense nebular spectra.** Spectra of BCDs are dominated by nebular emission lines. In most cases, strong recombination lines of H and He obliterate any stellar absorption lines which may be present. In addition, the nebular continuum combined with the relatively featureless continua of the O and B stars acts to effectively hide the presence of lines from the older stars of the host galaxy. Consequently, spectroscopy yields very little information regarding the stellar content of the BCD host galaxy. On the other hand, the nebular spectra *do* allow for the accurate determination of the abundances in the ionized gas.

• **Gas rich.** The typical BCD contains a large amount of HI gas. The mean value of M_{HI}/M_{tot} for a sample of 122 BCDs is 0.16 (Salzer et al. 1999a). Figure 1 shows the distribution of M_{HI}/L_B for this sample, plotted vs. absolute magnitude. The open symbols represent average values for spiral and irregular galaxies taken from the literature. A statistical correction to the luminosities and mass-to-light ratios of the BCDs has been applied to account for the fact that their luminosities are elevated by an average of 0.75 B magnitudes due to the starburst (see below). Thus, the figure shows where the **host galaxies of the BCDs** would lie in this diagram if the starburst component were removed. On average, the BCD hosts have a factor of ∼2 higher HI gas mass at a given absolute magnitude than do the more quiescent irregular galaxies with which they are compared. To be sure, there are some BCDs with low HI gas content, but on the whole, BCDs are quite gas rich.

The presence of many BCD hosts lying above the trend set by the comparison sample could be interpreted in two ways: either they have unusually high HI masses, or they have normal HI masses but unusually low luminosities. The presence of a population of galaxies in the upper portion of this diagram, with no counterpart in the comparison sample, suggests the possibility that these galaxies would be difficult to detect in typical galaxy surveys during their quiescent

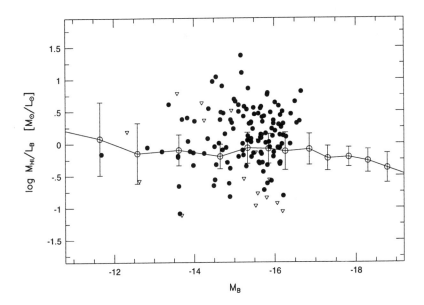

Figure 1. The HI content of BCDs. Here we plot the HI mass to blue-light ratio vs. absolute magnitude for a large sample of BCDs taken from Salzer et al. 1999a. The BCDs are shown as filled circles, while open triangles show upper limits for the BCDs not detected in HI. The connected open circles are the mean values of M_{HI}/L_B for a sample of spiral and irregular galaxies taken from the literature. After correcting the BCDs for the excess light due to their starburst, they are seen to mostly lie at or above the location of the normal galaxies in this diagram.

phases. This led us to suspect originally that the BCD hosts might in fact be LSB dwarfs.

- **Most not bursting for the first time.** The dominance of the starburst led a number of authors to suggest that the observed burst of star formation in some BCDs represents the first episode of star formation in these objects (e.g., Searle & Sargent 1972). However, more recent imaging studies have shown fairly convincingly that nearly all BCDs do possess an older, underlying population of stars (e.g., Papaderos et al. 1996a,b; Telles & Terlevich 1997). There are a few galaxies, such as I Zw 18 and SBS 0335-04, for which a strong case can be made that the current starburst represents the first major episode of star formation (e.g., Thuan et al. 1997), but these are the exception rather than the rule. Although it is virtually impossible to assign ages to the underlying population, one can say with some confidence that most BCDs began forming stars long ago.

- **Burst strengths not extreme.** Because of the optical dominance of the starburst component in BCDs, it is often assumed that the current burst involves a large fraction of the mass of the galaxy, and that it has elevated the brightness of the galaxy by a large amount (estimated at 3–5 magnitudes by some authors). However, this turns out to be an overestimate. Modeling of the starbursts in over a dozen extreme BCDs shows that, on average, only a few percent of the available HI gas is being used in the current starburst (Salzer et al. 1999b). Further, this work shows that the average B-band luminosity enhancement due to the starburst (both stellar and nebular contributions) is only 0.75 magnitudes. Thus, the "bursts" in BCDs are not such extreme events as one might think.

3. Some Recent Key Results

Three recent results have played a significant role in reshaping our view of the nature of BCD host galaxies (or at least in how we might interpret the available data).

The first of these are theoretical studies which attempt to account for the fate of the ISM in dwarf galaxies which experience a major starburst. Early work suggested that even modest numbers of supernovae were enough to completely remove the ambient gas in small galaxies. One implication of this was that BCDs would lose all of their gas following the starburst event, and after 1–2 Gyr resemble dwarf ellipticals. However, more recent studies (DeYoung & Heckman 1994; MacLow & Ferrarra 1998; Brighenti & D'Ercole 1999) have come to the opposite conclusion: only for extremely low-mass galaxies does a starburst remove all of the gas. These new simulations suggest that the hot SN ejecta (including most of the metals produced in the high mass stars) will escape, while the bulk of the colder ISM will remain. If correct, these new studies change drastically our view of post-BCDs. This result is also consistent with the picture that BCDs have had previous star formation. If dwarf galaxies lost their gas easily due to SN outflows, then they would have trouble creating additional generations of stars. Since most BCDs are known to possess at least two generations of stars (and perhaps many more), the ability to retain their gas is obviously crucial.

Another recent finding that may play a major role in our understanding of the BCD phenomenon is that BCDs not only have more HI gas than comparable-sized dIs, but the gas is *more centrally concentrated*. In a recent paper by van Zee et al. (1998), the azimuthally-averaged HI distributions for 8 BCDs were compared to those for a similar number of dIs. The BCDs tend to have strongly peaked gas distributions, i.e., a large reservoir of HI in the central portions of the galaxies. We believe that this is related to the presence of the starbursts in the BCDs, and in fact may be a necessary condition for the occurrence of a strong, sustained star-formation episode like that seen in BCDs.

The third result which has had a major impact on our view of BCD hosts, and which as the primary motivation for carrying out the current study, is the recent work by Papaderos et al. (1996a,b) and Telles & Terlevich (1997). These studies investigate the surface brightness distributions of BCDs, and shed new light on the nature of the host galaxies of BCDs. In particular, Papaderos et al. utilized surface photometry of BCDs out to faint surface brightness levels which allowed them to study both the distribution of the starburst light as well as the light from the underlying host galaxy. Among their results was the suggestion that the host galaxies of BCDs have significantly different characteristics than other, more normal, dwarf galaxies. Their success gave us the incentive to carry out a similar analysis on a large number of existing BCD images obtained previously for other purposes (Salzer & Elston 1992; Salzer et al. 1999b).

4. Structural Parameters of BCD Host Galaxies

We have carried out detailed surface photometry using deep B-band CCD images for a sample of 18 BCDs and 11 dIs. The BCDs were selected from a variety of survey lists, and were chosen to represent the subsample of dwarf star-forming galaxies with the most extreme properties (i.e., the most intense star-formation events). The dI galaxies were analyzed as a comparison sample. These galaxies are part of a separate study of the properties of nearby dwarf galaxies known to exhibit numerous holes in their HI distributions (see Rhode et al. 1999, this volume). Additional comparison dIs were taken from the study by Patterson & Thuan (1996). Both the BCD and dI samples were limited to galaxies with $M_B > -17$.

Since the BCDs are dominated by a (usually) central starburst, isophotal fitting is a tricky business. In general, it is not trivial to distinguish between light from the underlying host galaxy and light from the starburst. However, since we were not particularly interested in the starburst region, we adopted the following simple approach. First, we allowed our isophote-fitting software to fit the entire galaxy with no preset restrictions (i.e., we did not exclude the starburst region). We then used the radial brightness profiles derived from the isophote fitting to determine the structural parameters of the galaxy: central surface brightness (μ_o) and exponential scale length (α). We assumed that the underlying host galaxy could be represented by a single exponential profile. The exponential disk profile was fit *only in the outer portions of the galaxy, well outside the radius occupied by the starburst*. We used Hα images of each galaxy to define where the star formation was occurring. In this way, we can be fairly confident that the derived parameters represent those of the underlying galaxy, and are not affected by the intense starburst. For consistency, we carried out fits to the radial brightness profiles of the comparison dIs in the exact same way. Only the outer portions of the dI profiles were used to determine the structural fits. Complete details of our analysis are given in Norton & Salzer (1999).

The results of our structural parameter determinations are shown in Figures 2 and 3. Figure 2 plots the extrapolated central surface brightness vs. absolute magnitude, while Figure 3 shows disk scale length plotted against absolute magnitude. Also plotted are similar quantities for BCDs from Papaderos et al., and additional dIs from Patterson & Thuan. *The host galaxies of BCDs lie at the extremes of the dwarf irregulars in both plots*, in the sense that the BCDs have systematically higher central surface brightnesses and smaller disk scale

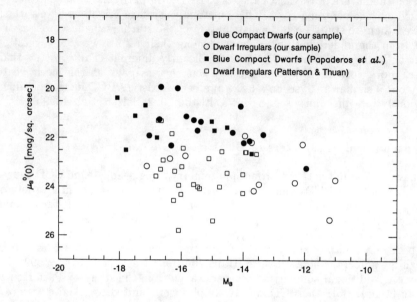

Figure 2. Plot of central surface brightness vs. absolute magnitude for our sample of BCDs and dIs. Also plotted are BCDs from Papaderos et al. and dIs from Patterson & Thuan. The BCDs have systematically higher central surface brightnesses than do the dIs.

lengths. That is, the hosts of BCDs are significantly more compact, and possess higher central mass densities than do more quiescent dIs. We stress that our measurements for the BCDs are done in a way that ignores the light from the starburst – the quantities plotted are for the underlying host galaxies only. If the light from the BCD were included, the difference between the BCDs and dIs would be even greater.

5. Discussion: Implications for Dwarf Galaxy Evolution

The results of our surface photometry, when combined with the recent studies mentioned in Section 3, can now be used to address the question posed in the introduction: are BCDs just typical low-surface-brightness dI galaxies currently in a bursting state? We believe that the answer to that question is a definite NO!

The typical BCD appears to be hosted by a gas-rich galaxy with structrual parameters at the extreme end of those exhibited by dwarfs. On average, BCDs are hosted by centrally concentrated systems, with very small scale lengths and high central mass densities. If there is a continuum of values for the scale length

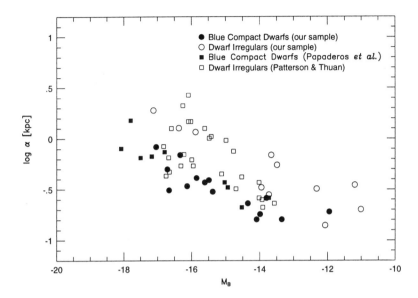

Figure 3. Plot of the exponential disk scale length (α) vs. absolute magnitude for our sample of BCDs and dIs. Also plotted are BCDs from Papaderos et al. and dIs from Patterson & Thuan. The BCDs tend to possess smaller disk scale lengths than the dIs, indicating a more compact distribution for the underlying stars.

and central surface brightness for dwarf galaxies, BCD hosts occupy the extreme end of the distribution. If this picture is correct, then BCDs cannot be hosted by just any dwarf galaxy. Rather, they are preferentially found in the most centrally concentrated systems. Further, they appear to occur in systems with the highest central HI gas densities. This makes perfect sense, since it would be exactly these systems which favor repeated star-formation bursts, and which could build up enough gas mass in a restricted area to generate a star-formation episode of sufficient magnitude to be called a BCD.

This scenario has a number of important implications for our views on how dwarf galaxies evolve. If BCDs and other dwarfs do not lose most of their gas mass as a result of a starburst, as suggested by the simulations of MacLow & Ferrara (1998), they will not evolve into gas-poor dE galaxies, as some authors have contended. Further, since little mass is ejected, there will be no significant dynamical relaxation of the system. In other words, the structural parameters will not change significantly after the starburst is over. Rather, the post-burst BCDs will remain compact. By remaining in such a configuration, with high central mass densities and high HI gas content, the period of time between bursts for a BCD could be quite short, probably on the order of the time scale for the cold ambient gas to settle down after the effects of the last burst are over (perhaps on a time scale of ∼100 Myr). If correct, this implies a very high duty cycle for BCDs. It also helps to explain why we don't see large numbers of compact, post-burst dwarf galaxies. First of all, most dwarfs cannot become BCDs in the first place, and those that do might spend a large fraction of their time in a bursting phase.

However one decides to interpret the current results, it seems clear that the old picture of dwarf galaxy evolution, where most or all gas-rich dwarfs go through one or more BCD-like episodes, is not correct. Rather, only a special subsample of the dwarf galaxy population can host a starburst of the magnitude that would qualify it as a BCD. This simple result requires substantial rethinking of our view of how dwarfs galaxies evolve, since in this scenario most galaxies do not evolve through the BCD phase. Thus transitions from gas-rich dIs to gas-poor dEs seem much less likely, and the galaxies which are currently seen as BCDs will not evolve into either normal dIs or gas-poor dEs after the starburst phase is over. A corollary to this is that galaxies currently seen as low-surface-brightness dwarfs have most likely never gone through a BCD-like phase, and probably never will.

Acknowledgments. We are grateful to our many collaborators on our various dwarf galaxy projects for input and suggestions. In particular, thanks to Liese van Zee, Katherine Rhode, David Westpfahl, and David Sudarsky. JJS gratefully acknowledges financial support from the National Science Foundation.

References

Brighenti, F., & D'Ercole, A. 1999, in preparation
De Young, D.S., & Heckman, T.M. 1994, ApJ, 431, 598
MacLow, M.-M., & Ferrara, A. 1998, ApJ, in press
Norton, S.A., & Salzer, J.J. 1999, in preparation
Papaderos, P., Loose, H.-H., Fricke, K.J., & Thuan, T.X. 1996a, A&A, 314, 59
Papaderos, P., Loose, H.-H., Thuan, T.X., & Fricke, K.J. 1996a, A&AS, 120, 207
Patterson, R.J., & Thuan, T.X. 1996, ApJS, 107, 103
Rhode, K.L., Salzer, J.J., & Westpfahl, D.J. 1999, this volume
Salzer, J. J., & Elston, R. 1992, in *I.A.U. Symposium No. 149*, ed. B. Barbuy and A. Renzini (Dordrecht: Kluwer), p. 482
Salzer, J.J., Rosenberg, J.L., Weisstein, E.W., Mazzarella, J.M., & Bothun, G.D. 1999a, in preparation
Salzer, J. J., Sudarsky, D.L., & Elston, R. 1999b, in preparation
Searle, L. & Sargent, W.L.W. 1972, ApJ, 173, 25
Telles, E., & Terlevich, R. 1997, MNRAS, 286, 183
Thuan, T.X., Izotov, Y.I., & Lipovetsky, V.A. 1997, ApJ, 477, 661
Thuan, T.X., & Martin, G.E. 1983, ApJ, 247, 823
van Zee, L., Skillman, E.D., & Salzer, J.J. 1998, AJ, 116, 1186

The Low Surface Brightness Universe, IAU Col. 171
ASP Conference Series, Vol. 170, 1999
J. I. Davies, C. Impey and S. Phillipps, eds.

Morphological aspects of star formation in dwarf galaxies

Noah Brosch, Ana Heller and Elchanan Almoznino
*The Wise Observatory and the School of Physics and Astronomy
Tel Aviv University, Tel Aviv 69978, Israel*

Abstract.
We studied the morphology of star formation in dwarf irregular galaxies and found that, in general, this takes place on one side of a galaxy and far from the center. This is mainly true for low surface brightness galaxies; in high surface brightness dwarf irregulars the star formation tends to be more centrally concentrated, as well as being more intense. We discuss possible star formation triggers in dwarf irregular galaxies, and evaluate the reasons for the peculiar distribution of star forming regions of these galaxies. Stochastic star formation, interactions with external gas, and tidal interactions appear to be ruled out as responsible for the asymmetric pattern of star formation. We conclude that asymmetry of a dark matter halo or of an asymmetric underlying stellar distribution may trigger an asymmetric pattern of star formation.

Keywords: star formation, dwarf galaxies, HII regions, morphology

1. Introduction

What triggers the star formation in galaxies ? How does it proceed once triggered ? Although much has been written on these subjects they are by no means much clearer today than they were two or three decades ago, when the field of galaxy evolution was just beginning to emerge as a branch of astrophysics.

In principle, the star formation (SF) can be characterized by two boundary condition parameters: the initial mass function (IMF) and the star formation rate (SFR). The IMF was originally described by Salpeter (1955) as a power law. It describes the number of stars formed per unit stellar mass and the exponent in the Salpeter formulation is -2.35. A number of possible modifications of the IMF have been proposed over the years; we shall not review these in detail. It suffices to mention modifications by Miller & Scalo (1979), Kennicutt (1983), and Scalo (1986), some of which may be represented as piecewise power laws. Parenthetically, we also mention the possibility of a metallicity dependence of the IMF (*e.g.*, Terlevich & Melnick 1983).

The SFR has been proposed to be a power law of the gas density (Schmidt 1959), but other formulations have also been put up. Some other possibilities were a dependence on the gas **surface** density, or on this density combined with a dynamical parameter (*e.g.*, Silk 1997). However, these relations may not always represent the most significant dependencies. For instance, in a sample of spiral galaxies the SFR, as measured by the strength of the Hα emission, was found to correlate with the blue surface brightness (Phillipps & Disney 1985). A similar correlation with starlight for a dwarf galaxy sample was found by Hunter et al. (1998).

There have also been numerous suggestions of possible SF triggers. For example, Larson (1986) proposed that large scale gravitational instabilities, cloud compression by spiral density waves, compression by shear forces in a rotating disk, and random cloud collisions in the ISM are the major SF triggers. To these one may add shock waves from stellar winds and SNe, tidal interactions, collisions with other galaxies, ISM stripping, and cooling flow accretion in specific galaxies. SF triggering mechanisms were reviewed by Elmegreen (1998).

Because of the proliferation of potential SF triggers, it is difficult to understand the process in large, well-established galaxies. For this reason we decided to concentrate our efforts in studying SF among the dwarf galaxies (DGs). These are devoid of differentially-rotating disks, thus the shear forces which act in such disks will not be counted among potential trigger mechanisms for DGs. In addition, dwarf irregular galaxies can hardly be classified as "spirals", thus an additional potential trigger may be ruled out: the density waves.

By careful selection of the sample galaxies one may probably discount tidal triggering of SF if the objects are selected to be far from other galaxies. One caveat related to this issue is the possibility of delayed star formation, following a soft (distant) tidal encounter, if this somehow triggers a rain of gas clouds onto the disk (Vázquez & Scalo 1989). Another option is to select the sample from a well-defined galaxy environment, where tidal triggering should average out among the galaxies of the sample. Selection within a cluster of galaxies would conform to this requirement but would not yield many dwarf irregulars with star formation, as these tend to avoid high galaxy density environments.

We selected our initial sample of galaxies from the Virgo cluster. The advantage is that the environment is well-defined, the objects are (relatively) nearby, and the Virgo sky region is accessible from both north and south hemispheres. The disadvantage is that the galaxies are more distant than those in the Local Group, thus one does not expect to detect much detail or individual stars. We concentrated in learning about the SF through integrated stellar populations from broad-band and Hα CCD imaging.

Our sample was selected from the Virgo Cluster Catalog (Binggeli et al. 1985, VCC) with the further proviso that the galaxies would have integrated HI non-zero measurements from Hoffman et al. (1987, 1989) and their heliocentric velocity would be less than 3,000 km s^{-1}. The galaxies form two sub-samples, selected according to their surface brightness as reflected by the morphological classification in the VCC. The high surface brightness (HSB) sub-sample is comprised of objects with the BCD classifier (mixed classifications, such as Scd/BCD, are accepted) and is described in Almoznino & Brosch (1998). The low surface brightness (LSB) sub-sample is comprised of ImIV and ImV objects (mixed classifications, such as dE/ImIV, are accepted) and is described in Heller, Almoznino & Brosch (1998). With an apparent magnitude threshold of 17.5 the LSB sub-sample is complete, while the BCD sub-sample is representative and contains ∼40% of the candidates in the VCC. Our original sample is, therefore, representative of the dwarf irregular galaxies (DIGs) in the VC.

2. Observational data and their interpretation

The HSB galaxies were observed at the Wise Observatory (WO) in Mizpe-Ramon from 1990 to 1997, with CCD imaging through the B, V, R, and I broad bands, and narrow Hα bandpasses in the rest frame of each galaxy. The LSB galaxies were observed mostly at the WO, with a few images obtained at the 6.0-m BTA telescope in Russia. The data set used in the analysis described here is derived exclusively from Hα line and off-line continuum images flux-calibrated against spectrophotometric standards. The data processing is described in detail by Heller et al. (1998 and these proceedings).

We estimate the SFR from the Hα flux using the formalism of Kennicutt et al. (1994) for an 18 Mpc distance adopted to the VC: SFR=$2.93\ 10^{11}$ F(Hα), where F(Hα) is the total line flux in erg s^{-1} cm^{-2} and the SFR is in $M_\odot$ yr^{-1}. Our main finding, explained in more detail by Heller et al. (this volume), is that SF takes place both in the HSB and LSB samples. The main difference between the two types of galaxies is the intensity of the phenomenon; HSB objects have SFRs higher by one order of magnitude than LSBs.

We searched for correlations between the SFR and other observable parameters, in order to understand what determines the SFR in DIGs. The search is described in detail in Brosch et al. (1998a). The relevant result is that the most significant correlation of the SFR, expressed as the SFR per unit solid angle to compensate for a measure of ignorance of the right distance, was not with the surface gas density nor with a surface gas density and dynamical parameter similar to that put forward by Silk (1997), but simply with the blue surface brightness. In other words, what seems to regulate the SFR in DIGs is the local pre-existing stellar population, as found for large spiral galaxies by Phillipps & Disney (1985).

We confirmed this in Heller et al. (1998), where we checked specifically for the link between the Hα flux of an individual HII region and the red continuum flux underneath this HII region; this showed a very strong correlation, supporting our previous result derived with the integrated properties of single galaxies. We conclude that in dwarf irregular galaxies the SFR is regulated by the underlying stellar distribution.

3. Morphology of star formation

A cursory perusal of the net Hα images collected for our entire DIG sample in the VC showed that the HII regions are mostly not centrally located on the galaxy image as shown by the red continuum image. In many cases the HII regions appear right at the edge of a galaxy, mostly to one side. In order to quantify this impression and to eliminate possible biases, due to specific details of our sample galaxies or of our observational procedures, we collected comparison samples of DIGs with Hα and red continuum intensity distribution information from the literature and analyzed them in exactly the same manner as used for the VC DIGs. The comparison samples and their analysis are described in Brosch et al. (1998b).

The quantitative analysis required the definition of two morphological indices. The first describes the degree of concentration in the distribution of HII regions and the second represents their measure of asymmetry. In both cases the reference is the red continuum intensity distribution in the galaxy image and the indices are derived using number counts of HII regions. We first determined the center and approximate extent of the red continuum image of each galaxy. This was done by eye-fitting an ellipse to the outermost visible contour of the galaxy's red continuum image and transposing this to the net-Hα image. We then counted the number of HII regions in the inner part of a galaxy, which we defined as the ellipse with the same center and axial ratio as the outer contour ellipse but with axes half the size of those of the outer contour ellipse. The number of HII regions in the inner part of the galaxy was divided by one-third (to compensate for the larger area) of the number of HII regions in the outer annulus, the space between the inner ellipse and the outer one. This ratio we call CI=concentration index and it can range between 0 and ∞. A value of CI=0 indicates a galaxy with HII regions exclusively in the outer part while CI=∞ indicates an object with exclusively central Hα emission.

A second morphological index, representing the asymmetry in the distribution of HII regions, was constructed by counting the number of HII regions on two sides of a bisecting line traced through the center of the ellipse representing the distribution of the red continuum light, which was transposed to the net-Hα image. The position angle of this bisecting line was set so as to maximize the "contrast" in the number of HII regions between the two halves of the image. The asymmetry index AI was then obtained as the ratio between the smaller number and the larger. AI can range between 0 and 1, with a zero value indicating an extremely asymmetric distribution having all HII regions to one side of the bisector line and a value of 1 representing a fully symmetric number distribution.

The distribution of galaxies in the AI-CI plane is shown in Figure 1. It is clear that most objects concentrate at low-AI and low-CI values, indicating that the HII regions are arranged near the edge of a DIG and mostly to one side of it. The points at very low CI are those with CI=0, i.e., galaxies with HII regions only in their outer parts. The concentration at (CI=100, AI=0) represents points with exclusively nuclear Hα emission; these are essentially BCDs and have CI=∞.

The tendency of DIGs to have asymmetric distributions of HII regions is emphasized by Figure 2, which is a histogram of the distribution of the AI index. Far from being a symmetric Gaussian around AI=0.5, as one could expect for a random distribution of spotty HII regions over the galaxies, the figure shows most objects with AI$\leq$0.5. If we eliminate the bin with AI$\approx$0, which represents those objects with $\sim$one HII region, we have slightly more galaxies in the AI$\geq$0.5 part of the distribution but the difference is not significant; the strong indication of asymmetry is thus driven by objects with few or single HII regions where the SF is generally not centered on the red light distribution. Objects which are not "extreme", i.e., AI$\neq$0 and CI$\neq$ ∞, seem to define a tendency of a more symmetric SF pattern the more centrally concentrated the SF is.

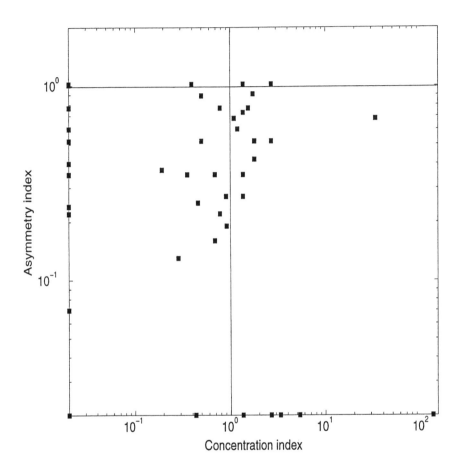

Figure 1. Distribution of the asymmetry and concentration indices. We indicate AI=1 (i.e., a fully symmetric distribution of HII regions) by a horizontal line. The vertical line indicates CI=1 (i.e., an equal number density of HII regions between the inner and outer parts of a galaxy). Most of the objects concentrate in the low AI bins, indicating a preference for asymmetric distribution of HII regions.

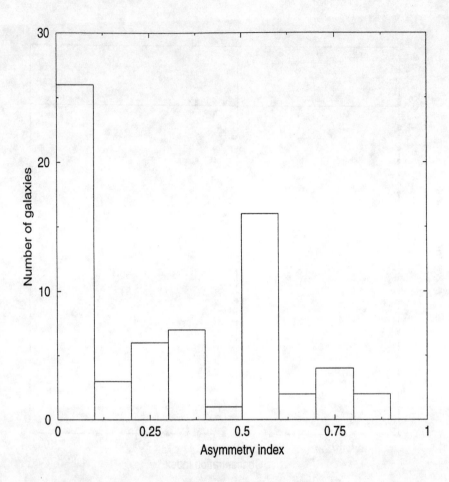

Figure 2. Histogram of the asymmetry index for all the sample galaxies, indicating that the majority have AI≤0.5. This behavior is driven by the large number of objects with a few, or just one, HII region. In such cases, the HII regions have a higher probability of not being located exactly in a symmetric location with respect to the red light distribution.

4. Discussion

We have demonstrated that there is inherent asymmetry in the distribution of star-forming regions in DIGs, and that these regions tend to reside in the outer parts of these galaxies. This is **not** a new discovery, as Hodge (1969) already remarked on the asymmetrical distribution of HII in a small number of DIGs. His seven objects were selected to be very near and could be analyzed on well-resolved photographic images. He checked the asymmetry by counting HII regions on different sides of a galaxy, where the reference mid-galaxy bisector was that of the HII region distribution itself. However, Hodge did not discuss the possible origin of this asymmetry.

The asymmetry and degree of concentration of light in galaxies have been used as morphological indicators for the barely resolved galaxies of the HST Medium Deep Survey (MDS: Abraham *et al.* 1996). There are differences between these indices and those used by us: firstly, the analyzed MDS images are broad-band I (representing the distribution of evolved stars), while we use net-Hα images (i.e., the pattern of newly formed stars). Secondly, our indices are based on eyeball number counts of HII regions, while those of Abraham *et al.* are calculated by "impartial" algorithms from fluxed images. The comparison sample at low redshift, from which Abraham *et al.* derived their calibration of the A & C indices against morphological types, lacks a good representation of irregular galaxies. Nevertheles, they concluded that the late-type galaxies (Sdm and Irr) in the MDS are concentrated at low A & C values, just as we find for our star-forming DIGs. This, they proposed, is evidence for the evolution of irregular galaxies.

Gerola & Seiden (1978) proposed the stochastic self-propagating SF (SSPSF) as a possible regulatory mechanism. Their simulations, as well as those by Jungwiert & Palous (1994), produce preferentially flocculent or grand-design spirals. There are no specific calculations for DIGs which show "snapshots" of the SF proceeding with time through the galaxy. If the galaxy is small, and a number of SNs explode off its center, it may be possible for a compression wave to form stars in suitable location while escaping from the galaxy in places where the ISM is thin or altogether absent. This could, in principle, give rise to an asymmetric pattern of SF and can be examined in a nearby DIG (Ho II: Puche *et al.* 1992). The Hα and off-line images show that Ho II forms stars near its center. The Hα emission originates either at the interfaces between large holes in the HI distribution or in small HI holes. Thus this case does not support the SSPSF forming stars asymmetrically or at the edges of a galaxy.

O'Neil, Bothun & Schombert (1998) argued that the SF trigger in LSB disks is distant tidal interactions. Specifically, they combined arguments from Vázquez & Scalo (1989), that tidally-induced starbursts may be considerably delayed after a gas density enhancement, with simulations from Mihos, de Block & McGaugh (1997) who showed that collisions of LSBs with compact galaxies produce long-lived disturbances in the LSB disks. Although this may operate in large LSB galaxies, serving there as SF triggers, the mechanism is probably not relevant for SF in DIGs. This is because disky DIGs tend to be more stable than large galaxies. Also, the simulations of Mihos *et al.* (1997) showed that the tidally-induced distortions in LSBs form mostly spiral arms and inner rings;

it is probable that any subsequent SF event will retain the same geometry which is a form of density wave and is not the asymmetry which we observe in DIGs.

Icke (1985) proposed a mechanism by which distant tidal interactions may trigger SF through shocks in the ISM of gas-rich disks. His mechanism is probably not relevant for the case of DIGs, because the real strength of the interaction is likely to be much weaker than he estimated. The reason is that the geometrical factor **g** which Icke uses to account for the configuration of the interaction is most likely <0.05, and not 0.39 as fitted by Icke. The factor **g** drives the strength of the tidal interaction and to keep this at the required level one needs to correct upward another parameter in the relation. This is probably very unlikely, making Icke's work not relevant to SF triggers in DIGs. These all argue against tidal interactions as being SF triggers in DIGs, as Heller et al mention here.

Rudnick & Rix (1998) discussed the asymmetry of early-type disk galaxies. They used a different algorithm than ours or that of Abraham *et al.* (1996) to detect asymmetry; the amplitude of the m=1 azimuthal Fourier coefficient of the surface brightness in the R-band. Their claim is that the R-band samples stellar populations older than 1 Gyr, thus the asymmetry found in these disks must be an inherent property of the stellar mass distribution. However, this asymmetry in the mass is not followed by them in the SF properties.

A similar form of asymmetry, lopsidednes of a galactic disk, has also been studied by Jog (1997). Its cause is the asymmetric motion of particles in a lopsided halo. Jog also showed that the gravitational coupling of stars and gas would tend to make the gas more unstable in such a situation than if the gas should be self-gravitating (Jog 1996, 1998). This effect would enhance the asymmetry observed in the young stars in comparison with any asymmetry shown by the old star component. Another explanation for the asymmetry of disks has been proposed by Levine & Sparke (1998). Their scenario has the disk orbiting in an off-center location in the galaxy's DM halo and spinnig in a retrograde sense to its orbital motion. It is not clear what implication do these models have on the asymmetry of SF in DIGs; they refer mostly to collisionless particle simulations of disks while the behavior we witness in DIGs is manifested by the dissipative component of a galaxy, its ISM. However, if the asymmetry of disks or halos implies a similar asymmetry of the gravitational potential, this could serve as a trigger for the asymmetric SF we observe in DIGs.

5. Conclusions

We have shown that star formation takes place in both HSB and LSB DIGs, and that the difference between the two flavors of DIGs is the intensity of the phenomenon. A consideration of a large sample of DIGs has demonstrated that these galaxies tend to form stars in an asymmetric pattern at their outer edge. We could not identify the mechanism responsible for this behavior, but suggest that some asymmetry in the gravitational potential of the galaxies may be the cause.

Acknowledgments. NB is grateful for continued support of the Austrian Friends of Tel Aviv University. EA is supported by a special grant from the Ministry of Science to develop TAUVEX, a UV imaging experiment. AH acknowledges support from the US-Israel Binational Science Foundation. Astronomical research at Tel Aviv University is partly supported by a grant from the Israel Science Foundation. Discussions with Federico Ferrini and Simon Pustilnik on the subject of star formation in DIGs are greatly appreciated.

References

Abraham, R.G., van den Bergh, S., Glazebrook, K., Ellis, R.S., Santiago, B.X., Surma, P. & Griffiths, R.E. 1996, ApJS, 107, 1
Almoznino, E. 1996, PhD thesis, Tel Aviv University
Almoznino, E. & Brosch, N. 1998, MNRAS, 298, 920
Binggeli, B., Sandage, A. & Tammann, G.A. 1985, AJ, 90, 1681 (VCC)
Brosch, N., Heller, A. & Almoznino, E. 1998a, ApJ, 504, 720
Brosch, N., Heller, A. & Almoznino, E. 1998b, MNRAS, in press
de Blok, W.J.G., van der Hulst, J.M. & Bothun, G.D. 1995, MNRAS, 274, 235
Elmegreen, B.G. 1998, in *Origins of Galaxies, Stars, Planets and Life* (C.E. Woodward, H.A. Thronson, & M. Shull, eds.), ASP series, in press
Gerola, H. & Seiden, P.E. 1978, ApJ, 223, 129
Heller, A., Almoznino, E. & Brosch, N. 1998, MNRAS, in press
Hodge, P. 1969, ApJ, 156, 847
Hoffman, G.L., Helou, G., Salpeter, E.E., Glosson, J. & Sandage, A. 1987, ApJS, 63, 247
Hoffman, G.L., Williams, H.L., Salpeter, E.E., Sandage, A., Binggeli, B. 1989, ApJS, 71, 701
Hunter, D.A., Elmegreen, B.G. & Baker, A.L. 1998, ApJ, 493, 595
Icke, V. 1985, A&A, 144, 115
Jog, C.J. 1996, MNRAS, 278, 209
Jog, C.J. 1997, ApJ, 488, 642
Jog, C.J. 1998, private communication
Jungwiert, B. & Palous, J. 1994, A&A, 287, 55
Kennicutt, R.C. 1983, ApJ, 272, 54
Kennicutt, R.C. 1989, ApJ, 344, 685
Kennicutt, R.C., Tamblyn, P. & Congdon, C.W. 1994, ApJ, 435, 22
Larson, R.B. 1986, MNRAS, 218, 409
Levine, S.E. & Sparke, L.S. 1998, ApJ, 496, L13
Loose, H. H. & Thuan, T.X. 1986 in *Star forming dwarf galaxies and related objects* (D. Kunth, T.X. Thuan & J. Tran Thanh Van, eds.), Gif sur Yvette: Editions Frontières, p. 73
Mihos, C., de Blok, W. & McGaugh, S. 1997, ApJ, 477, L79
Miller, G.E. & Scalo, J.M. 1979, ApJS, 41, 513
O'Neil, K., Bothun, G.D. & Schombert, J. 1998, astro-ph/9808359
Patterson, R.J. & Thuan, T.X. 1996, ApJS, 107, 103

Phillipps, S. & Disney, M. 1985, MNRAS, 217, 435
Rudnick, G. & Rix, H.-W. 1998, AJ, 116, 1163
Salpeter, E.E. 1955, ApJ, 121, 161
Scalo, J.M. 1986, Found. Cosmic Phys., 11, 1
Schmidt, M. 1959, ApJ, 129, 243
Silk, J. 1997, ApJ, 481, 703
Terlevich, R. & Melnick, J. 1983, ESO preprint no. 264
Vázquez, E. & Scalo, J. 1989, ApJ, 343, 644

The Low Surface Brightness Universe, IAU Col. 171
ASP Conference Series, Vol. 170, 1999
J. I. Davies, C. Impey and S. Phillipps, eds.

The Connection Between dE and dI Galaxies

Bryan W. Miller

Leiden Observatory, P.O. 9513, 2300 RA Leiden, The Netherlands

Abstract. We combine specific globular cluster frequencies (S_N) with newly measured surface brightness profiles to identify dEs that may be stripped dIs. Luminous dEs generally have higher surface brightnesses and steep central light profiles. Conversely, fainter dEs have low surface brightnesses and flatter central light profiles. The most likely candidates for stripped dIs have low S_N and low central surface brightnesses.

1. Introduction

There is little consensus over how dwarf galaxies form or whether there is an evolutionary connection between gas rich dwarf irregular galaxies (dIs) and gas poor dwarf ellipticals (dEs). Yet, several factors point to such a connection. Within the Virgo and Fornax clusters, the non-nucleated dEs appear to form an extended population, with a spatial distribution similar to the spirals and irregulars rather than the bright ellipticals and the rest of the dEs (Ferguson & Sandage 1989). These non-nucleated dEs also have flattenings more like dIs than the rounder, nucleated dEs. A possible explanation for these differences is that the non-nucleated dEs are stripped dwarf irregular (dI) galaxies (e.g. Lin & Faber 1983). The most popular alternative scenario for dE formation (e.g. Dekel & Silk 1986) is that they formed in one monolithic collapse, and subsequently ejected their ISM via supernova winds.

Miller *et al.* (1998) have recently addressed these issues by studying the globular cluster systems of dE galaxies with the Hubble Space Telescope. The specific globular cluster frequency, $S_N = N_c 10^{0.4(M_V+15)}$, is a useful quantity for testing whether dEs are more like giant ellipticals or spirals and irregulars. They confirm the high mean value of specific frequency, $S_N \sim 5$, suggested by earlier studies and show that S_N in dE,Ns is about twice that in dE,noNs. Therefore, it seems that most bright dEs are more like giant ellipticals, which have $S_N \approx 2-6$, than spirals or irregulars with $S_N < 1$ (Harris 1991).

Yet, several of the non-nucleated dEs in Miller *et al.*'s sample have $S_N < 2$ and they could be stripped and faded dIs. If this is the case, then these galaxies should have structural parameters like dIs. Therefore, we have obtained ground-based V and I imaging of the 17 Leo and Virgo dEs in the HST sample. The new data allow us to check the HST surface photometry and yield more accurate total magnitudes for the lowest surface brightness galaxies. In addition, we obtained B-band imaging of some of the brighter galaxies in order to look for bluer nuclei.

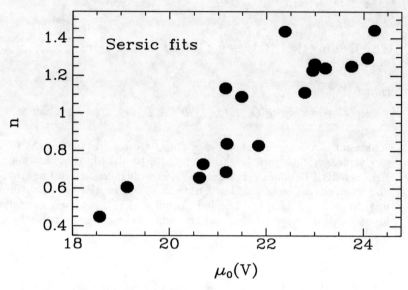

Figure 1. The correlation between μ_0 and n from Sérsic fits to the V light profiles of dE galaxies. Higher surface brightness galaxies also tend to be more luminous.

2. Observations and Procedure

The data were obtained 1–5 April, 1997 at the Las Campanas 40-in. telescope during photometric conditions. Calibration using Landolt (1992) standards shows a RMS scatter of < 0.03 mag. Surface photometry was performed with the GALPHOT routines kindly supplied by Wolfram Freudling. These routines make use of the ISOPHOTE ellipse-fitting package in STSDAS. We fit the light distributions to modified exponentials, or Sérsic profiles, of the form

$$I = I_0 \exp[-(r/r_0)^n]$$

Functions of this form have been shown to fit the light profiles of dEs more generally than simple exponential or $r^{1/4}$ laws (see Durrell 1997).

3. Results

Figure 1 shows the correlation between central surface brightness and exponent n from the Sérsic fits. More luminous galaxies have higher central surface brightnesses and $n < 1$. That is, their profiles are more "cuspy" than an exponential ($n = 1$). Lower luminosity galaxies have lower μ_0 and are flatter than exponentials ($n > 1$). An interesting exception is VCC 1876, which is the third most luminous galaxy yet has $n > 1$ and $\mu_0(V) = 21.15$. The faintest seven galaxies would be considered LSBs since they have $\mu_0(V) > 22.5$ ($\mu_0(B) > 23$).

Figure 2 shows how central surface brightness correlates with S_N. S_N increases with increasing μ_0 for the nucleated galaxies. For non-nucleated galaxies

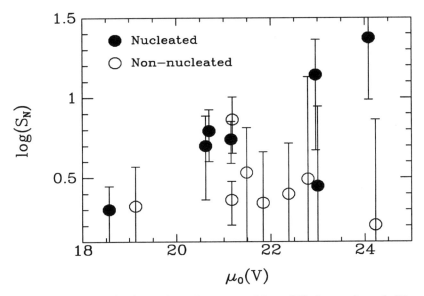

Figure 2. $\log(S_N)$ tends to increase with $\mu_0(V)$ for nucleated dEs. S_N is low and roughly constant for non-nucleated dEs.

S_N is roughly constant and relatively low. The exception is VCC 1577, which has a high S_N but whose most luminous cluster lies within 35 pc of the center. The orbit of this cluster may be decaying by dynamical friction so that the cluster will eventually be the nucleus (Miller *et al.* 1998). Galaxies with $S_N \approx 0$ and $\mu_0(V) > 22.5$ are the best candidates for being stripped dIs. Thus, galaxies like VCC 118, VCC 1651, and VCC 2029 are the best examples. VCC 503 has a very low value of S_N for a nucleated galaxy and has a low surface brightness, so it would be a useful comparison object.

More work is necessary to determine the early star formation histories of dEs and establish any connection with present-day dIs or BCDs. The combination of near-IR and optical colors may help us determine the ages of the dominant stellar populations. Stripped dIs may still show net rotation or residual HI gas. Also, velocity dispersions of the globular systems and the galaxies themselves are needed to constrain the intrinsic shapes and dark matter contents of dEs.

References

Dekel, A., & Silk, J. 1986, ApJ, 303, 39
Durrell, P. R. 1997, AJ, 113, 531
Ferguson, H. C., & Sandage, A. 1989, ApJ, 346, L53
Harris, W. E. 1991, ARAA, 29, 543
Lin, D.N.C., & Faber, S. M. 1983, ApJ, 266, L21
Miller, B. W., Lotz, J., Ferguson, H. C., Stiavelli, M., & Whitmore, B. C. 1998, ApJ, in press

The Low Surface Brightness Universe, IAU Col. 171
ASP Conference Series, Vol. 170, 1999
J. I. Davies, C. Impey and S. Phillipps, eds.

Star Formation Thresholds in Low Surface Brightness Dwarf Galaxies

Liese van Zee

National Radio Astronomy Observatory, Socorro, NM 87801

Abstract. The results of a comparative study of star formation thresholds in gas-rich, low surface brightness, dwarf galaxies are presented. Approximately half the galaxies in the study were "high M_H/L_B" galaxies, which appear to have inefficient star formation properties. The comparison sample comprised of otherwise "normal" dwarf galaxies, with moderate current star formation rates. In all systems, sites of active star formation were associated with local peaks in the HI column density. For both types of galaxies, the azimuthally averaged gas column density is low. Similar to other LSB systems, the global gas densities are well below the critical threshold for star formation throughout the system. Thus, star formation is inhibited globally, but occurs locally in these gas-rich dwarf galaxies.

1. Introduction

One of the intriguing questions today is what mechanisms inhibit or assist star formation activity in low surface brightness (LSB) galaxies. Most LSB galaxies are gas-rich (e.g., Sprayberry et al. 1995), yet their low stellar densities imply that the star formation process has been inefficient in converting the available gas into stars. Several studies of the gas distribution in luminous LSB galaxies (e.g., van der Hulst et al. 1993; de Blok et al. 1996; Hunter et al. 1998) indicate that while these systems are gas-rich, their neutral gas surface densities fall well below the Toomre (1964) instability threshold throughout their optical disks. This is in direct contrast to the results for high surface brightness (HSB) galaxies, where the gas surface density exceeds the Toomre (1964) instability level in the inner disk, and star formation generally occurs in regions where the gas density exceeds 2/3 of the Toomre criterion (Kennicutt 1989). Thus, the low gas surface density inherent in giant LSB galaxies may explain their low star formation rates and low stellar densities.

Several studies have recently applied the same star formation threshold arguments to low luminosity systems (e.g., Hunter & Plummer 1996; van Zee et al. 1996; van Zee et al. 1997c; Hunter et al. 1998). One difficulty with this type of analysis for low mass systems is that their entire optical disks fall within the region of solid body rotation. Some debate has centered around whether the Toomre criterion is even valid in the limit of zero rotational shear; however, it should be valid for all types of single-fluid rotating disks since it is derived through perturbation analysis. Other threshold criteria, such as those derived from pressure balance (e.g., Elmegreen & Parravano 1994) or those from two-fluid rotating disks (e.g., Hunter et al. 1998), have also been explored.

There are two additional caveats about the application of such criteria to observed gas distributions. First, most studies (with the notable exception of Kennicutt 1989) only include the distribution of the atomic gas since the molecular gas component is difficult to determine (or even detect!) (e.g., Wilson 1995; Verter & Hodge 1995). Second, these criteria are applied to the azimuthally averaged gas surface density, which has the tendency to smooth over regions of high gas density/star formation activity. Thus, it is also important to investigate the correlation (or lack thereof) between the local gas density and sites of active star formation. Here, I summarize the results of a comparative study of global and local star formation thresholds in LSB and "normal" dwarf galaxies. Full details of the observations and results can be found in van Zee et al. (1997a,b,c).

2. Sample Selection

The primary sample of gas-rich low surface brightness dwarf galaxies were selected on the basis of unusually high values of M_H/L_B (> 5, based on catalogued values). Although low surface brightness was not initially a selection criterion, the high M_H/L_B systems tend to be low surface brightness objects. In these systems, a significant fraction of their total luminosity comes from regions outside of D_{25} and thus their catalogued magnitudes were severe underestimates of their true luminosity. As discussed by Patterson & Thuan (1996) and van Zee et al. (1997a), the catalogued magnitudes underestimate the true magnitudes by $\sim 1.5^m$ for low surface brightness dwarf galaxies. Thus, systems which appear in the catalogues with extremely high M_H/L_B are typically low surface brightness systems for which the optical magnitudes have been incorrectly estimated. However, even with revised luminosities derived from deep optical imaging, these systems still have values of M_H/L_B which are typically factors of two or more higher than normal for dwarf galaxies.

A comparison sample of "normal" dwarf galaxies were also observed. These systems have similar global properties to the LSB dwarfs, but have values of $M_H/L_B \sim 1$. In particular, these objects are *not* blue compact dwarf galaxies. Rather, they are low luminosity dwarf galaxies which have approximately the same linear optical sizes as the LSB systems but slightly higher central surface brightnesses and star formation rates. Note that both the LSB and the normal dwarf galaxies in this sample are rotation dominated systems, with typical maximum rotation velocities of 40–100 km s^{-1}. Finally, both the normal and LSB dwarf samples were selected to be isolated systems, based on inspection of the Palomar Observatory Sky Survey (POSS) prints.

The current star formation rates for both the LSB and normal dwarfs are quite low and span a wide range, from 0.0015 to 0.35 M$_\odot$ yr^{-1}. Qualitatively, the Hα emission from the normal dwarfs tends to be concentrated in discrete knots near the center of the system, while several of the LSB dwarfs have a significant contribution of diffuse Hα emission. An example of the extremely low star formation rates found in the LSB dwarf galaxies is shown in Figure 1b.

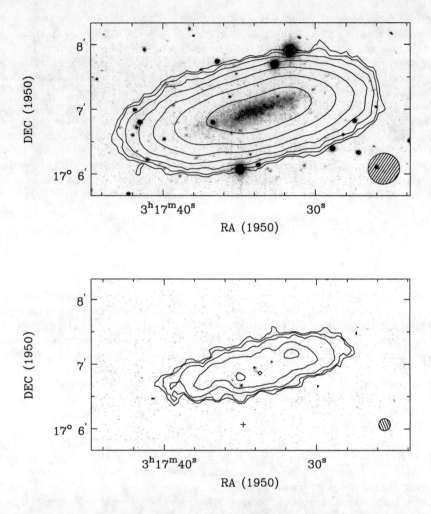

Figure 1. The HI column density distribution of UGC 2684, a low surface brightness dwarf galaxy. The upper panel shows a B–band image of UGC 2684 with a low resolution (23.7″× 22.0″) HI map superposed. The HI contours are logarithmic, with the lowest and highest contours corresponding to densities of 2.0×10^{19} and 1.3×10^{21} atoms cm^{-2}, respectively. The lower panel shows a continuum subtracted Hα image of UGC 2684 with a higher resolution (11.1″× 11.1″) HI map superposed. The HI contours are logarithmic, with the lowest and highest contours corresponding to densities of 1.6×10^{20} and 2.6×10^{21} atoms cm^{-2}, respectively. Note the coincidence between sites of active star formation and local peaks in the HI surface density.

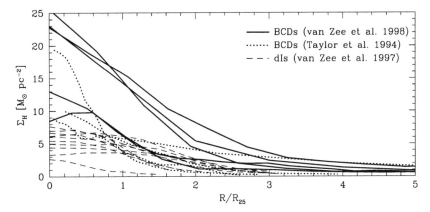

Figure 2. Radial HI surface densities for BCDs and dIs. The BCDs (Taylor et al. 1994; van Zee et al. 1998) have much higher central surface densities than comparable dIs (van Zee et al. 1997c).

3. Star Formation Thresholds

Azimuthally averaged gas surface density profiles of the normal and LSB dwarf galaxies are shown in Figure 2. Also included in this figure are the results of HI studies of star-bursting blue compact dwarf (BCD) galaxies (Taylor et al. 1994; van Zee et al. 1998). A sharp contrast is found between the more quiescent dwarf irregular galaxies (both LSB and "normal") and the star-bursting BCD galaxies. Not only are the HI distributions of the BCDs centrally concentrated, their central surface densities are significantly higher than those of the dIs. Thus, it appears that star-bursting BCDs have inherently different gas mass distributions than dIs, which may be a clue to the origin of strong star formation activity in BCDs.

The ratio between the azimuthally averaged gas surface density and the Toomre (1964) instability criterion for the LSB and normal dwarf galaxy samples are shown in Figure 3 (upper and lower panels, respectively). Of particular interest is the fact that none of the LSB dwarf galaxies have gas densities which exceed the Toomre instability level. Again, this is in direct contrast with results for blue compact dwarf galaxies, where the Toomre criterion is exceeded throughout the star forming regions (Taylor et al. 1994). However, little difference is seen when the LSB and normal dwarf galaxies (objects with similar scale lengths, but slightly higher surface brightnesses) are compared. Similar results were found by Hunter et al. (1998) in their study of gas density thresholds for dwarf galaxies. Thus, it appears that star formation is inhibited for both the LSB and normal dwarf galaxies; in both types of systems, the global gas density appears to be too low to support percolation of star formation.

So, how do these systems form stars at all if their gas density is too low? One important consideration is the fact that star formation is a local process in dwarf irregular systems. Unlike the massive spiral galaxies (both HSB and

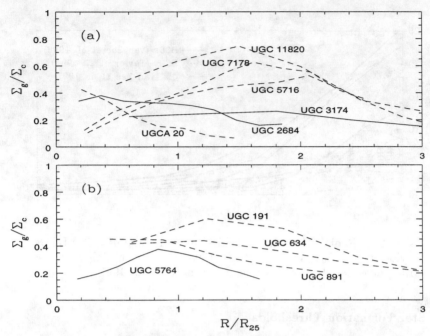

Figure 3. The ratio between the observed neutral hydrogen surface density and the Toomre instability threshold density. (a) The low surface brightness dwarf galaxies. (b) The "normal" dwarfs. The dashed lines denote galaxies with optical scale lengths greater than 1.5 kpc. Note that all of the galaxies have gas densities lower than the predicted instability threshold, but that the intrinsically larger systems do have slightly higher surface densities.

LSB) where spiral density waves introduce global patterns in the star formation activity, the onset of star formation is primarily stochastic in low mass systems. Thus, the azimuthal averages shown in Figures 2 and 3 are not representative of the local gas density in regions of active star formation. As seen in Figure 1b, there is a good correlation between sites of star formation activity and high column density regions on the local level. Thus, while star formation may be inhibited globally in both the LSB and normal dwarf galaxies, it is permitted locally.

4. Are Low Surface Brightness Dwarf Galaxies Young?

The high ratios of M_H/L_B found for the LSB dwarf galaxies suggest that much of the gas in these systems has not yet been processed into stars. Two possible explanations for this are that either the time averaged star formation rate is extremely low in these objects, or that star formation has been initiated recently. Under the first hypothesis, inefficient star formation would slowly convert gas into stars over a Hubble time, resulting in a mixed stellar population at the

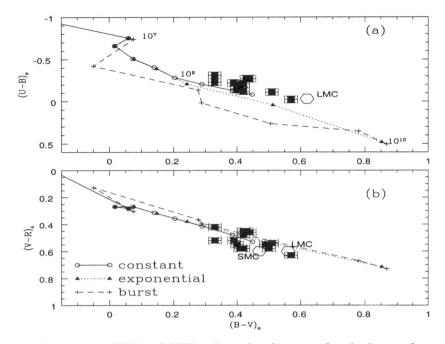

Figure 4. UBV and BVR color–color diagrams for the low surface brightness dwarf galaxies. The color evolution for 3 different star formation rate laws are shown (Bruzual & Charlot 1993). The colors for a galaxy with a Salpeter IMF and constant star formation rate is shown with a solid line. The evolution of a 10^7 year burst is shown with a dashed line. An exponentially decreasing star formation rate with an e–folding time of 1 Gyr is shown with a short dashed line. The time range for each model is marked from 10^7 to 10^{10} years every 0.5 dex.

present epoch. Under the second hypothesis, the recent onset of star formation would result in a predominately young stellar population at the present epoch. Further, in this scenario, the recent star formation rate would have to be relatively high to have generated the observed luminosities. Thus, the observed colors and current star formation rates could distinguish between these two possible evolutionary histories.

The results of UBVR broad band imaging of the low surface brightness dwarf galaxies are shown in Figure 4 (van Zee et al. 1997b). Also plotted in Figure 4 are the color evolution tracks of Bruzual & Charlot (1993). These models assume solar metallicity, which is clearly an incorrect assumption for most of these systems. Thus, the ages derived below are lower limits due to the unknown level of the line blanketing effect.

Age determinations from color evolution models require two major assumptions: (1) the IMF, and (2) the star formation rate history. We have assumed an IMF of Salpeter form and experimented with three different star formation rate histories: (1) the evolution of a single burst of 10^7 year duration, (2) an exponentially decreasing star formation rate with an e–folding time of 1 Gyr, and (3) a constant star formation rate.

In the BVR plane, all three star formation rate histories are degenerate. In this plane it is impossible to distinguish between an aging burst population of 1 Gyr, a 3 Gyr old system with an exponentially decreasing star formation rate, and a 15 Gyr old system with a constant star formation rate.

In the UBV plane, an aging burst population is too red to fit any of the observed colors. This is not unexpected since these systems are fairly large and isolated; it is thus unlikely that the entire system has undergone a simultaneous burst of star formation. The exponentially decreasing star formation rate and constant star formation rates come close to fitting the observed colors (the remaining offset is probably due to using solar metallicity models for subsolar metallicity systems). Both of these star formation rate histories indicate that the dominant stellar populations of the LSB dwarf galaxies were formed within the last 2–4 Gyr. Furthermore, it is not unreasonable to assume that older stellar populations are also present in the LSB dwarf galaxies.

Thus, based on the observed colors and current star formation rates, it appears that one reasonable explanation for the observed enhanced M_H/L_B is that the star formation rate has been extremely low in these objects. It is highly unlikely that these objects are truly "young" systems undergoing their first episode of star formation. Rather, it appears that the star formation process is simply inefficient.

5. Conclusions

In conclusion, we find that the global gas surface density in both the LSB and the normal dwarf galaxies is significantly below the critical surface density at the present epoch. In fact, the two samples appear remarkably similar in HI, despite differences in their optical properties (surface brightness, M_H/L_B, etc.). Further, in both types of galaxies the sites of active star formation are correlated with local peaks in the gas surface density, which approach the instability criterion. Thus, star formation appears to be inhibited globally, but permitted locally in low mass galaxies.

The initiation of current star formation in these objects is probably due to self–regulating feedback, where local regions approach or exceed the critical density to form new HI clouds. These clouds collapse to form stars, once again decreasing the local surface density. Propagation of star formation through the disk may occur if the neighboring gas is compressed by the subsequent evolution of massive stars (e.g., stellar winds, ionizing flux, and supernovae explosions). Clearly, it is the latter step which inhibits star formation if the global surface density is too low. Thus, one possible explanation for the slightly higher surface brightness of the normal dwarfs is the initial gas density distribution. Perhaps the normal dwarfs started with a higher gas density distribution, but then processed their "excess" gas into stars until star formation became too inefficient (i.e., the present epoch). If so, this could also explain why the normal dwarfs have a lower ratio of M_H/L_B than the LSB dwarf galaxies, since they have converted a larger fraction of their initial gas into stars.

Finally, despite their blue colors and abundant supply of relatively unprocess gas, the low surface brightness dwarf galaxies do not appear to be young systems. Rather, their dominant stellar populations are 2–4 Gyr old, and star formation has probably been underway for a large fraction of a Hubble time.

Acknowledgments. I thank my collaborators, Martha Haynes, John Salzer and Adrick Broeils, for many useful conversations. The National Radio Astronomy Observatory is a facility of the National Science Foundation, operated under a cooperative agreement by Associated Universities Inc.

References

Bruzual, G., & Charlot, S. 1993, ApJ, 405, 538
de Blok, W. J. G., McGaugh, S. S., & van der Hulst, J. M. 1996, MNRAS, 283, 18
Elmegreen, B. G., & Parravano, A. 1994, ApJ, 435, L121
Hunter, D. A., Elmegreen, B. G., & Baker, A. L. 1998, ApJ, 493, 595
Hunter, D. A., & Plummer, J. D. 1996, ApJ, 462, 732
Kennicutt, R. C. 1989, ApJ, 344, 685
Patterson, R. J., & Thuan, T. X. 1996, ApJS, 107, 103
Sprayberry, D., Bernstein, G. M., & Impey, C. D. 1995, ApJ, 438, 72
Taylor, C. L., Brinks, E., Pogge, R. W., & Skillman, E. D. 1994, AJ, 107, 971
Toomre, A. 1964, ApJ, 139, 1217
van der Hulst, J. M., Skillman, E. D., Smith, T. R., Bothun, G. D., McGaugh, S. S, & de Blok, W. J. G. 1993, AJ, 106, 548
van Zee, L., Haynes, M. P., & Salzer, J. J. 1997a, AJ, 114, 2479
van Zee, L., Haynes, M. P., & Salzer, J. J. 1997b, AJ, 114, 2497
van Zee, L., Haynes, M. P., Salzer, J. J., & Broeils, A. H. 1996, AJ, 112, 129
van Zee, L., Haynes, M. P., Salzer, J. J., & Broeils, A. H. 1997c, AJ, 113, 1618
van Zee, L., Skillman, E. D., & Salzer, J. J. 1998, AJ, 116, 1186
Verter, F., & Hodge, P. W. 1995, ApJ, 446, 616
Wilson, C. D. 1995, ApJ, 448, L97

Testing environmental influences on star formation with a sample of Low Surface Brightness dwarf galaxies in the Vigo cluster

Ana Heller, Elchanan Almoznino & Noah Brosch

The Wise Observatory and the School of Physics and Astronomy
Tel Aviv University, Tel Aviv 69978, Israel

Abstract. We analyze the star formation activity of an homogeneous LSB dwarf galaxy sample in the Virgo cluster, as a function of the radial velocity relative to the cluster mean velocity and the projected distance from the center of the cluster, using CCD images obtained at the Wise Observatory. The localized Hα emission in the HII regions of this sample is compared to that of an isolated gas-rich sample of LSB dwarf galaxies and that of a representative sample of Blue Compact Dwarf (BCD) galaxies in the cluster. We report preliminary results on the LSB dwarf star formation histories obtained from surface color distribution.

Keywords: LSB dwarf galaxies, environment, star formation

1. Introduction

There is an endless debate in the literature regarding the influence of the neighborhood over the physical processes that govern the evolution and the relative abundance of types in dwarf galaxies (Koopmann 1997). Futhermore, recent investigations of distant galaxies suggest that while rich cluster galaxies have reduced star formation compared to field galaxies of the same central concentration of light, the star formation rate (SFR) is quite sensitive to local galaxy density, both inside and outside of clusters, but the highest levels of star formation are encountered in the intermediate density environment (Hashimoto *et al.* 1997). Therefore, we must consider not only the surface brightness of galaxies in each class, but also their environments, to see if they influence the star formation properties.

The high Galactic latitude of the Virgo cluster (VC), and therefore very small foreground extinction, coupled with the relative nearness of the cluster, allows one to study the effect of the environment on the star formation histories of low surface brightness (LSB) dwarf galaxies. Their low mean density and gravitational binding energies (Bothun *et al.* 1985) make them susceptible to dynamical processes that potentially operate in the cluster, which has a mean galaxy density $\sim$10 galaxies Mpc^{-3}. However, the VC is not a relaxed, virialized cluster; it shows a complex structure in which the LSB dwarf population has no preferential velocity distribution. A kinematic structure map, showing the position on the sky of the various groups and clouds within the VC cluster area, has been published by Hoffman *et al.* (1989).

External forces produced by tidal interactions or tidal shocks should depend strongly on the mass of the perturber/perturbed galaxy, the relative speed, and the minimal distance between them. Other effects of the surroundings, such as ram pressure stripping, evaporation, or turbulent viscous stripping, are believed to be active in gas removal, with estimated stripping time-scale $\sim 10^9$ yr, and should be most efficient in the hot and dense central part of the VC, within $\sim$300kpc of M87 (Ferguson & Binggeli 1994). The removal of gas by the cluster environment is also believed to be responsible for the dependence of chemical abundance properties of HII regions on cluster location in late-type spiral galaxies in VC (Skillman et al. 1995). Note that none of the LSB dwarf galaxies included in our sample (see below) is located in the central region of the VC, and all of them are more distant than $\sim$2.5 degrees ($\sim$1Mpc, projected) from M87. Therefore, the processes mentioned before may not be affecting these galaxies. This is supported by VLA HI maps of the more intense sources of our sample of galaxies (Skillman & Bothun 1986, Skillman et al. 1987)) and Arecibo HI maps (Hoffman et al. 1996, Salpeter & Hoffman 1996). We found that all mapped galaxies support the previous finding of Skillman et al. (1987); at specific projected distances from the cluster center (R_{M87}), the derived HI to optical diameter ratios (D_{HI}/D_{opt}) are larger than those of a sample of VC spiral galaxies.

However, long before passing near the cluster core region, the pressure of the intracluster medium (ICM) may induce star formation in a gas-rich galaxy, due to processes such as compression of gas clouds, density enhancements, accumulation of gas in clouds that later collapse gravitationally, or cloud-cloud collisions in the interstellar medium (ISM) (Elmegreen 1997). Pressure confinement may also prevent or reduce the outflow of gas driven by supernovae or winds from OB stars (Babul & Rees 1992). If these dynamical effects are capable to significantly drive the star population and evolution of LSB dwarf galaxies in the VC, we should then expect to see systematic differences in their SFR related to the location within the cluster.

2. The sample

In the Virgo Cluster Catalog (VCC, Binggeli, Sandage & Tammann 1985), the surface brightness serves as a luminosity class indicator for late-type galaxies. The highest surface brightness objects are assigned to class III while those with the lowest surface brightness belong to class V. There are 31 galaxies in the VCC with certain classification ImIV and ImV; these are LSB galaxies with mean surface brightness fainter than 24-25 mag arcsec^{-2}. We rejected two galaxies from the 24 ImIV to ImV galaxies with non-zero HI measurements (Hoffman et al. 1987): a small, faint 17.5 magnitude galaxy which fell below our threshold, and another with very high heliocentric radial velocity, which violated our membership VC criterion of $v_\odot <$ 3000 km sec^{-1}. This $v_\odot$ restriction arises because of the void behind the VC, between the W and M clouds, from 2800 km/sec to 3500 km/sec, where no galaxies are detected (Binggeli et al. 1993). Here we consider all members in the various clouds as part of the VC. The mean heliocentric radial velocities of the galaxies is 1200 km/sec, close to the mean velocity of dE galaxies, 1139±67 km/sec, though slightly higher than the mean velocity of the cluster (1050±35km/sec; Binggeli et al. 1993).

Our final sample sample comprises of 27 galaxies; it includes 22 galaxies of type ImIV to ImV, four of uncertain classification ImV/dE, ImV? or Im:, and one Im III-IVpec. The limiting magnitude for inclusion in our sample is $m_B = 17.5$; the galaxies are small and their major-axes range from 16 to 120 arcsec at the 25 mag arcsec^{-2} isophote.

3. Observations, Results and Analysis

In order to study the on-going star formation we observed the galaxies with a narrow-band filter in the red continuum near Hα (Hα-off) and with a set of narrow-band filters centered on the rest-frame Hα line (Hα-on) of the galaxies. Deep images in the broad-band filters U, B,V, R and I provided some constraints on the older stellar populations. The narrow-band Hα images were taken during the observing runs of 1996 and 1997 at the Wise Observatory (WO). Two of the faintest galaxies were observed with narrow-band Hα filters at the SAO 6m telescope. The Hα-on images were calibrated with observations of spectrophotometric standards, co-added, and the Hα-off image was subtracted from each one to produce final net-Hα images. The Hα images of the sampled galaxies are shown in Heller et al. (1998). The typical limiting Hα flux is 10^{-16} erg sec^{-1} cm^{-2}; the limiting spatial resolution is 300-400 pc. All UBVRI images were collected at the WO during 1997 and 1998 and standard stars (Landolt 1973, 1992) were used for calibration.

We detected Hα emission in 62% of the sample galaxies. The detection rate is higher when considering only those galaxies with certain ImIV to ImV classification (68%). Three of the galaxies in which we did not detect Hα emission were those with uncertain classification (dE2 or ImIV, Im?) in the catalog. Their HI fluxes put them in low HI content group with a line flux integral ≤776 mJy Km/sec. The high HI content group, with an equal number of galaxies, had a line flux integral ≥950 mJy Km/sec.

We calculated the total SFR of the galaxies as in Kennicutt et al. (1994), SFR=2.93 10^{11} F(Hα), where F(Hα) is the total line flux in cgs units and the SFR is in M$_\odot$yr^{-1}. We adopted a common distance of 18 Mpc for all the galaxies of our sample. This may not be exactly true, but allows a comparison with the values calculated in the same way for BCD galaxies in the VC (Almoznino 1996). The typical SFR of our LSB sample is 0.007 M$_\odot$yr^{-1} a factor 10 lower than BCDs. The typical Hα equivalent width (EW) of the LSB galaxies is ∼30Å peaking at 100Å, a factor 2 lower than the BCDs.

Some of the LSB galaxies in the sample have low heliocentric velocities; it is possible that these are objects falling into the VC from its distant side (Tully & Shaya 1984). We tested the possibility that these galaxies may have enhanced star formation because of interaction with the cluster gas, but this did not prove out. Likewise, we did not find any clear dependence between the net-Hα emission, the EW, the relative velocity of a galaxy with respect to the mean cluster velocity, the angular distance to the VC center, subclustering, or the HI flux of a galaxy.

A number of tests were performed on the level of individual HII regions. We first compared the range of net-Hα fluxes of individual HII regions of our LSB sample to those of isolated, gas rich LSB galaxies studied by van Zee (1998) and found that they are extremely similar (Figure 1).

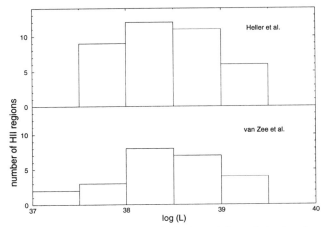

Figure 1. Distribution of the logarithm of the luminosity for individual HII regions, Virgo Cluster LSB dwarf sample (Heller *et al.* 1998) (up), isolated gas rich LSB sample (van Zee 1998) (down).

Both samples have approximately the same number of HII regions per galaxy and in both the luminosity peaks at $\sim 3 \; 10^{39}$ erg s^{-1}. The results indicate no differences between galaxies in the cluster and isolated ones as long as they are dwarfs LSBs, although it should be noted that the covering factor of a galaxy by HII regions is much lower in the isolated sample, originally selected to have extended HI envelopes.

In Figure 2 we plotted the line and continuum fluxes of the HII regions of each galaxy, including the HII regions of the BCD sample. The dotted lines indicate EW=1000, 100, 10 and 1Å respectively. The dearth of HII regions with low EW is easy to understand. It is the result of our detection technique, where low EW HII regions can hardly be distinguished against the red continuum background. On the other hand, it is not clear what limits the high end of the EW distribution. It is very interesting to note that both samples align with approximately the same EW for their HII regions between EW=10 and 100Å.

In a model in which the interaction between the ISM and the ICM is the dominant star formation trigger, the spin of the galaxy moves the HII region from the recent SF burst, which is the locus of the interaction with the ICM, away from the galaxy leading edge. As a burst evolves, the Hα EW should decrease. This is because of the reduction of the line emission (a few Myrs after the disappearance of the ionizing flux) at the same time as the continuum increases due to the net increase in the number of low mass stars. Note that this is true for a star burst which takes place in a pre-existing old population. Our results for individual HII regions in each galaxy (Figure 3) do not support this scenario.

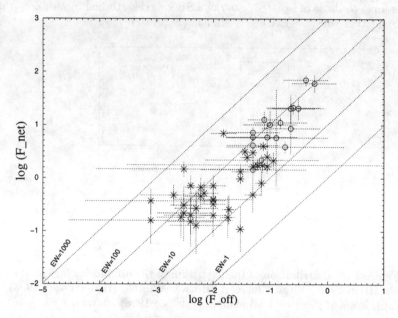

Figure 2. Correlation between the logarithms of the Hα emission and of the red continuum light for individual HII regions in LSB galaxies. The data for HII regions of BCDs is plotted as circles and that for the HII regions of LSBs is represented by stars. The units are 10^{-14} erg cm^{-2} s^{-1} for the line flux and 10^{-14} erg cm^{-2} s^{-1} Å^{-1} for the continuum flux density. The dotted lines indicate EW=1000, 100, 10 and 1Å respectively.

An analysis of the number distribution of the star formation regions for these galaxies (and in other star-forming dwarfs), based on the visual inspection of the net-Hα images, results in asymmetric indices (AI, ratio of the number of HII regions counted in the poor to the rich area) $AI < 0.5$ (Brosch et al. 1998a and these proceedings). In other words, the HII regions are located predominantly on one side of the galaxy, but are not the result of a bow-shock induced SF.

In case that the bursts are the result of both external and internal agents, we should expect to see a broad range of EW values due to the different locations of the orbiting galaxies in the cluster. For most of the objects this is not the case (Figure 4), and the EW of individual HII regions changes very smoothly in a galaxy, with no correlation with the angular distance to the cluster center (M87) nor with the heliocentric radial velocity of the galaxy.

Lacking a clear understanding of external triggers which may be relevant to explain the star formation in Virgo LSB dwarfs, we turned to internal effects which might influence this process. We searched for correlations between the SFR and other individual observable parameters of the galaxies in the sample.

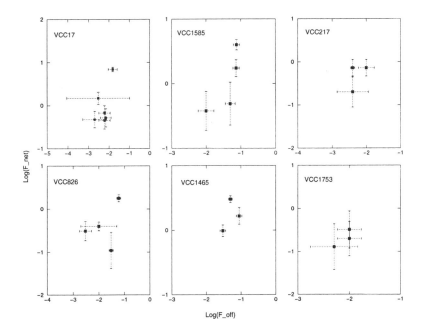

Figure 3. Correlation between the logarithms of the Hα emission and of the red continuum light for individual HII regions in LSB galaxies with multiple HII regions.

The strongest correlation was found between the line and continuum fluxes of the individual HII regions; for more intense line emission the red continuum is also more intense. A linear regression between these variables indicates a correlation coefficient of 0.62. The same calculation for the sample of 17 BCDs in the VC leads to a correlation coefficient of 0.82 (Heller et al. 1998). The result is reminiscent of the correlation found between the SFR and the mean blue surface brightness of late-type dwarf galaxies (Brosch et al. 1998b). We conclude that the star formation of most LSB dwarf galaxies in VC depends on internal processes; a self regulating heating-cooling mechanism modulated by the local volume density of stars has to be at work, probably limiting the SFR of these types of galaxies.

In order to constraint the star formation histories we analyzed the surface color distribution of the galaxies. At present we applied this method to a single object from our sample. We mapped the color indices (U-R), (U-B) and (B-V) for VCC826 and we used the monochromatic Hα-off magnitude as R. We find that the area under the HII region is bluer than the rest of the galaxy. This indicates that there are many young stars under the HII region, and it is not just a lop-sided IMF which creates only massive stars. Therefore a truncated IMF with only stars with mass higher than 10 $M_\odot$ cannot be considered a possible explanation for the low luminosity of LSB galaxies. The color difference between

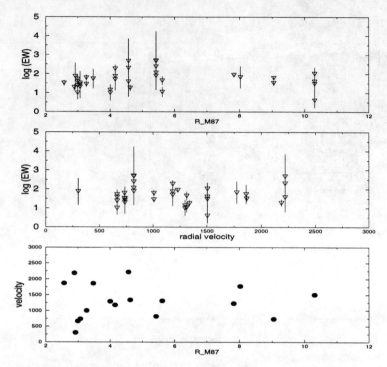

Figure 4. Log(EW) vs. the projected angular distance in degrees from cluster center (R_{M87}) (up). Log(EW) vs. heliocentric radial velocity in km/sec ($v_\odot$) (middle). Heliocentric radial velocity ($v_\odot$) vs. R_{M87} (down).

the area under the strong HII region of VCC826 and the rest of the galaxy is only $\delta(U-R) \sim 0.8$; this indicates that the stellar population of the rest of the galaxy cannot be very old. Although we have not yet attempted a full evolutionary model calculation, possible combination which yield this color difference could have typically B3V stars under the HII regions and A0V in the rest of the galaxy, or O6V under the HII regions and B7V in the rest of the galaxy. This is similar to the interpretation of broad-band colors for BCDs as due to at least two stellar populations, reached by Almoznino (1996). Note that we cannot rule out the existence of even older stellar generations on the basis of the present results.

4. Conclusions

We found no strong evidence for environmental influences in the SF activity of LSB dwarf galaxies in the Virgo Cluster. The SF process seems to be mostly local and regulated by the local population of stars. The SF histories of LSB dwarfs and BCD are probably similar but the bursts differ in intensity. The method of local color analysis proves to be very significant in constraining possible evolutionary scenarios.

Acknowledgments. AH acknowledges support from the US-Israel Binational Science Foundation and thanks Liese Van Zee for kindly providing comparison images. EA is supported by a special grant from the Ministry of Science and the Arts to develop TAUVEX, a UV space imaging experiment. NB is grateful for the continued support of the Austrian Friends of Tel Aviv University. Astronomical research at Tel Aviv University is partly supported by a Center of Excellence Award from the Israel Academy of Sciences.

References

Almoznino, E. 1996, PhD thesis, Tel Aviv University.

Babul, A. & Rees, M.J. 1992, MNRAS, 255, 346.

Binggeli, B., Sandage, A. & Tammann, G.A. 1985, AJ, 90, 1681.

Binggeli, B., Popescu, C.C. & Tammann, G.A. 1993, A&AS, 98, 275.

Bothun, G.D., Mould, J.R , Wirth, A. & Caldwell, N. 1985, ApJ, 90, 697.

Brosch, N., Heller, A.B. & Almoznino, E. 1998a, MNRAS, in press.

Brosch, N., Heller, A.B. & Almoznino, E. 1998b, ApJ, 504, 720.

Elmegreen, B.G. 1998, in "Origins of Galaxies, Star, Planets and Life" (C. E. Woodward, H. A. Thronson & M. Shull, eds.), ASP series, in press.

Ferguson, H. & Binggeli, B. 1994, A&A Rev, 6, 67.

Hashimoto, Y., Oemler, A., Lin, H. & Tucker, D.L. 1998, ApJ, 499, 589.

Heller, A.B., Almoznino, E. & Brosch, N. 1988, MNRAS, in press.

Hoffman, G.L., Helou, G., Salpeter, E.E., Glosson, J. & Sandage, A. 1987, ApJS, 63, 247.

Hoffman, G.L., Helou, G., Salpeter, E.E. & Lewis, B. M. 1989, ApJ, 339, 812.

Hoffman, G.L., Salpeter, E.E., Farhat, B., Roos, T., Williams, H. & Helou, G. 1996, ApJS, 105, 269.

Kennicutt, R.C., Tamblyn, P. & Congdon, C.W. 1994, ApJ, 435, 22.

Koopmann, R.A. 1997, PhD thesis, Yale University.

Landolt, A.U. 1973, A. J 78, 958.

Landolt, A.U. 1992, A. J 104, 340.

Salpeter, E.E. & Hoffman, G.L. 1996, ApJ, 465, 5958.

Shields, G. A., Skillman, E.D., Kennicutt, R.C. & Zaritsky, D. 1995, RMexAA, 3,149.

Skillman, E.D, Bothun, G.D., Murray, M.A. & Warmels, R.H. 1987, A&A, 185, 61.

Tully, R. B. & Shaya, E. 1984, ApJ, 281, 31.

van Zee, L. & Haynes, M.P. 1998, private communication.

Searching for LSB - V

It's much further away than that, I have seen it in a vision.

The Parkes Multibeam Blind HI Survey

R.L. Webster, V. Kilborn and J.C. O'Brien

School of Physics, University of Melbourne, Parkville, Vic, Australia, 3052

L. Staveley-Smith

Australia Telescope National Facility, PO Box 76, Epping, NSW, Australia, 2121

M.E. Putman

Mount Stromlo and Siding Springs Observatory, Australian National University, Weston Creek PO., ACT, Australia, 2611

G. Banks

Department of Physics and Astronomy, University of Wales, Cardiff, PO Box 913, Cardiff, CF2 3YB, Wales, UK

Abstract. A thirteen-beam HI receiver has been constructed for the Parkes radio telescope. When this instrument is used in active scanning mode, it can rapidly survey large areas of sky, with a relatively uniform sensitivity. The Multibeam Working Group, comprising about 30 astronomers from more than a dozen institutions, is undertaking a blind HI survey of the entire southern sky. The status of the survey is described, with some of the first scientific results.

1. The Survey

Surveys of large areas of sky in HI are time-consuming, and require either a dedicated telescope (e.g. Dwingeloo Obscured Galaxy Survey (Henning et al. 1998)) or the multiplexing advantage of a multibeam system. The multibeam receiver at Parkes is a purpose-built instrument with 13 receivers, arranged in a hexagonal grid (Staveley-Smith 1997). The beam centres are separated by about two beam widths. With an appropriate orientation, the full array will Nyquist sample as the telescope scans the sky.

Two surveys are underway: the HI All-sky Survey (HIPASS) which will survey the region $\delta < 0°$ (though a northern extension is being discussed), and a more sensitive Zone-of-Avoidance survey (ZOA), within 5° of the galactic plane. Details of the HIPASS survey will be described. For a more complete description of the ZOA survey see Henning et al (1998). The HIPASS survey uses active scanning where the telescope is driven at 1° per minute in declination strips, recording a spectrum every 5 sec. Each region of sky is scanned 5 times, with small offsets between each set of scans. Data is band-pass corrected online, and both the raw and the preprocessed data are archived (Barnes et al. 1998). On the completion of the full set of scans, data is gridded onto the sky in cubes of $8° \times 8°$, with $4'$ pixels.

In mid-1998, the survey was $\sim 40\%$ complete. Extensive modelling of the gridding procedure has resulted in a fast robust gridder (Barnes 1998). The survey parameters are given in the Table 1.

Table 1. Survey parameters, updated from Staveley-Smith (1997)

Parameter	Value
Declination range	$\delta < 0°$
Equivalent Integration time	500 sec
Velocity Range	-1200 to $12,700$ km sec^{-1}
Channel Resolution	13.2 km sec^{-1}
Velocity Resolution	18.0 km sec^{-1}
Positional Accuracy (3σ)	$\sim 2 - 5'$
Detection Limit (3σ)	40 mJy per beam
HI Mass Limit (3σ)	$1.4 \times 10^6 d^2_{Mpc} M_\odot$
Angular Resolution	$\theta \gtrsim 7'$
Limiting Column Density (3σ)	7×10^{17} HI cm^{-2} per velocity channel
Number of Data Cubes	388

At the present time, galaxy finding is by eye. Cubes are scanned in velocity-position space for regions of enhanced HI emission. This process is both time-consuming and statistically unquantifiable. The Multibeam Working Group is investigating automated galaxy-finding methods which will increase the speed of this process and provide quantifiable detection limits. Nearly all galaxy detections are unresolved on the sky. Thus we are searching for point sources which are distributed in velocity space. Our most promising techniques are based on wavelet transforms, using a point source profile with variable filtering in velocity space. The current flux limit for detection by eye is ~ 4 Jy km sec^{-1}, which is equivalent to $\sim 9\sigma$ for a velocity width of ~ 70 km sec^{-1}. The principle issue for the galaxy finder is to efficiently find galaxies in data with non-gaussian noise. However we are also investigating new methods of filtering noise from the data. An automated finder will provide a powerful tool with which to explore the data cubes and will be necessary before an all-sky catalogue of galaxies can be published.

In order to fully realise the scientific program associated with the survey, a complimentary followup program is required. Synthesised images of selected HI detections are being obtained at the ATCA. These images not only provide more accurate positions for optical identifications, but also resolve the HIPASS detections both spatially and in velocity (Kilborn et al. 1998). Optical B and R-band images are being obtained on 1-metre telescopes at Siding Springs and

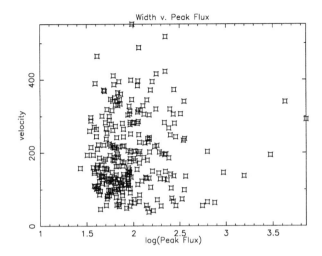

Figure 1. Plot of velocity width against peak intensity for 274 galaxies detected in the southern in the galactic pole region.

Las Campanas. Our data will also be combined with data from other all-sky surveys at different wavelengths, eg. near-IR.

2. HI Mass Function

The HI mass function has been determined for two regions of sky which have been fully scanned. The −74 region is an 8°-wide declination band centred on $\delta = -74$. A total of 99 galaxies have been detected in 0.242 steradians of sky or about 4% of the southern sky. Confused sources have been omitted from the catalogue. Four different people have searched the cubes for galaxies. Figure 1 plots the peak flux of detected galaxies as a function of measured velocity width at 20% of the peak intensity. A clear cutoff is shown at $v \sim 50 \, \text{km sec}^{-1}$. This cutoff is simply the velocity resolution of the survey. The cutoff in peak intensity is also relatively constant, and may be due to the efficiency of the eye in detecting relatively high peaks against a noisy background. There is a concentration of galaxies in the lower left-hand corner of the plot, suggesting that galaxies in this region will be missed due to the detection limits. High resolution observations from the ATCA are being used to quantify how these observational cutoffs limit the parameter space (inclination, velocity width, peak flux, etc) in which galaxies are observed.

Of the galaxies detected in the −74 region, 14% are new detections and 36% have new redshifts. There are a total of $\sim 50\%$ new HI detections. Only those galaxies with a total flux greater than $4 \, \text{Jy km sec}^{-1}$ are included in the determination of the HI mass function. The $\Sigma \frac{1}{V_{Max}}$ method is used to determine the mass function in bins of 0.5 in the logarithm of the mass. Distances are

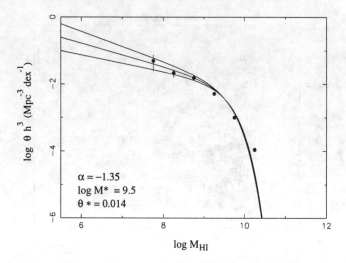

Figure 2. HI mass function 99 galaxies detected in the -74 region.

determined using the Hubble law and a value of $H_0 = 100 \text{ km sec}^{-1}\text{Mpc}^{-1}$. A Schechter function has been fitted by eye. The functional form is

$$\theta(x)\, dx = \theta_* x^\alpha e^{-x}\, dx \qquad (1)$$

where $\theta(x)$ is the number of galaxies per decade of mass per Mpc^3, $x = M_{HI}/M_*$ where M_* is the characteristic mass defining the knee of the mass function, and α is the low mass slope of the mass function.

The low mass end of the mass function is still poorly determined. There are only a couple of galaxies in the lowest mass bin, and the completeness of the sample has not been determined. In addition, the $\Sigma \frac{1}{V_{Max}}$ method is sensitive to clustering, whereas a more robust method, such as a maximum likelihood method is not. The fitted values of the Schechter funtction are $\theta_* = 0.014$ galaxies per Mpc^3 per decade of mass, $log_{10}M_* = 9.5$ and $\alpha = -1.35$. The HI mass function is plotted in Figure 2. These values can be compared with other recent determinations. For example, Zwann et al. (1997) find $\alpha = -1.2$ and $log_{10}M_* = 9.55$.

Finally, and importantly, so far all of the HI detections have an optical counterpart. Thus we are not detecting HI galaxies which do not contain stars, to the limiting HI surface density of the survey. This is an important result for theories of star formation.

3. Cen A Group of Galaxies

The Cen A group of galaxies was extensively surveyed by Côté, Freeman and Quinn (1997). Optical candidates were selected from UK Schmidt plates, and followed up with pointed observations at the Parkes radio telescope. This

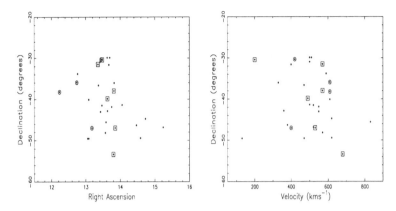

Figure 3. Members in the Cen A group, where the new HIPASS detections are the larger symbols. Both a spatial plot and a plot in velocity space are shown.

work provides a good basis to test the multibeam's survey capabilities. A total of 29 galaxies were detected in the HIPASS data in the veolcity range $200 - 900 \,\mathrm{km\,sec^{-1}}$. Of these, 18 had previously been detected, principally by Côté et al. 10 new galaxies were found with fluxes greater than our fiducial cutoff of $4\,\mathrm{Jy\,km\,sec^{-1}}$. Of these 5 were previously catalogued, and 5 were not. Thus the HIPASS observations have increased the number of galaxies detected in the group by $\sim 50\%$. Figure 3 shows the spatial distribution of galaxies in the Cen A group, with the new members indicated by the larger symbols. Figure 3 also plots the group members as a function of velocity. From this figure it appears that putative members at the extreme ends of the velocity range may not belong to the gravitationally relaxed core of the group. Indeed if traditional cluster member algorithms are applied (Yahil and Vidal 1977), then only those galaxies in the velocity range $300 - 700 \,\mathrm{km\,sec^{-1}}$ satisfy the 3σ cutoff criterion. A Schechter function can be fitted to the mass distribution for the galaxies in the group, assuming that they are all located at a distance of 3.5 Mpc. Values of $log_{10}M_* = 9.3$ and $\alpha = 1.3$ are obtained.

In addition, the region around all the optically verified members of the group was exhaustively searched by eye, resulting in a detection of ESO 272-G025 with a flux of $1.2\,\mathrm{Jy\,km\,sec^{-1}}$, which is a $\sim 3\sigma$ detection. Figure 4 shows the HIPASS profile of this galaxy. A blind eye search would not have found this galaxy, however this detection provides a benchmark with which automated finding algorithms can be tested.

4. Galaxy Formation

Our preliminary datasets all give a slope for the low mass end of the mass function of $\alpha \sim -1.3$, irrespective of whether we search in the field or in a

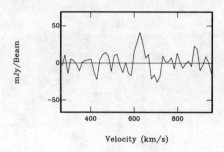

Figure 4. HIPASS profile for ESO272-G025, an optically detected member of the Cen A group. This is a 3σ detection.

denser galaxy environment. Improved galaxy detection algorithms may increase the detections at the low mass end and a more sophisticated treatment of the incompleteness will certainly provide a more robust measurement of α. Thomas (1997) has predicted the shape of the mass function under a range of cosmological assumptions. In a standard CDM cosmology, if the baryons are located in CDM halos, then using the Press-Schechter formalism, a low mass slope of $\alpha = -1.8$ is calculated for the mass function. The optical galaxy luminosity function has a faint end slope of $\alpha = -1.25$ and this function can be matched to the CDM halo mass function at the knee if a $M/L = 15h\,M_\odot/L_\odot$ is assumed for normal galaxies. At the high mass end, the dynamical timescale is greater than the cooling time. Thus we would not expect massive optically luminous galaxies to form. At the low mass end, the CDM halos have a very different distribution from the optically catalogued galaxies. This would result if there was a natural bias of light with respect to the dark matter. In addition, if the low mass end of the HI mass function has a different slope to the faint end of the luminosity function, this would support the idea that the star formation rate in HI halos is not simply a function of total mass. However a full exploration of the HI parameters of detected galaxies is required before strong conclusions can be drawn from these results.

5. Dynamics of the Local Group

With the completion of the first of the five scans, a full mosaic of the southern galactic cap ($\delta \lesssim 62°$) was generated (Putman et al. 1998). In order to improve the detection of HI structure near the Milky Way, the raw data was reprocessed using a modified bandpass correction (Barnes 1998).

The origin of the Magellanic Stream has remained controversial since its discovery by Matthewson et al. (1977). Currently two models for the stream are considered viable: tidal distortion which predicts a leading arm in conjuction with the trailling Magellanic Stream, and ram pressure stripping by gas in the halo of the Milky Way. Figure 5 shows the full mosaic of the HI distribution. A leading arm is clearly seen stretching from the bridge between the Large

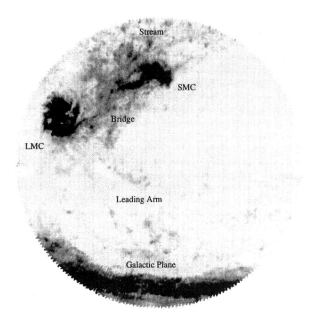

Figure 5. The southern galactic cap as viewed by HIPASS. Different features are labelled.

Magellanic Cloud and the Small Magellanic Cloud towards the Milky Way. This new image is a substantial piece of evidence in favour of the tidal theory for the formation of the Magellanic Stream. This model of the interaction of the LMC/SMC system with the Milky Way supports the idea of Moore et al. (1998) that small disk galaxies are tidally harassed by the larger neighbours in their local environment.

Interestingly, if we also look at the image of the Large Magellanic Cloud in HI, it appears to be a regular spiral galaxy. In optical light, only the bulge is readily visible, and the galaxy is classified as a Irregular. Thus the galaxy morphology based on HI is different from the optical morphology. In addition there is little evidence of tidal distortion in HI distribution in the Large Magellanic Cloud.

6. Discussion

The HIPASS survey will be available as a public database, with the first cubes released in late 1998. So far only about 4% of the southern sky has been completed, and the data searched albeit by eye. Our planned scientific program is wide-reaching: a HI mass function derived from > 4000 galaxies with masses > $10^6 M_\odot$; a DEEP survey, pushing the limits of detectability within the constraints of the scanning technique; and a bivariate brightness distribution, based on HI detections with followup optical imaging. The survey will provide a full inventory of the local HI distribution for regions with $\gtrsim 10^{18}\,\mathrm{HI\,cm^{-2}}$ – an ideal

laboratory for the study of galaxy formation. Progress on the survey can be found at the website:
http://www.atnf.csiro.au/research/multibeam/multibeam.html.

Acknowledgments. The HIPASS survey is the result of the collaboration of the Multibeam Working Group. The Parkes Observatory staff are thanked for their continued assistance in the survey.

References

Barnes, D.G. 1998, PhD thesis, University of Melbourne
Barnes, D.G., Staveley-Smith, L., Ye, T. and Oosterloo, T. 1998 ADASS VII (San Franciso)
Henning, P.A. et al, 1998, AJ in press
Henning, P.A., Staveley-Smith,L., Kraan-Korteweg, R.C. and Sadler, E.M., these proceedings
Kilborn, V., de Blok, E., Staveley-Smith, L. and Webster, R.L. 1998, these proceedings
Matthewson, D.S., Schwarz, M.P. and Murray, J.D. 1977 ApJ 217, L5
Moor, B., Lake, G. and Katz, N. 1998 ApJ 495, 139
Putman, M.E. et al. 1998, Nature 394, 752
Staveley-Smith, L. 1997, PASA, 14, 111
Thomas, P.A. 1997 PASA, 14, 25
Yahil, A and Vidal, N.V. 1977 ApJ 214, 347
Zwaan, M., Briggs, F. and Sprayberry, D. ApJ 490, 173

HI in Karachentsev Objects
Properties of new nearby Dwarf Galaxies

W. K. Huchtmeier

Max-Planck-Institut für Radioastronomie, Auf dem Hügel 69, 53121 Bonn, Germany

Abstract. This is a report on HI observations of newly discovered nearby dwarf galaxies, most of which of low surface brightness, from the first section of the Karachentsev catalog. Observations were performed using the 100-m radiotelescope at Effelsberg, the Nançay radiotelescope, and the compact array of the Australia Telescope. We observed 220 galaxies with a detection rate of about 60%. 35 of the observed galaxies are located within the Local Volume (i.e., within 10 Mpc). The smallest detected galaxies have diameters around 0.3 kpc and HI-masses of a few times 10^6 solar masses. We confirm a correlation between HI column density and surface brightness of gas-rich disk galaxies.

1. Introduction

The study of the faintest and smallest galaxies is limited to our local cosmic neighborhood, i.e., the Local Group. In order to get a reasonable sample of galaxies representing (nearly) all morphological types the volume searched needs to have a certain size. A value of the order of 10 Mpc has been choosen for catalogs of the Local Volume (Kraan-Korteweg and Tammann 1979 (KKT), Schmidt and Boller 1992, Karachentsev 1994).

As this sample covers the whole sky it will be incomplete by two reason, the Zone of Avoidance with its strong foreground extinction and by sensitivity. However, our knowledge of the Local Volume is increasing steadily from a membership of 179 objects in 1979 (KKT) to 303[1] in 1998 (Karachentsev 1998, priv. comm.). The latest improvement became possible with the availability of the Second Palomar Sky Survey (POSS-II) which is more sensitive compared to its predecessor. Karachentseva and Karachentsev started to search the POSS-II and the ESO/SERC films for weak galaxies to a limiting diameter of 0.5' instead of 1', as was used before for many galaxy catalogs based on the POSS-I. In a first step, Karachentseva and Karachentsev (1998) searched for candidates of nearby dwarf galaxies in the neighborhood (i.e., within 50 times the optical diameter D_{25}) of known galaxies and in known groups of galaxies within the Local Volume. This survey covers roughly 25% of the total sky. They found

---

[1] not counting the detection of two spheroidal companions of M31 reported in this conference; And V (Armandroff) and Peg B (Karachentsev)

245 objects from which 139 were not cataloged so far. A great majority of the galaxies in the Karachentsev catalog is of low surface brightness.

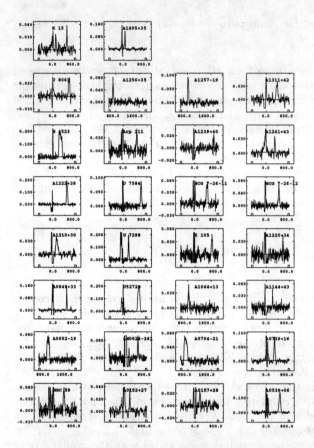

Figure 1. HI profiles of galaxies in the Local Volume as observed with the 100-m radio telescope at Effelsberg. The observed flux in Jansky [Jy] is plotted versus heliocentric radial velocity in [km s^{-1}].

In order to check the 'dwarf' nature of these objects we want to observe their radial velocities for a rough distance estimate. Gas rich dwarf galaxies seem to be relatively easy objects to detect in the 21-cm line of neutral hydrogen. Therefore we decided to search the list from Karachentseva and Karachentsev (1998) for HI emission. The result of this search is presented in this paper.

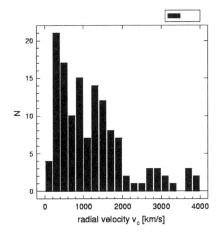

Figure 2. The distribution of corrected radial velocities (v_o) of our galaxy sample demonstrates that most galaxies are relatively nearby. This histogram shows the number of galaxies per velocity interval of $200\,\mathrm{km\,s^{-1}}$.

2. Observations

HI observations have been performed for 220 galaxies using three different radio telescopes for three different declination ranges. 165 galaxies north of declination $-30°$ have been observed using the 100-m radio telescope at Effelsberg which has a half power beam width (HPBW) of 9.3'; 15 galaxies in the declination range between $-30°$ and $-38°$ have been observed with the Nançay radio telescope which has a HPBW of 3.6' x 22' (in R.A. and Dec., respectively). In both cases the velocity coverage was ~ 4400 $\mathrm{km\,s^{-1}}$ with a channel separation of $5\,\mathrm{km\,s^{-1}}$, i.e., a velocity resolution of $6\,\mathrm{km\,s^{-1}}$ ($10\,\mathrm{km\,s^{-1}}$ after Hanning smoothing) using a 1024 channel autocorrelation receiver split into four banks of 256 channels each and a bandwidth of 6.25 MHz.

40 galaxies south of declination $-38°$ have been observed with the Compact Array of the Australia telescope (ATFN) in Culgoora (NSW). For the 'snapshot' observations (in the 750A configuration) we spent a total of 5 to 6 times 10 min per galaxy. The resulting synthesized beam was of the order of 1'. The velocity coverage was of the order of 3000 $\mathrm{km\,s^{-1}}$ yielding a channel separation of 6.6 $\mathrm{km\,s^{-1}}$ amd a velocity resolution of $7.9\,\mathrm{km\,s^{-1}}$. The overall detection rate was slightly above 60%. This is a good value in view of possible background objects which are not included within the bandwidths used in these observations. In addition we might have 'lost' galaxies in the velocity range of local neutral hydrogen in case of the single-dish observations.

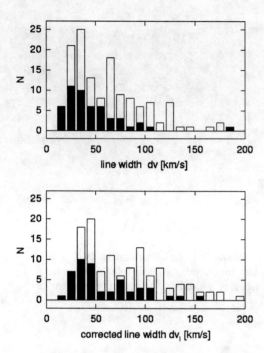

Figure 3. The distribution of line widths of our galaxies is given for the observed values (dv) in the upper panel and for the (for inclination) corrected values in the lower panel (dv_i). Galaxies within the Local Volume are shown by shaded areas.

In a number of cases HI features with negative radial velocities have been observed. All these features were found to be in regions where the Dwingeloo Galactic HI Survey (Hartman and Burton 1997) shows evidence for local HI emission in the corresponding velocity range. Therefore these features were considered to be high velocity clouds and not emission from the searched galaxies.

As an example for the HI profiles observed we selected for Fig. 1 the HI profiles (100-m telescope) of the galaxies within the Local Volume (i.e., within a distance of 10 Mpc). The single-dish observations were performed in the total power mode combining the ON-source observation with a reference field earlier in R.A. by 5 min. Hence, for the local HI emission only a residual from the ON - OFF procedure is left around a velocity of $0\,\mathrm{km\,s^{-1}}$. There have been only few cases of confusion as the galaxies in this sample tend to be isolated.

Most of the HI profiles are narrow (small rotational velocities). The ATNF observations show HI emission centered on the optical positions and an extent of the HI distributions of two to three times the optical diameter of the galaxies.

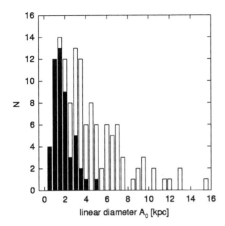

Figure 4. The distribution of the linear diameters of the whole sample is given in this figure. Galaxies within the distance limit of the Local Volume (i.e., within 10 Mpc) are shown by the shaded areas.

3. Discussion

For the discussion we will also include recent observations of nearby dwarf galaxies like the pilot observations for this project (Huchtmeier et al. 1997) and observations of new nearby galaxies (Kraan-Korteweg et al. 1994, Huchtmeier et al. 1995, Burton et al. 1996, Huchtmeier and van Driel 1997).

In Fig. 2 we show the radial velocity (v_o, corrected for the rotation of our galaxy) distribution of our galaxy sample; approximately 25 % are located within the Local Volume and most of the galaxies are within the Local Supercluster. Apart from a few large background galaxies (mostly LSB objects) most of the galaxies are rather small as can be seen from their small line widths (i.e., small rotational velocities) (Fig. 3) and their small linear dimensions (Fig. 4).

Two more global parameters are shown in Fig. 6. Here we show the range The line width distribution in Fig. 3 shows the observed line widths in the upper panel whereas the (for inclination) corrected line widths are given in the lower panel. The optical axial ratio has been used as a measure for the inclination. Galaxies within the Local Volume are indicated by shaded areas.

The distribution of the linear diameters (Fig. 4) of our galaxy sample extends from 0.2 kpc to 26 kpc, the galaxies within the limits of the Local Volume (closer than 10 Mpc) are given as shaded areas. It is evident that all new nearby galaxies are relatively small with 1 to 2 kpc on the average for the optical diameter (D_{25}).

How do the galaxies of our present sample compare to the galaxies of the Local Volume known before? In Fig. 5 we compare global parameters of these two samples. The total mass of neutral hydrogen (M_{HI}) is plotted versus the

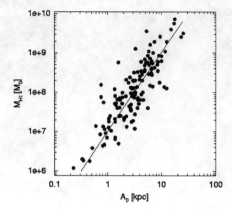

Figure 5. The total mass of neutral hydrogen of the galaxies in our present sample is plotted versus the linear extent (in kpc). The full line represents the regression line for the KKT sample (Huchtmeier and Richter 1988).

linear extent (A_0) for the present sample of galaxies. The full line represents the regression line for the galaxies of the Kraan-Korteweg-Tammann sample (Huchtmeier and Richter 1988). This regression line seems to be an excellent fit for the present sample, too.

Two more global parameters are shown in Fig. 6. Here we show the range of the pseudo HI surface density Σ_{HI} which is obtained by dividing the total HI mass (M_{HI}) by the disk area of the galaxy as defined by its optical diameter (D_{25}). This quantity is given in units of solar mass per square parsec and in the usual HI column density (N_{HI}) in $atoms\,cm^{-2}$ (on the right hand side of the figure).

This quantity is plotted versus the relative HI content in mass (M_{HI}/M_T). This diagram demonstrates that this sample fills the usual range of the defined quantities as observed for normal galaxies. To summarize this plot we can say that the present sample of galaxies is relatively rich in HI on average. Some of the scatter in this diagram is due to uncertainties in the observed quantities, especially the inclination which is used to correct the line width. This widths enters the total mass calculation by the square. The optical diameters get uncertain for galaxies at low galactic latitudes due to the high extinction, e.g., Cas 2, ESO 137-G27, BK 12, ESO 558-11.

Last but not least I show the correlation between HI surface density and optical surface brightness. The Karachentsev catalog contains a surface brightness (SB) class in four steps from high surface brightness to low, very low, and extremely low SB which are coded here in this sequence from 4 to 1 in steps of 1. The different values of the errors of the mean of each class depends essentially on the different population size of each SB class. There are relatively few objects

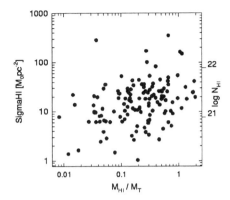

Figure 6. The pseudo column density of neutral hydrogen (Σ_{HI} in $M_\odot\, pc^{-2}$) of our present sample is plotted versus the relative HI content (M_{HI}/M_T).

in classes 1 and 4. However, the difference in HI column density over the range in SB is a factor of 2 to 3. This value has been quoted in this conference before (e.g. de Blok 1998).

Acknowledgments. V. E. Karachentseva, I.D. Karachentsev, and M. Ehle participated in this project.

The Australia Telescope is funded by the Commonwealth of Australia for operation as a National Facility managed by CSIRO.

The Nançay Radio Observatory is the Unitè Scientifique de Nançay of the Observatoire de Paris, associated as Unitè de Service et de Recherche (USR) N0. B704 to the French Centre National de Recherche Scientifique (CNRS). The Nançay Observatory also gratefully acknowledges the financial support of the Conseil Régional of the Région Centre in France.

This work has been partially supported by the Deutsche Forschungsgemeinschaft under Project no. 436 RUS 113/470/0 (R).

References

de Blok, E. 1998, this conference, page . . .

Burton, W.B., Verheijen, Kraan-Korteweg, R.C., Henning, P.A. 1996, A&A, 293, L33

Hartmann D., Burton W.B. 1997, Atlas of Galactic Neutral Hydrogen, Cambridge University Press, Cambridge

Huchtmeier, W.K., Karachentsev, I.D., Karachentseva, V.E. 1997, A&A, 322, 375

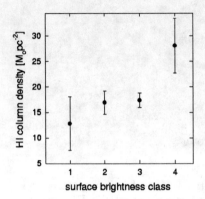

Figure 7. In this figure we show a correlataion between the pseudo HI column density with the optical surface brightness of the galaxies of our actual sample. The surface brightness class is taken from the Karachentsev catalog; 1 = extremely low SB, 2 = very low SB, 3 = low SB, 4 = high SB. The error bars correspond to twice the rms error of the mean of each SB class.

Huchtmeier, W.K., Lercher, G., Seeberger, R., Saurer, W., Weinberger, R. 1995, A&A, 293, L33
Huchtmeier, W.K., van Driel,W. 1996, A&A, 305, L25
Huchtmeier, W.K., Richter, O.-G. 1988, A&A, 203, 237
Karachentsev, I.D. 1994, Astron.Astrophys. Transactions, 6, 1
Karachentseva, V.E., Karachentsev, I.D. 1998, A&AS, 127, 409
Kraan-Korteweg, R.C., Loan, A.J., Burton, W.B. et al. 1994, Nature, 372, 77
Kraan-Korteweg, R.C., Tammann, G.A. 1979, Astron. Nachr., 300, 121
Schmidt, K.-H., Boller, T. 1992, Astron. Nachr., 313, 189

An HI Survey of LSB galaxies selected from the APM Survey

Stéphanie Côté

Dominion Astrophysical Observatory, Herzberg Institute of Astrophysics, National Research Council of Canada, 5071 W. Saanich Rd., Victoria, BC, V8X 4M6, Canada

Tom Broadhurst

Astronomy Department, University of California, Berkeley, CA94720, USA

Jon Loveday

Astronomy & Astrophysics Department, University of Chicago, 5640 S Ellis Ave, Chicago, IL60637, USA

Shannon Kolind

Department of Physics & Astronomy, University of Victoria, Victoria BC, V8X 4M6, Canada

Abstract. We present preliminary results of a neutral hydrogen (HI) redshift survey to find Low Surface Brightness (LSB) galaxies in the very nearby universe. Our sample consists of all galaxies in the APM catalog (Maddox et al. 1990) with a mean surface brightness of $\mu \geq 24$ mag/arcsec^2, down to a magnitude limit of $b_j \leq 17$. With the Parkes 64m radiotelescope 35 objects were detected at $v < 4300$ km s^{-1}. The resulting luminosity function, HI mass function, and for the first time total mass function are presented. It is found that LSBs make a negligible contribution to the overall integrated luminosity, HI mass, and total mass contained in galaxies.

1. Introduction

Redshift surveys of the general field galaxy population have revealed an interesting class of optically Low Surface Brightness (LSB) galaxy which refuse to yield a redshift by optical means, and which may represent however a significant population. Studies of nearby groups of galaxies have unveiled a large population of neutral hydrogen (HI) rich dwarfs at low redshifts (Côté et al. 1997). It is natural to suppose that more of these gas-rich dwarfs could be lurking in the field neighborhood, and could thus be relatively easily amenable to redshift detection in the radio. The following survey was designed to explore this possibility. Here we present preliminary results.

Our sample was extracted from the 'APM Galaxy Survey' catalog of Maddox et al. (1990), which contains about 2 million galaxies spread over 4300 square degrees covering the southern galactic cap. Because images were identified by the APM as connected groups of pixels (16 pixels minimum) with densities higher than a set threshold above the local sky, the resulting catalog is effectively diameter-limited, to $A \simeq 4$ arcsec^2, with a limiting surface brightness $\mu \simeq 24.5$ mag/arcsec^2. To generate a well-defined sample of LSB field galaxies we retained all galaxies satisfying the following criteria:
- mean surface brightness of $\mu \geq 24$ mag/arcsec^2
- magnitude of $b_j \leq 17$

This produced a candidate list of 88 galaxies.

2. Observations

All our candidates were observed in HI at the 64m Parkes Radiotelescope, using a 32 MHz bandwidth, which resulted in a channel separation of 6 km s^{-1} for a (usable) velocity range of -300 to 4300 km s^{-1}. This negative velocities coverage was necessary to ensure that unknown Local Group objects would be recovered. Our 3σ detection limit is $M_{HI} = 1.5 \times 10^8 M_\odot$ (at 4300 km s^{-1}, using H$_o$=75 km/s/Mpc throughout this study). This yielded 35 redshifts.

These nearby galaxies were then observed at the CTIO 1.5m with a Tek1024 CCD in B and R, with typical exposure times of 20 minutes. Surface photometry was performed, and luminosity profiles (down to $\mu_B \sim 26.5$mag/arcsec^2) were fitted to obtain the structural parameters, and integrated to get isophotal magnitudes, more reliable than the APM magnitudes.

This survey is therefore very much in the same flavor as the large APM survey of Impey et al. (1996) who obtained 332 redshifts in HI at Arecibo and in Hα. Their sample was not magnitude-limited though and their diameter limit was higher ($A \simeq 104$ arcsec^2), which means that their sample is more susceptible to observational selection bias, that needed to be corrected for (Sprayberry et al. 1996, 1997). By restricting ourselves to a relatively bright magnitude limit we avoided being affected by these biases (see Figure 1). Also our HI spectra have better resolution and sensitivity since our intended targets were nearby dwarfs.

3. Results

The absolute magnitudes of the objects recovered in this survey range from M_B =-18.5 down to -10.9, with disk scalelengths from 4 kpc to 0.15 kpc. The average central surface brightness is 23.4 mag/arcsec^2, but there is a wide range of central surface brightnesses at a given absolute magnitude. The median colour of our sample is $\langle B-R \rangle$=0.93, which is essentially the same as that expected from High Surface Brightness (HSB) galaxies: de Blok et al. (1995) have analysed a sample of galaxies extracted from the ESO-LV catalog (to use as a comparison sample for their LSB galaxies) and found $\langle B - R \rangle = 0.92$ for the HSB galaxies. A wide range of colours is recovered for our nearby galaxies, from $B - R$=0.55 to 1.22, but again this is typical of the range found in normal late-type galaxies.

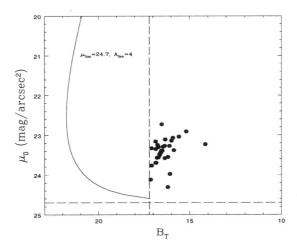

Figure 1. The curve shows the envelope of a selection function defined by a limiting isophote $\mu_{lim}=24.7$ and limiting angular size of 2.25". Because of the bright magnitude limit of our sample ($B \simeq b_j + 0.2 = 17.2$) and the small limiting size we avoid being affected by the usual selection bias.

No correlation is seen between the surface brightnesses and the colours which would not be the case if LSB galaxies were the faded remnants of HSB galaxies (see the discussion in O'Neil et al. 1997). The disk scalelengths in B and R agree within 20% for almost all the objects; combined with the fact that no correlation between colour and inclination is seen, and that only one object has been detected by IRAS, this all suggests that these LSBs have a relatively low dust content.

Figure 2 shows the distribution of morphological types in our sample. As expected the majority of them are irregulars. While spirals have more HI than irregulars (in terms of total gas mass in $M_\odot$), proportionally more irregulars are LSB galaxies, and with our HI detection limit we can recover 'typical' irregulars down to $M = -15$, according to the magnitude-HI mass relations derived by Tully (1988) and Rao & Briggs (1993). One elliptical is also found, since following these same relations from Wardle & Knapp (1986) and Rao & Briggs (1993), we should be HI sensitive to them down to a magnitude of $M_B = -18$. The shaded histogram in the Figure shows the numbers of barred galaxies amongst them. Contrary to previous claims the same proportion of barred objects is seen amongst LSB galaxies (about 34%) compared to normal galaxies, which are 30% barred (van den Bergh, 1998) (see also Knezek, this volume).

Amongst our 35 detections only 9 objects had already published redshifts (from the ESO catalog, the Southern Sky Redshift Survey Catalogue of da Costa et al. (1991), or Côté et al. (1997)). As far as very nearby dwarfs are concerned only 3 objects were detected at $V_\odot < 1000$ km s^{-1}: ESO305-G2 (already detected in da Costa et al. 1991), ESO473-G24 (already known from Côté

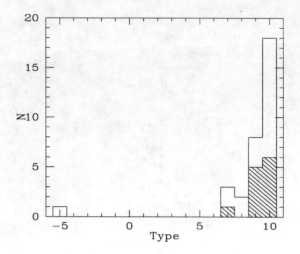

Figure 2. Distribution of morphological types. The hatched histogram shows the fraction of barred galaxies in the sample.

et al. 1997), and APM156-15-05 detected here for the fist time at $V_\odot = 230$ km s^{-1}.

4. Implications

• Luminosity function

After transforming our observed velocities to the Local Group frame (using $v_{cor} = v + 300 \sin l \cos b$), the luminosity function was estimated using Schmidt (1968) V/V_{max} method (Figure 3). None of our galaxy has $\langle V/V_{max}\rangle < 0.25$ and the average is $\langle V/V_{max}\rangle = 0.63$, meaning that the sample is not suffering from serious incompleteness. A Schechter fit yields $\alpha = -2.14$ and $M_* = -19.7$ (although with large uncertainties). In Figure 3 our results (where the error bars plotted are just poissonian) compare well with the luminosity function obtained by Sprayberry et al. (1997) with their larger sample of LSB galaxies, for which $\alpha = -1.42$ and $M_* = -18.34$. Note that their raw values were considerably boosted to correct mostly for surface brightness selection biases (Sprayberry et al. 1996), assuming -like many other authors- that scalelengths and central surface brightnesses of galaxies are uncorrelated, which is most likely not the case (de Jong this volume). Nevertheless these corrections do not appear to be so unreasonable because in the end their final values agree well with our raw ones (not corrected, since as stated above we selected our sample such as to minimise biases).

By integrating our luminosity function we derive an estimate of the total luminosity density of LSB galaxies of $1.6 \times 10^7 L_\odot$ Mpc^{-3}, while Marzke et al. (1994) finds $11 \pm 4 \times 10^7 L_\odot$ Mpc^{-3} over all galaxy types from the CFA survey, comparable to Loveday et al. (1992) value of $15 \pm 3 \times 10^7 L_\odot$ Mpc^{-3}. Clearly the luminosity contribution of LSB galaxies is only a small fraction of

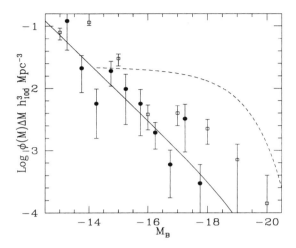

Figure 3. Luminosity function of our LSBs, compared to Sprayberry et al (1997) (squares), and Marzke et al (1994) LF for all galaxy types (dashed line)

the total luminosity in the universe, despite our steep faint-end slope, because these high number counts are for LSB galaxies of insignificant luminosities.

- HI mass function

The HI spectra can yield more info than just a redshift, by integrating the profiles one gets an estimate of the HI mass contained in LSB galaxies. The HI profiles of our LSB sample have widths typical of late-type spirals and dwarf irregulars (Schombert et al. 1992), with values for W_{50} ranging from ~ 20 to 203 km s^{-1}, about half of them exhibiting the familiar double-horned shape. As noted before by e.g. de Blok et al. (1996) and Sprayberry et al. (1995) LSB galaxies tend to be more gas-rich than HSB galaxies of the same luminosity. However in terms of total HI mass contribution they do not seem to tilt the balance in their favour: Figure 4 show our HI mass function, compared to the Schechter fit to the HI mass function of all galaxies detected in the Arecibo Strip Survey (Zwaan et al. 1997, see also Zwaan this volume). Galaxies with $10^9 M_\odot$ of HI will tend to be normal spirals, and only a small number of them are classified as LSB galaxies. But down at $10^8 M_\odot$ of HI this is typically a late-type or an irregular, and most of them are LSB galaxies (de Blok et al. 1995) find an average effective brightness of 24.25 mag/arcsec^2 for Sd's), which explains why our survey basically 'catches up' with the Zwaan HI function eventually for these low HI masses. But the majority of the HI mass in the universe is not in LSB galaxies but in galaxies of about $10^9 M_\odot$ of HI (Zwaan et al. 1997).

- Mass function

But besides an estimate of the total HI mass of the object, an HI profile also reveals, from its width, something about the kinematics of the galaxy. From the observed $W_{20\%}$ we estimated for each galaxy its maximum rotation velocity V_{max}, by using its derived inclination from our photometry and the Tully-Fouqué

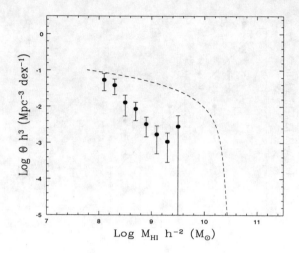

Figure 4. HI Mass function of our LSBs (blue dots), compared to the schechter fit to the Arecibo survey data of Zwaan *et al.* (1997) (dashed line)

(1985) corrections for random motions. This is still a reasonable thing to do for dwarfs down at $M_B = -13.25$ like in our sample, because they are known to be still mainly rotationally-supported at that luminosity (see DDO154, Carignan & Freeman 1988, also Côté *et al.* 1997, van Zee *et al.* 1997). This is the only way (other than fully mapping them in HI with aperture synthesis) to estimate their V_{max} because dwarfs deviate from the Tully-Fisher relation defined by spirals (Carignan & Freeman 1988, see also Freeman, this volume). With this V_{max} one can then obtain an indicative dynamical mass with a relation of the form $M_{dyn} = RV_{max}^2/G$, where we will use $R = 7\alpha^{-1}$, with the scalelengths derived from our photometry, because HI rotation curves for dwarfs typically reach at least $7\alpha^{-1}$ (Broeils 1992). This first field mass function is presented in Figure 6, showing the number of galaxies of a particular mass per mass decade per Mpc^3. The rise at the faint-end is much steeper than for the mass function calculated by Ashman *et al.* (1993) who converted the luminosity function of Efstathiou *et al.* (1988) using the variation of the mass-to-light ratio of Salucci *et al.* (1991). But it's still not steep enough for small galaxies to dominate the mass in galaxies in the universe. With our limited sample of only 35 redshifts so far this figure should be taken just as an illustration of the interesting potential of HI surveys, compared to the conventional deep imaging surveys of LSB galaxies.

Acknowledgments. Many thanks to the organisers for such a pleasant and stimulating workshop. We also thank the Parkes and CTIO TACs for all the observing time we needed. No thanks to big corporations for polluting our radiobands which make HI surveys like this one more and more difficult each year (see *Science*, vol. 282, no. 5386, p.34, for more details).

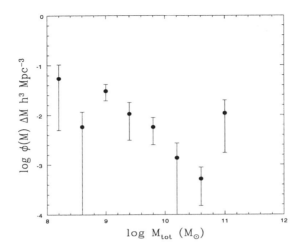

Figure 5. Total Mass function of our LSBs (the errorbars are from Poisson statistics)

References

Ashman, K.M., Salucci, P., Persic, M. 1993, MNRAS, 260, 610
Broeils, A., 1992. PhD thesis, Rijksuniversiteit Groningen
Carignan, C., Freeman, K, 1988, ApJ, 332, 33
Côté, S., Freeman, K., Carignan, C. 1996. In "Dark and Visible matter in galaxies", ASP 117, eds. M.Persic,P.Salucci, 52
Côté, S., Freeman, K., Carignan, C., Quinn, P. 1997. AJ, 114, 1313
da Costa, L., Pellegrini, S., Davis, M et al 1991, ApJS, 75, 935
de Blok, W., McGaugh, S., van der Hulst, J. 1996, MNRAS, 283, 18
de Blok, W., van der Hulst, J., Bothin, G. 1995, MNRAS, 274, 235
Efstathiou, G., Ellis, R.S., Peterson, B.A. 1988, MNRAS, 232, 431
Impey, C., Sprayberry, D., Irwin, M., Bothun, G., 1996, ApJS, 105,2091
Loveday, J., Peterson, B., Efstathiou, G., Maddox, S. 1992, ApJ, 390, 338
Maddox, S., Sutherland, W., EWfstathiou, G., Loveday, J. 1990, MNRAS, 243, 692
Marzke, R., Geller, M., Huchra, J., Corwin, H. 1994, AJ, 108, 437
O'Neil, K., Bothun, G., Schombert, J., Cornell, M., Impey, C. 1997, AJ, 114, 2448
Rao, S., Briggs, F. 1993, ApJ419, 515
Salucci, P., Ashman, K.M., Persic, M., 1991, ApJ, 379, 89
Schmidt, M. 1968, ApJ, 151, 393
Schombert, J., Bothun, G., Schneider, S., McGaugh, S. 1992, AJ, 103, 1107
Sprayberry, D., Impey, C., Irwin, M. 1996, ApJ, 463, 535

Sprayberry, D., Impey, C., Irwin, M., Bothun, G. 1997, ApJ, 481, 104
Tully, B., 1988, Nearby Galaxies Catalog, cambridge univ. press
Tully, B., Fouqué, P. 1985, ApJS, 58, 67
van den Bergh, S. 1998, "Galaxy morphology and classification", cambridge univ.press
van Zee, L., Haynes, M., Salzer, J., Broeils, A. 1997, AJ, 113, 1618
Wardle, M., Knapp, G. 1986, AJ, 91, 23
Zwaan, M., Briggs, F., Sprayberry, D., Sorar, E. 1997, ApJ, 490

Using weak Mg II lines to chart Low Surface Brightness Galaxies

V. Le Brun
IGRAP/LAS du CNRS, BP 8, 13376 Marseille CEDEX 12

C.W. Churchill
Penn. State University, University Park, PA 16802

Abstract. We report the detection, based on HST and Keck data, of two peculiar absorbers in the Lyα forest of the quasar PKS 0454+039. These clouds, at redshifts $z = 0.6248$ and 0.9315 respectively, display both Mg II and Fe II absorption lines in addition to the Lyα line. Based upon photoionization models, these are inferred to be photoionized by the intergalactic UV background, and to have H I column densities in the range $15.8 \leq N(\mathrm{H\,I}) \leq 16.8$. Furthermore, if one supposes that the relative abundances of heavy elements is similar to that of depleted clouds of our galaxy, the abundances of these two absorbers are greater than the solar value, which is a unique case for absorbers which are not associated to the quasar. We tentatively suggest that these absorbers may select giant low surface brightness galaxies.

1. Introduction

Quasars, beside the interest of their internal mechanisms, are unique tools for the study of the gaseous content of the Universe. In fact, any single gaseous cloud (down to a surface density of a few 10^{12} cm^{-2}) placed on the sightline to one of these bright and distant point sources can leave an imprint, in the form of one or several absorption lines, on the spectrum of the latter.

Absorption line systems (a system is a set of several lines of different ionization states of different elements) are classified following their content in neutral hydrogen :

- The so-called Damped Lyα systems (DLAS), because their Lyα line is located on the damped part of the curve of growth, are characteristic of large neutral hydrogen column densities ($N(\mathrm{H\,I}) \geq 2\ 10^{21}$ cm^{-2}). Their similarity with Galactic clouds in term of physical conditions (ionization degree, temperature ...), led people to associate them to distant spiral galaxies (Wolfe et al. 1986). In fact, they were recently confirmed, thanks to HST imaging and spectroscopy, to be associated with galaxies of any type (Le Brun et al. 1997),

- The Mg II systems display less neutral hydrogen than DLAS, even if they are still optically thick ($N(\mathrm{H\,I}) \geq 10^{17}$ cm^{-2}). The gas of these clouds is of low ionization level, and they have been shown to be associated with large ($R \sim 90h_{50}^{-1}$ kpc) gaseous halos of bright field galaxies (Bergeron & Boissé 1991, Steidel 1995).

- At last, the absorbers of the Lyα forest outnumber the other classes by several decades, each system displaying only the Lyα line (at least with an average spectroscopic set-up), $N(\text{H\,\textsc{i}})$ being in the range $10^{12} - 10^{16.5}$. Photoionization models show that the gas of these clouds is highly ionized, with some of them displaying ionic lines from C\,\textsc{iv}, N\,\textsc{v} or O\,\textsc{vi}. The abundances are quite low : $Z \simeq 10^{-3} - 10^{-2} Z_\odot$. These clouds could trace both the external parts of the galactic halos and the intergalactic gas spread along the large scale structure of the galaxy distribution (Le Brun & Bergeron 1998).

In this paper, we present the complete study of two absorbers of a new kind, the "weak Mg\,\textsc{ii} absorbers". Section 2 presents the state-of-art about these absorbers and Sect. 3 the peculiar sightline to the quasar PKS 0454+039, with all data we obtained on it. At last, Sect. 4 presents our discussion and conclusions about these objects.

2. The weak Mg II absorbers

The Mg\,\textsc{ii} systems were extensively studied in the 80's thanks to systematic spectroscopic surveys of several tens of quasars at signal to noise ratio about 10, with limiting rest equivalent width of about 0.3 Å. The most complete of those was published in 1992 (Steidel & Sargent). These surveys have shown that the number density, either in value or in evolution with redshift, was fully compatible with these absorbers being linked with field galaxies. This hypothesis was confirmed in the same time by the the first identifications of absorbing galaxies (Bergeron & Boissé 1991). It is only with the advent of the Keck Telescope that Churchill et al. (1998) could initiate a survey for "weak" Mg\,\textsc{ii} absorbers, that is with rest equivalent width down to 0.02 Å. The survey was made with HIRES (Vogt et al. 1994), all details are given in Churchill et al. (1998). As can be seen on Fig 1, the weak Mg\,\textsc{ii} absorber outnumber the strong ones by a factor of 2 to 3, and there is no lower cutoff in the distribution of the rest equivalent width down to 0.02 Å. Also, the evolution of the number density of weak absorbers show that it is compatible with a non evolving (in number) population (Fig. 2). If these absorbers are of the same nature as the strong ones, by comparing the density of absorbers and of galaxies, it requires galaxies to be surrounded by halos of radius $R \sim 120 h_{50}^{-1}$ kpc. However, this value can be lowered if a fraction of the weak absorbers is perhaps of different origin for example Low Surface Brightness (LSB) galaxies.

3. The sightline to PKS 0454+039 : data and analysis

To ascertain the nature of weak Mg\,\textsc{ii} absorbers, we have focused on the sightline toward the quasar PKS 0454+039. This object is part of the Keck/HIRES survey, and a high resolution ($R \sim 45,000$) high signal to noise ratio (~ 50) has

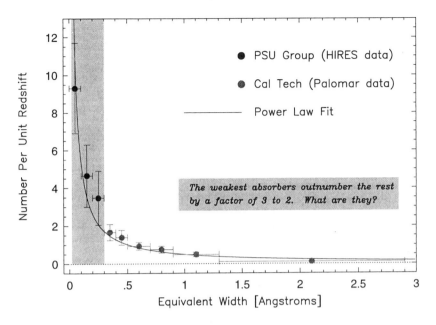

Figure 1. Rest equivalent width distribution of the weak Mg II absorbers (in the shadened zone) as compared to strong ones (on the right of the diagram)

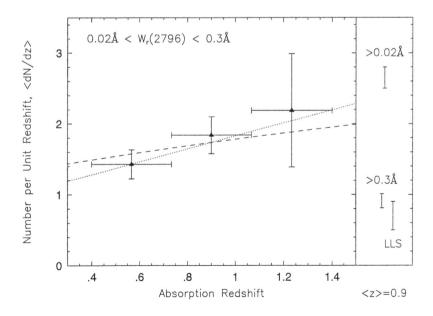

Figure 2. Redshift evolution of the weak Mg II absorbers population.

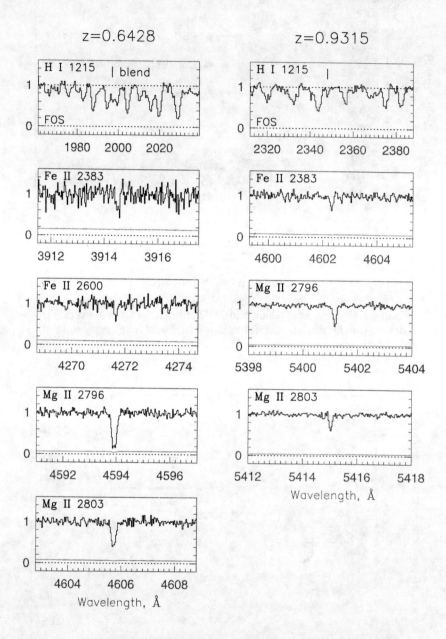

Figure 3. FOS/HST (top panel) and Keck/HIRES (lower panels) spectra of the two absorbers. Note the very different wavelength scales between the optical and UV spectra

been obtained in the visible domain. Beside of this, the HST/FOS UV spectrum has been extensively studied by Boissé et al. (1998) for the characterization of the $z = 0.8596$ DLAS present in front of the quasar. The spectral resolution is $R \simeq 1300$, and the signal to noise ratio is about 10. The limiting equivalent width for a 3σ detection is about 0.3 Å. At last, we have obtained deep CFHT and HST/WFPC2 R band images of the field surrounding this quasar (Le Brun et al. 1997).

The analysis of the HIRES spectrum shows that two faint Mg II absorbers are present at redshifts $z = 0.6248$ and $z = 0.9315$, together with Fe II lines at the same redshift derived from Fig. 3 displays both the Lyα line, as present in the HST/FOS spectrum (top panels), and the metal lines (lower panels). As can be seen, the Lyα lines are very faint, and were not even included in the 3σ limited sample of absorption lines listed in Boissé et al. (1998). Their rest equivalent width are 0.33 and 0.15 Å respectively.

We have thereafter tried to use the standard analysis methods to derive the physical properties of the gas. However, since, even at the HIRES resolution, the Mg II and Fe II lines are barely resolved, we have used Monte Carlo simulations to determine the best values for the column densities and dispersion parameters, using the doublet ratio. We obtain that, for both systems, $N(\text{Fe II}) \sim N(\text{Mg II}) \sim 10^{12.5}$, while the b parameter are 5 and 2 km s^{-1} for the $z = 0.6428$ and 0.9315 systems respectively.

Unfortunately, the FOS spectrum is of poorer quality, and a Voigt profile fitting was impossible, so that we could not derive directly the properties of the H I gas. We have therefore proceeded in several steps that are summarized below (see Churchill & Le Brun 1998 for a detailed description of this work):

1. We have introduced the turbulence parameter, $f = b_{\text{turb}}/b_{\text{tot}}$, which can have values between 0 and 1. If $f = 0$, the gas is thermally excited, and the b parameter for different elements scales as the square root of the mass ratio. On the contrary, if $f = 1$, the gas is fully collisionally ionized, and all the lines of all elements have the same b value. Of course, all intermediate situations are possible. The variation of $b(\text{H I})$ as a function of the f value is shown on Fig. 4 for the two systems.

2. From this range of possible variations for the b parameter, we thus can derive, using the curve of growth analysis, the range of possible values for the neutral hydrogen column densities : it covers nearly 3 decades from $\sim 10^{14}$ to $\sim 10^{17}$ cm^{-2}.

3. The latter result makes it impossible to derive any hints on the physical states of these absorbers just from the data. We thus have used CLOUDY (Ferland 1996), to make some simulations of the absorbing gas. For each value of $N(\text{H I})$ between 10^{14} and 10^{17}, by step of 0.5 in log, we have run CLOUDY in 'optimized' mode, so that the simulation converges toward the observed values of $N(\text{Fe II})$ and $N(\text{Mg II})$. The other inputs are i) the UV ionizing external radiation field : it could either have a galactic-shaped spectrum, or an intergalactic UV background shape, as given by Haardt & Madau (1996), and ii) the abundance pattern, that is the relative abundances of heavy elements : solar, H II region, that is including depletion by dust, or enhanced abundances of α elements. The output of CLOUDY

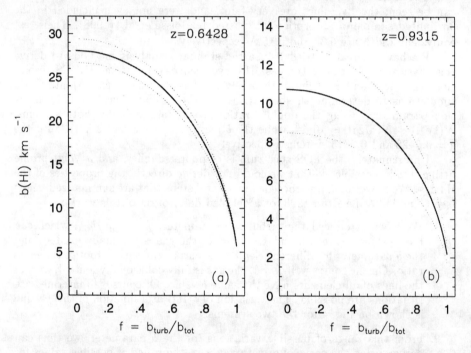

Figure 4. Variation of the H I broadening value as a function of the turbulence parameter for the two absorbers

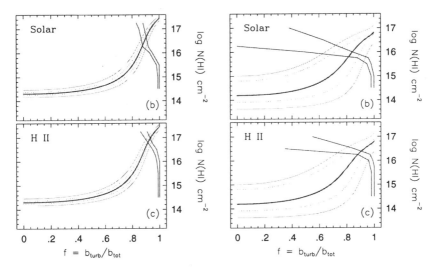

Figure 5. Properties of the two absorbers ($z = 0.6428$ on the left, $z = 0.9315$ on the right). Thick solid lines give the value of $N(\text{H\,\textsc{i}})$ derived from observed $b_{\text{tot}}(\text{Mg\,\textsc{ii}})$ and $W_r(\text{H\,\textsc{i}})$. Thin curves give the uncertainties from the measurements. The curves that originate in the lower-right corner and rise upwards and then to the left are the allowed locus of f for a cloud model of a given $N(\text{H\,\textsc{i}})$

is the full physical state of the gas, including temperature, from which we could derive the f parameter value. Thereafter, we only had to compare the output of the simulation to the observational constraints to get the possible values of $N(\text{H\,\textsc{i}})$.

4. Discussion and conclusions

These simulations allowed us to eliminate some hypothesis : the α-enhanced abundance pattern fails to produce coherent models, as well as the galactic-shaped ionizing flux, which requires unrealistic spatial densities of stars to reproduce the physical quantities of the gas. Thus, Fig. 5 displays the domain that is allowed in the $f - N(\text{H\,\textsc{i}})$ plane for the two absorbers : it is the intersection of the band coming from lower-left to upper right, which displays the range allowed by data (with the errors), and the band going from lower-right to upper-right, which reflects the uncertainties in the Fe\,\textsc{ii} and Mg\,\textsc{ii} column densities. As can be seen, only a small domain is allowed for each absorber, which covers less than a decade in $N(\text{H\,\textsc{i}})$.

As a result of this, we can now get estimates of the metallicity of these clouds, which are surprisingly high : the $z = 0.6428$ absorber has abundances $Z \geq 0.2 Z_\odot$ if the abundance pattern is solar, and $Z \geq 1.6 Z_\odot$ if the gas has abundances similar to the Galactic H\,\textsc{ii} regions, i.e with depletion on dust grains.

For the $z = 0.9315$ absorber, the abundances are above the solar value, whatever the abundance pattern.

Their very high abundances make these absorbers very peculiar, and in any case different from the strong Mg II absorbers, which have metallicities $Z \sim 0.01 Z_\odot$, and at least a part of the weak Mg II absorbers thus seems not to originate in field galaxy halos. Furthermore, we have searched the deep CFHT and HST images of the field around the quasar, and there is no galaxy close enough to the sightline that could give rise to these absorption systems, when one takes into account the galaxies that are likely to host the four already know "normal" metallic absorption systems.

There is however a class of galaxies which present the same characteristics as our absorbers : The Giant Low Surface Brightness Galaxies. Their abundances, as measured in H II regions, are above the solar value (McGaugh 1994, Pickering & Impey 1995), and the H I gas velocity dispersion is quite low, in the range $10 - 30$ km s^{-1}, thus similar to the value derived in our absorbers. These two similarities, which can not be found in any other class of galaxies, leads us to suggest that a least a fraction of the weak Mg II absorbers is due to these giant LSBGs. The spatial densities of absorbers and galaxies cannot be compared yet, since the samples of both are to small to derive useful statistics.

The immediate follow-up of this work should go in two direction : first, more detailed UV spectroscopy is necessary, to derive better constraints on both the Lyα profile and other ions absorption lines (C IV, O III, O VI, ...), and also by more imaging and spectroscopy in the field, to try to identify the absorbing galaxies, or companions of them. These developments will need large aperture space and ground-based telescopes.

References

Bergeron J., Boissé P., 1991, A& A 243, 344
Boissé P., Le Brun V., Bergeron J., Deharveng J.M., 1998, A&A 333, 841
Churchill C.W., Le Brun V., 1998, ApJ 499, 677
Churchill C.W., Rigby J.R., Charlton J.C., Vogt S.S. 1999, ApJ Supp 120, in press (Jan. 99) astro-ph/9807131
Haardt F., Madau P.,1996, ApJ 461, 20
Le Brun V., Bergeron J., Boissé P., Deharveng J.M., 1997, A&A 321, 733
Le Brun V., Bergeron J., 1998, A&A 332,814
McGaugh S., 1994, ApJ426, 135
Pickering T.E., Impey C.D., 1995, BAAS 186, 39.07
Quillen A.C., Pickering T.E., 1997, AJ 113, 2075
Steidel C.C., 1995, in ESO Astrophysics Symp., QSO Absorption Lines; ed G. Meylan (Springer, Heidelberg), p. 139
Steidel C.C., Sargent W.L.W., 1992, ApJS 80, 1
Wolfe A.M., Turnshek D.A., Smith H.E., Cohen R.D., 1986, ApJS 61, 249

Simulations of Lyα Absorption from Low Surface Brightness Galaxies

Suzanne M. Linder
Pennsylvania State University, University Park, PA 16802 USA;
slinder@astro.psu.edu

Abstract. Using simulations of the low redshift galaxy population based upon galaxy observations, it is shown (Linder 1998) that the majority of Lyα absorbers at low redshift could arise in low surface brightness (LSB) galaxies. The contribution to absorption from LSB galaxies is large for any galaxy surface brightness distribution which is currently supported by observations. Lyα absorbers should become powerful tools for studying the properties and evolution of galaxies, but first it will be necessary to establish observationally the nature of the Lyα absorbers at low redshift. Further simulations, in which the absorbing galaxy population is 'observed' with some selection criteria, are used to explore how easy it is for an observer to test for a scenario in which LSB galaxies give rise to most of the Lyα absorbers. It is shown that absorption arising in LSB galaxies is often likely to be attributed to high surface brightness galaxies at larger impact parameters from the quasar line of sight.

1. Introduction

Lyα absorption at low redshift is a powerful probe of gas in or around galaxies. With Lyα absorber observations it is possible to detect neutral hydrogen column densities which are several orders of magnitude lower than those typically seen with 21 cm observations. For example, the HST Key Absorption Line project (Bahcall et al. 1996) is complete to approximately $N_{HI} \sim 10^{14.3}$ cm^{-2}. Observations of Lyα absorption should therefore be capable of detecting neutral gas which is known to exist in galaxy disks as well as more diffuse gas or neutral components of highly ionized gas surrounding galaxies. Given that Lyα absorption lines can be observed even more easily at higher redshifts, they should become especially useful tools for studying galaxy evolution in the future.

Some stronger Lyα absorption lines are known to arise in lines of sight thorough galaxies, but the nature of the more common, weak forest absorbers is more difficult to establish. Supposing that these absorbers arise in gas associated with galaxies, the galaxies may be located at large impact parameters from a quasar line of sight. Therefore it becomes impossible to be sure that any particular galaxy is actually causing the absorption, so that it is difficult to make a direct test of any model in which galaxies cause absorption.

Testing absorber-galaxy models becomes even more complicated if absorption arises in unidentified galaxies. Given that low surface brightness (LSB)

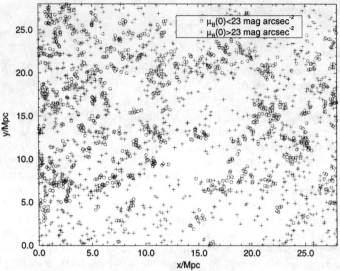

Figure 1. A slice of the cluster simulation which is 5 Mpc thick is shown, illustrating the positions of LSB and HSB galaxies. LSB galaxies are made to be clustered more weakly than HSB galaxies.

galaxies are common relative to high surface brightness (HSB) galaxies, and that LSB galaxies are often found to be rich in gas, they *must* make an important contribution to Lyα absorption. LSB galaxies are typically bigger than HSB galaxies at a given luminosity. If LSB galaxies also have larger sizes compared to HSB galaxies in Lyα absorption, then they may easily dominate the cross section for Lyα absorption. Lyα absorbers may allow for an important way to study the properties and evolution of galaxies, especially LSB galaxies, but first it will be necessary to find a way to test absorber-galaxy models and to establish what types of galaxies could actually cause most of the Lyα absorption.

2. Simulations of Absorbing Galaxies

Simulations are used (Linder 1998) to find out how many absorbers could be associated with galaxies. A population of galaxies is defined according to observed distributions of galaxy parameters, and the galaxies are placed in a box. Random lines of sight are chosen to go through the box, and absorption occurs where the lines of sight intersect the galaxies. In these simulations it is assumed that absorption arises in galaxy disks and their ionized outer extensions (Charlton, Salpeter & Linder 1994). It is shown that Lyα absorber counts at low redshift can easily be explained by galaxies when LSB galaxies are included. The model parameters are tuned to be consistent with observed luminosity functions for HSB and LSB galaxies, standard nucleosynthesis predictions of the baryon

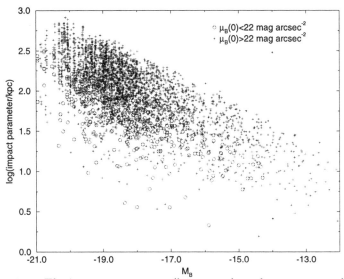

Figure 2. The impact parameter (between the galaxy center and the quasar line of sight) is plotted versus M_B for the actual absorbing ($> 10^{14.3}$ cm^{-2}) galaxies in the simulations. More luminous galaxies have larger absorption radii, and can thus cause absorption at larger impact parameters from the line of sight. LSB galaxies have larger absorption radii than HSB galaxies at a given luminosity.

density, and Lyman limit absorber counts. Preferred scenarios are found to have moderate galaxy absorption cross sections and moderate numbers of LSB galaxies.

The galaxies in the earlier simulations are placed randomly in the box. More recently, clustered galaxies have been simulated using a fractal-type method as in Soniera & Peebles (1978) where eight levels of clustering are used. The LSB galaxies are moved further out from the centers of the clusters at the second largest level, so that they are made to be clustered more weakly than HSB galaxies, as shown in a slice of the box in Fig. 1.

More luminous galaxies have larger absorption radii in the simulations, so that they are able to cause absorption when they are at larger impact parameters from quasar lines of sight, as shown in Fig. 2. The Holmberg relation ($R \sim L^{0.5}$) and the typical absorbing radius of a luminous HSB galaxy are similar to those reported in some observational studies (Chen et al. 1998; Lanzetta, this *Proceedings*). It can also be seen from Fig. 2 that LSB galaxies have larger absorption radii than HSB galaxies at a given galaxy luminosity. Thus the majority of Lyα absorbers arise in lines of sight through LSB galaxies, as seen in Fig. 3. It can also be seen in Fig. 3 that the Lyα absorption is generally

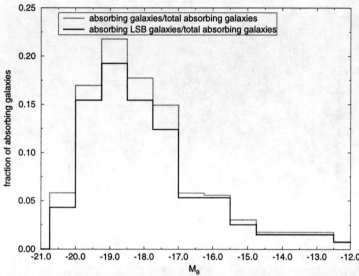

Figure 3. A distribution of M_B is shown for the actual absorbing ($> 10^{15}$ cm^{-2}) LSB (> 22 B mag arcsec^{-2}) galaxies and all galaxies, both normalized to the total number of absorbing galaxies. Moderately luminous LSB galaxies are shown to cause most of the Lyα absorption.

caused by moderately luminous, 'normal' LSB galaxies and not by extremely luminous Malin-type objects.

A flat central surface brightness distribution (McGaugh 1996) at a given scale length was assumed for most of the simulations. Preliminary simulations have also been done using a surface brightness distribution which is lognormal at a given luminosity (de Jong, this *Proceedings*). While this surface brightness distribution allows for fewer extremely luminous LSB galaxies, there are still numerous moderately luminous LSB galaxies which still give rise to most of the Lyα absorption.

3. 'Observing' the Simulations

Given some scenario where LSB galaxies make a significant contribution to Lyα absorption, such as the one described above, is it possible for an observer to test for such a scenario? Suppose we can 'observe' the galaxies simulated above using some selection criteria. Here the nearest galaxy to a line of sight is found within a velocity difference of 400 km s^{-1} to the absorption line, where the galaxy has $M_B < -16$ and $\mu_B(0) < 23$ mag arcsec^{-2}.

A plot of M_B versus the impact parameter of the 'observed' absorbing galaxy is shown in Fig. 4, using the same scale as the actual plot shown in Fig. 2. It is likely that an observer could find a way of determining that the

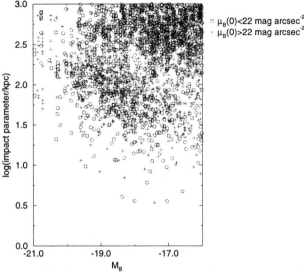

Figure 4. The impact parameter versus M_B are shown as if they were obtained by 'observing' the simulated galaxies, in order to identify an absorbing galaxy, according to the selection criteria given in the text. While the same simulated galaxies shown in Fig. 2 are 'observed' here, it is no longer obvious that LSB galaxies have larger absorption cross sections.

points in the upper right-hand corner for Fig. 4 are unphysical absorber-galaxy associations. For the remaining points, a correlation between luminosity and impact parameter can still be seen, although the slope of the 'observed' Holmberg relation may change. However, absorption arising in LSB galaxies is frequently 'observed' as arising in HSB galaxies at typically larger impact parameters from the quasar line of sight. Thus it is no longer possible to verify from the 'observed' plot that LSB galaxies have larger absorption cross sections, as assumed in the simulation.

In any reasonable absorber-galaxy model, it is expected that, on average, the neutral hydrogen column density should fall off with galaxy radius. Thus it is likely that absorbing galaxies at larger impact parameters from the quasar line of sight should produce lower column densities. An anticorrelation between impact parameter and neutral column density is seen for the simulated actual absorbing galaxies (Linder 1998). However, a large amount of scatter exists in such plots, which is caused mostly by variations in the properties of absorbing galaxies. (Varying disk inclinations produce comparatively little scatter, so that a spherical halo-type model for absorbing galaxies should produce a similar plot.)

The anticorrelation between neutral column density and impact parameter can still be seen for the galaxies 'observed' with the selection criteria above, as shown in Fig. 5. Note that if more reasonable selection effects were used to

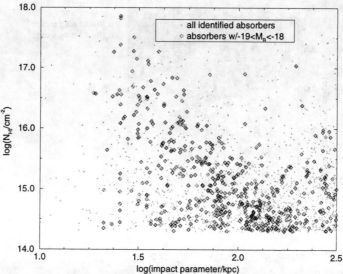

Figure 5. The neutral column density for Lyα absorption is plotted versus the impact parameter between galaxy and line of sight for the 'observed' (see text) absorbing galaxy. An anticorrelation between neutral column density and impact parameter is still seen even though the actual absorbing galaxy is unlikely to be identified (particularly for the weaker absorption lines).

'observe' the simulations, compact HSB dwarf galaxies would be excluded, so that fewer points would be seen toward the lower left-hand corner. Thus, when also considering incompleteness for observing galaxies at large impact parameters, Fig. 5 should bear even more resemblance to the plot shown in Chen et al. (1998). LSB absorbers are frequently misidentified in the simulation shown. In fact, many LSB absorbers have no possible 'observed' galaxy which satisfies the selection criteria, especially when the LSB galaxies are clustered more weakly than the HSB galaxies. Thus an observer would be likely to conclude that some fraction of absorbers cannot be explained by galaxies. Less scatter is seen in Fig. 5 for galaxies with a narrow range of luminosities. However, it is more difficult to reduce the scatter by limiting the range of 'observed' absorbing galaxy surface brightnesses. Thus from this plot an observer is likely to conclude again that galaxy absorption cross sections are correlated with galaxy luminosity but not with galaxy surface brightness.

Another plot frequently made by observers (for example, Bowen, Blades, & Pettini 1996), that illustrates the absorption covering factor, or the fraction of galaxies found to cause absorption as a function of impact parameter, is shown in Fig. 6. 'Observed' galaxies appear to cause absorption at large impact parameters compared to those seen in Fig. 2. Again it can be seen that many

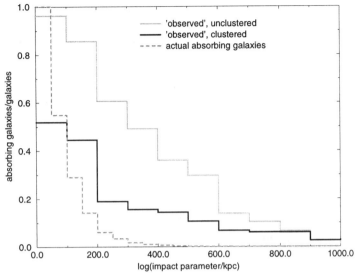

Figure 6. A 'covering factor' plot shows the fraction of galaxies found at some impact parameter from a line of sight which cause absorption ($> 10^{14.3}$ cm^{-2}). Galaxy absorption radii are likely to be overestimated when the galaxies are observed with strong selection effects against LSB galaxies. Galaxy clustering also causes misidentification of the actual absorbing galaxy to occur more frequently.

absorbers arising in LSB galaxies are attributed to HSB galaxies at larger impact parameters from the quasar line of sight. It can also be seen that clustered galaxies are more frequently misidentified. It will be difficult to use an observed covering factor plot to test an absorber-galaxy model since its appearance will be sensitive to the actual absorbing properties of the galaxies, the clustering behavior of the galaxies, and the observational selection criteria.

4. Conclusions

It has been shown that Lyα absorbers at low redshift can easily be explained by galaxies when LSB galaxies are included. The majority of absorbers are likely to arise in LSB galaxies, even if extremely luminous LSB galaxies are rare. Lyα absorber observations will be useful for constraining properties of galaxies, such as the gaseous extent of galaxies, and for studying evolution in such galaxy properties. Lyα absorbers will be especially powerful tools for studying galaxies which are LSB at any evolutionary stage.

In order to take advantage of the information from Lyα absorption, it will be necessary to establish observationally specifically what kinds of galaxies give rise to Lyα absorption. Observational studies which attempt to match absorption lines with possible absorbing galaxies will be crucial to establishing the nature of the Lyα absorbers. However, as shown above, it is also necessary to use simulations in order to understand the absorbing properties of galaxies. For example, an observer may overestimate the absorbing radius of a galaxy by looking directly at observations which are subject to severe selection effects. Selection effects will always be present in observational studies, but it should be possible to define the selection criteria more rigorously so that they can be simulated in order to make more realistic tests of absorber-galaxy models.

Acknowledgments. I am grateful to J. Charlton and C. Churchill for helpful discussions, R. de Jong for supplying a surface brightness distribution before publication, WISE CIC and NASA GSRP for travel funding, and R. Gunesch and A. Panaitescu for assistance with figures.

References

Bahcall, J. N. et al. 1996, ApJ, 457, 19
Bowen, D. V., Blades, J. C., & Pettini, M. 1996, ApJ, 464, 141
Charlton, J. C., Salpeter, E. E., & Linder, S. M. 1994, ApJ, 430, L29
Chen, H, -W., Lanzetta, K. M., Webb, J. K., Barcons, X. 1998, ApJ, 498, 77
Linder, S. M. 1998, ApJ, 495, 637
McGaugh, S. S. 1996, MNRAS, 280, 337
Soniera, R. M. & Peebles P. J. E. 1978, AJ, 83, 845

First Results from the HI Parkes Zone of Avoidance Survey

P. A. Henning

Institute for Astrophysics, University of New Mexico, Albuquerque, NM, 87131

L. Staveley-Smith

Australia Telescope National Facility, CSIRO, P.O. Box 76, Epping, NSW 1710, Australia

R. C. Kraan-Korteweg

Departamento de Astronomia, Universidad de Guanajuato, Guanajuato, Gto. CP 36000, Mexico

E. M. Sadler

School of Physics, University of Sydney, NSW 2006, Australia

Abstract. The HI Parkes Zone of Avoidance Survey is a 21 cm blind search with the multibeam receiver on the 64-m radiotelescope, looking for galaxies hidden behind the southern Milky Way. The first phase of the survey has uncovered 107 galaxies, two-thirds of which were previously unknown. The addition of these galaxies to existing extragalactic catalogs allows the connectivity of large-scale structures across the Zone of Avoidance within 3500 $km\,s^{-1}$ to become evident. No nearby hidden "Andromeda-class" object was uncovered. Our census of the most dynamically important HI-bearing nearby galaxies is now complete, at least for those objects whose HI profiles are not totally buried in the Galactic HI signal. The full survey is ongoing, and is expected to produce a catalog of thousands of objects when it is finished.

1. Introduction

The dust and high stellar density of the Milky Way obscures up to 25% of the optical extragalactic sky, creating a Zone of Avoidance (ZOA). The resulting incomplete coverage of surveys of external galaxies leaves open the possibility that dynamically important structures, or even nearby massive galaxies, remain undiscovered.

Careful searches in the optical and infrared wave bands can narrow the ZOA, (see Kraan-Korteweg, this volume) but in the regions of highest obscuration and infrared confusion, only radio surveys can find galaxies. The 21 cm line of neutral hydrogen (HI) passes readily through the obscuration, so galaxies with sufficient HI can be found through detection of their 21 cm emission. Of course, this method will miss HI-poor, early-type galaxies, and cannot discriminate HI galaxies with redshifts near zero velocity from Galactic HI.

Here we describe an HI blind survey for galaxies in the southern ZOA conducted with the new multibeam receiver on the 64-m Parkes telescope. A survey of HI galaxies in the northern ZOA is underway with the Dwingeloo radiotelescope (Henning et al. 1998; Rivers et al. this volume).

2. The Shallow Survey Observations and Data Analysis

The HI Parkes ZOA survey covers the southern ZOA ($212° \leq l \leq 36°; |b| \leq 5°$) over the velocity range $(cz) = -1200$ to 12700 km s^{-1}. The multibeam receiver is a focal plane array with 13 beams arranged in an hexagonal grid. The spacing between adjacent beams is about two beamwidths, each beamwidth being 14 arcmin. The survey is comprised of 23 contiguous rectangular fields which are scanned parallel to the galactic equator. Eventually, each patch will be observed 25 times, with scans offset by about 1.5 arcmin in latitude. The shallow survey discussed here consists of two scans in longitude separated by $\Delta b = 17$ arcmin, resulting in an rms noise of about 15 mJy, equivalent to a 5σ HI mass detection limit of 4×10^6 d$^2_{\text{Mpc}}$ M$_\odot$ (for a galaxy with the typical linewidth of 200 km s^{-1}).

After calibration, baseline-subtraction, and creation of data cubes, all done with specially developed routines based on aips++ (Barnes et al. 1998, Barnes 1998) the data are examined by eye using the visualization package Karma (http://www.atnf.csiro.au/karma/). The data are first displayed as right ascension – velocity planes, in strips of constant declination. Data cubes are then rotated, and right ascension – declination planes are checked for any suspected galaxies (eg. Figure 1).

3. Galaxies Found by the Shallow Survey

The shallow 21-cm survey of the southern ZOA has been completed, and 107 galaxies with peak HI flux densities $\geq$ about 80 mJy have been cataloged. Refinement of the measurement of their HI characteristics is ongoing, but the objects seem to be normal galaxies. Most of the galaxies are within 4000 km s^{-1}, which is about the redshift limit for detection of normal spirals of this shallow phase of the survey. As the deep survey continues, spirals at higher velocities will be recovered. Of the 107 objects, 28 have counterparts in the NASA/IPAC Extragalactic Database (NED) with matching positions and redshifts. Optical absorption, estimated from the Galactic dust data of Schlegel et al. (1998), ranges from $A_B = 1$ to more than 60 mag at the positions of the 107 galaxies, and is patchy over the survey area. No objects lying behind more than about 6 mag of obscuration have confirmed counterparts in NED, as expected.

The connection of these objects with known large-scale structures in optically unobscured regions is discussed by Kraan-Korteweg in this volume, and the reader is referred to her paper for graphical depiction. The shallow multibeam HI survey connects structures all the way across the ZOA within v < 3500 km s^{-1}. The ongoing, deep ZOA survey will have sufficient sensitivity to connect structures at higher redshifts.

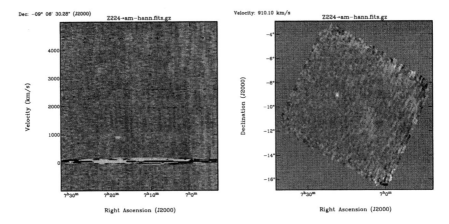

Figure 1. Left panel shows a right ascension – velocity slice to 5000 km s^{-1} which includes a galaxy discovered by the survey. Galactic HI appears as the strong horizontal feature at zero velocity. Note the extragalactic HI signal at 7^h18^m, 900 km s^{-1}. Right panel shows the right ascension – declination plane at the velocity of the suspected signal. The galaxy is evident at 7^h18^m, $-9°$

While 14 of the 107 galaxies lie within 1000 km s^{-1} and are therefore fairly nearby, all of the newly-discovered objects have peak HI flux densities an order of magnitude or more lower than the Circinus galaxy . Thus, it seems our census of the most dynamically important, HI-bearing nearby galaxies is now complete, at least for those objects with velocities offset from Galactic HI. Simulations are currently being conducted to investigate our sensitivity to HI galaxies whose signals lie within the frequency range of the Milky Way's HI.

Acknowledgments. We thank HIPASS ZOA collaborators R. D. Ekers, A. J. Green, R. F. Haynes, S. Juraszek, M. J. Kesteven, B. S. Koribalski, R. M. Price, and A. Schröder. This research has made use of the NASA/IPAC Extragalactic Database (NED) which is operated by the Jet Propulsion Laboratory, Caltech, under contract with the National Aeronautics and Space Administration. The research of P.H. is supported by NSF Faculty Early Career Development (CAREER) Program award AST 95-02268.

References

Barnes, D.G. 1998, in ADASS VII, eds. Albrecht, R., Hook, R.N., & Bushouse, H.A., San Francisco: ASP

Barnes, D.G., Staveley-Smith, L, Ye, T., & Osterloo, T. 1998, in ADASS VII, eds. Albrecht, R., Hook, R.N., & Bushouse, H.A., San Francisco: ASP

Henning, P.A., Kraan-Korteweg, R.C., Rivers, A.J., Loan, A.J., Lahav, O., & Burton, W.B. 1998, AJ, 115, 584

Schlegel, D.J., Finkbeiner, D.P., & Davis, M. 1998, ApJ, 500, 525

Results from the Dwingeloo Obscured Galaxies Survey

A. J. Rivers, P. A. Henning

University of New Mexico, Dept. of Physics and Astronomy, Albuquerque, NM 87131

R. C. Kraan-Korteweg

Dept. de Astronomia, Universidad de Guanajuato, Mexico

O. Lahav

Institute of Astronomy, Madingley Road, Cambridge CB3 0HA, UK

W. B. Burton

Sterrewacht Leiden, P.O. Box 9513, 2300 RA Leiden, The Netherlands

Abstract. Approximately 25% of the extragalactic sky is obscured by dust in our own Milky Way galaxy. Diligent optical and infrared surveys are successful at detecting galaxies through moderate Galactic dust extinction, but in the most heavily obscured regions near the Galactic plane, only radio surveys are effective.

The Dwingeloo Obscured Galaxies Survey (DOGS) is a 21-cm blind survey out to 4000 km s^{-1} in the northern "Zone of Avoidance" (ZOA). The DOGS project is designed to reveal hidden dynamically important nearby galaxies and to help "fill in the blanks" in the local large scale structure.

1. Introduction

To begin filling in the gaps in our knowledge of the distribution of nearby galaxies, a full survey of the northern ZOA is in progress utilizing the 25 m radiotelescope of the Netherlands Foundation for Research in Astronomy, in Dwingeloo. We report here on the initial results and the implications for our knowledge of the local large scale structure.

Some of the early survey results are reported by Henning et al. (1998, paper I) which also gives a detailed scientific justification and a more thorough discussion of the survey strategy and goals.

2. Telescope and Survey Background

The 25 m Dwingeloo telescope operating at 21 cm has a half-power-beamwidth (HPBW) of 0°.6 which may be thought of as the survey resolution. A DAS-1000

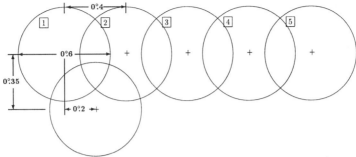

Figure 1. Distribution of survey grid points. Galactic longitude spacing is $\Delta \ell = 0°\!.4$ while the successive rows of constant Galactic latitude are separated by $\Delta b = 0°\!.35$. Each grid point is marked with a + surrounded by a circle indicating the HPBW of $0°\!.6$.

channel autocorrelator spectrometer is utilized in the telescope backend; the rms noise per channel is typically σ_{ch}=40 mJy for a 1 hr integration.

The completed survey will incorporate 15,000 partially overlapping pointings covering a spatial range of $30° \leq \ell \leq 220°$; $|b| \leq 5°\!.25$. A 20 MHz bandwidth is used, covering a velocity range of $0 \leq V_{LSR} \leq 4000$ km s^{-1}, resulting in a velocity resolution of 4 km s^{-1}. Under this arrangement, the narrowest linewidth galaxies are covered by several channels.

Each DOGS observation consists of a sequence of 5 contiguous pointings at constant Galactic latitude (figure 1). Five On-Off pairs created from the sequence ensure that a real galaxy will appear twice, once as a positive signal, and again as a negative one, referenced against two independent scans. Overlapping the constant latitude grids to form a honeycombed coverage of the sky allows for detection of galaxies in adjacent pointings and facilitates a more accurate determination of their positions.

3. Results

With approximately 60% of the survey complete, 36 galaxies have been detected, 23 of which were previously unknown (no NASA/IPAC Extragalactic Database (NED) counterpart). The number of galaxies registered so far is in agreement with calculations based on survey sensitivity and an HI mass function (Zwaan et al. 1997) which predict 50 to 100 detections in the survey range. Currently, we expect the completion of this survey by the summer of 1999.

Five of the 36 sources were originally identified by the shallow survey (5 min integration per pointing, cf. paper I) including Dwingeloo 1, a member of the nearby Maffei / IC 342 group of galaxies (Kraan-Korteweg et al. 1994). Two other Maffei group galaxies were detected (Maffei 2 and MB 1, cf. McCall & Buta 1995) and two possible members await confirmation observations.

The most significant nearby, previously unknown galaxy identified by DOGS was Dwingeloo 1. Given the 80% coverage of the survey region by the shallow survey, chances are low that a massive nearby spiral was missed, since nearby galaxies appear in many adjacent pointings.

In addition to the Maffei / IC 342 detections, 11 galaxies were discovered in the Supergalactic plane crossing region; 6 of these sources are noted in NED. Known structures appear continuous across the Galactic plane ($\ell \sim 140°$ in figure 2).

The most interesting sources uncovered by DOGS are located near NGC 6946 ($\ell=95°.72$, b=11°.67). Dw095.0+1.0, originally recorded as a high velocity HI cloud (Wakker, 1990), is probably a nearby dwarf galaxy. Two additional $\sim 10^7$ M$_\odot$ sources, one of which was previously known, were also found in the vicinity. All have recessional velocities $V_{LSR} \leq 250$ km s^{-1}. If these dwarf galaxies signify a new nearby galaxy group, this group would lie some 40° off the Supergalactic plane, considerably more than any other known group in the local universe.

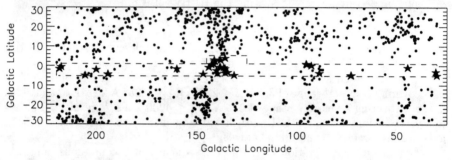

Figure 2. Spatial distribution of DOGS sources (indicated by $\star$). combined with LEDA galaxies out to $V_{LSR} \leq 4000$ km s^{-1} ($\bullet$). The survey coverage to date is shown by the dashed line.

Acknowledgments. We are grateful to A. Foley and D. Moorrees for their supervision of the day-to-day telescope operations, and to T. Hess for the development of data translation routines. The Dwingeloo 25 m radio telescope is supported by the Netherlands Foundation for Scientific Research (NWO). This research has made use of the NASA/IPAC Extragalactic Database (NED) and the Lyon-Meudon Extragalactic Database (LEDA). The research of P.A. Henning is supported by NSF Faculty Early Career Development (CAREER) Program award AST 95-02268.

References

Henning, P. A., Kraan-Korteweg, R. C., Rivers, A.J., Loan, A.J, Lahav, O., & Burton, W.B. 1998, AJ, 115, 584, (paper I)

Kraan-Korteweg, R. C., Loan, A. J., Burton, W. B., Lahav, O., Ferguson, H. C., Henning, P. A., & Lynden-Bell, D. 1994, Nature, 372, 77

McCall, M. .L, & Buta, R. J., 1995, AJ, 109, 2460

Wakker, B. P., 1990, Ph.D. thesis, Groningen Univ.

Zwaan, M., Briggs, F., & Sprayberry, D. 1997, PASA, 14, 126

HI Properties of LSB Dwarf and Blue Compact Dwarf Galaxies

C.L. Taylor[1], E. Brinks[2], E.D. Skillman[3]

[1] Astronomisches Institut, Ruhr-Universität Bochum, Germany
[2] Depto. de Astronomía, Universidad de Guanajuato, Mexico
[3] Dept. of Astronomy, University of Minnesota, Minneapolis, USA

Abstract.
We present high resolution ($\sim 15''$) VLA HI observations of one LSB dwarf galaxy and 5 Blue Compact Dwarfs. Previous works have emphasized the difference in the peak column density of the HI between the two types of galaxy. However, the peak column density is typically only a factor of 2—4 greater in BCDs than in LSB dwarfs, suggesting that the difference between the two is one of degree, not quality. The HI morphologies of the two types are often quite similar.

1. Introduction

Gas rich dwarf galaxies can be divided into two groups based upon their optical properties. Dwarf irregulars tend to be of low surface brightness, and generally have low current star formation rates. Blue Compact Dwarfs (BCDs), in contrast, tend to have higher surface brightnesses, their optical morphologies being dominated by a large burst of star formation. The differences in the global properties of their gaseous components, as traced by HI, are not so immediately apparent. Previous work (e.g. Taylor et al. 1994, van Zee et al. 1998) has shown that the two types of galaxy often have similar HI masses, average column densities, and rotation curves, but that the BCDs have peak column densities of HI higher by a factor of $\sim$ 2—4 than LSB dwarf irregulars. BCDs also often show signs of kinematically irregular gas – detached clouds, or warped disks, for example. Here we present HI data which, along with data from the literature, will be used to compare LSB dwarf galaxies with BCDs in an upcoming paper.

2. Observations

21-cm line HI observations of 7 BCDs and one LSB dwarf galaxy were conducted with the VLA in the most compact (D) and second most compact (C) configurations, with the correlator configured to provide a velocity resolution of 5.2 km s^{-1}. The angular resolution of the resulting data cubes ranged between $13''$ to $22''$. The rms noise in a channel map ranged from 0.6 mJy beam^{-1} to 1.1 mJy beam^{-1} Figure 1 shows for six dwarf galaxies an HI column density map (left) and a velocity field map (right).

The Challenge: The HI distributions of the galaxies are quite similar. Can you pick out which galaxy is the LSB dwarf, based upon the HI distribution alone? Answers will appear at the end.

3. LSB Dwarfs, BCDs, and Interactions

The HI distributions of LSB dwarfs and BCDs can appear similar, with central concentrations and diffuse outer regions occurring often in both types of galaxy. Thus the difference between them may be more a matter how high the central gas density reachs, rather than a qualitative distinction. If the gas density explains the difference in star formation properties between the two types, the question then becomes, what is the cause of this difference in gas density. One possibility is that gas rich dwarfs naturally exhibit a range in central gas density, and the fraction with the highest density become BCDs. Another possibility is that the increased gas density at the center is created by the inflow of gas from the outer regions of the galaxies. Galaxy interactions are known to drive this sort of radial gas motion in spiral galaxies, and may play a similar role for dwarfs.

Taylor (1997) compared the frequency with which HI rich companions are found near LSB dwarfs with that for BCDs. The LSB dwarfs had such companions at less than half the rate as did the BCDs, suggesting that interactions between galaxies can play a role in triggering star bursts in dwarf galaxies. But interactions are not the whole story. Not all of the LSB dwarfs were isolated, and not all of the BCDs had close neighbors. To investigate the differences between the two types due to differences in the degree of interaction, we selected galaxies from the surveys of Taylor et al. (1995, 1996) to include isolated examples, cases of distant interaction, and cases of close interaction.

The global properties (e.g. mass, *average* column density) of the HI in LSB dwarf galaxies and BCDs are very similar. We will use our data to compare small scale properties (e.g. peak column densities), kinematics, and interaction properties to determine which of these elements are most important in determining the star formation properties in these galaxies. For example, in the case of massive galaxies, it is well known that prograde interactions result in much stronger disturbances (and more stronger effects on star formation rates) than do retrograde interactions. Data such as ours are essential to understanding the star formation properties in dwarf galaxies.

Answers
a: F495-V1 – LSB dwarf; b: Haro 21 – BCD; c: UM372 = UGC1297 – BCD; d: UM500 = UGC7531B – BCD; e: UM456 – BCD; f: UM501 – BCD.

References

Taylor, C.L. 1997 ApJ, 480, 524
Taylor, C.L., Brinks, E., Pogge, R.W., & Skillman, E.D. 1994 AJ, 107, 971
Taylor, C.L., Brinks, E., Grashuis, R.M. & Skillman, E.D. 1995 ApJS, 99, 427: erratum 102, 189
Taylor, C.L., Thomas, D.L., Brinks, E. & Skillman, D. 1996 ApJS, 107, 143
van Zee, L., Skillman, E.D. & Salzer, J.J. 1998 AJ, 116, 1186

HI properties of LSB dwarfs 339

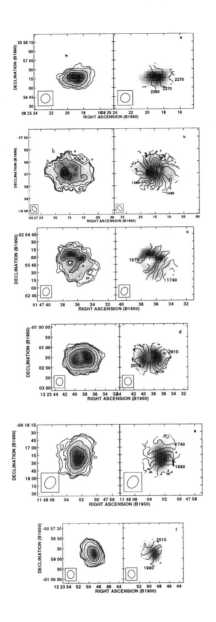

Figure 1. HI column density (left) and velocity field (right) maps for 5 BCDs and one LSB dwarf galaxy. In each case the greyscale shows the column density, and represents a different range of values for each galaxy. The column density contours, however, are the same for every galaxy: 0.5, 1, 2, 4, 8 and 16×10^{20} cm^{-2}.

Searching for LSB - VI

OK, we are all agreed, we will need a bus.

The HST/LCO Measurement of the Optical Extragalactic Background Light

R.A. Bernstein

Carnegie Observatories, 813 Santa Barbara St., Pasadena, CA 91101

Abstract.
We present the first detection of the mean flux of the optical extragalactic background light (EBL) at 3000, 5500, and 8000Å, derived from coordinated data sets from *HST* and Las Campanas Observatory. To isolate the extragalactic component, we have measured and subtracted the flux from foreground sources explicitly. In addition to detections in all three bands, we identify the minimum surface brightness contributed by resolved galaxies ($23 < V < 28$ AB mag) using a non-standard method of aperture photometry to which these data are uniquely suited. Individually resolved galaxies account for $\sim 30\%$ of the mean EBL coming from galaxies fainter than $V = 23$ AB mag. Taking into account the effective surface brightness detection limits of the deepest galaxy counts, and the results of LSB surveys at low redshift, the EBL we detect can be explained by galaxy populations already cataloged.

1. Introduction

The Extragalactic Background Light (EBL) is the integrated light from all extragalactic sources, both resolved and unresolved. The cosmological significance of the EBL was first appreciated in the 1700's, when expectations of a infinite, static Universe, uniformly filled with stars, led astronomers to puzzle over the fact that the nighttime sky is dark. The apparent conflict posed by the darkness of the night sky became known as Olbers' paradox. While it is now easily explained by the expansion of the Universe, the finite speed of light, and, most importantly, the finite lifetimes of stars, Olbers' paradox stands as an excellent illustration of the power of background measurements to test our model of the Universe. With new 10m–class telescopes and HST, the limits of resolved–source detection are being extended to ever fainter levels; however, a measurement of the EBL remains an invaluable complement to the source–count approach. Low surface brightness objects at low redshift and the majority of the galaxy luminosity function at high redshift are all easily missed both in surface brightness limited galaxy counts and in redshift surveys. In addition, photometry and even identification of faint galaxies becomes uncertain near detection limits. The EBL is immune to these surface brightness selection effects.

On a practical note, because the *HST*/WFPC2 field of view is only 5 arcmin^2, galaxies brighter than $V = 23$ AB mag are not statistically well sampled in these data. Our measurement of the EBL is therefore defined as the flux

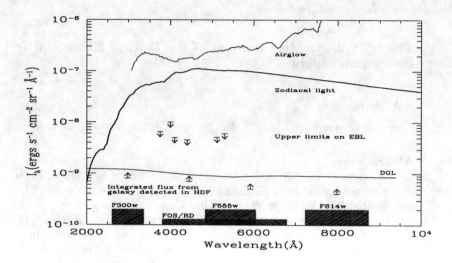

Figure 1. The relative surface brightness of airglow, zodiacal light, and diffuse galactic light (DGL) compared to upper limits for the EBL from previous investigations and the cumulative surface brightness of sources with $V > 23$ AB mag in the Hubble Deep Field (Williams et al. 1996). All foregrounds are shown at the flux levels we identify in this work. Hatched regions show the *HST*/WFPC2 and FOS band-passes.

from resolved and unresolved galaxies fainter than $V = 23$ AB mag. We use the abbreviation EBL23 as a reminder of this bright magnitude cut-off. This result can be combined with ground-based counts at brighter magnitudes to obtain the total EBL.

2. Foreground Subtraction and Data Sets

The key to any background measurement is the successful subtraction of foreground contamination. Figure 1 shows the surface brightnesses of foreground sources along lines of sight where they are faintest — at Galactic latitudes greater than 65 degrees and ecliptic latitudes greater than 30 degrees. For comparison, upper limits from previous attempts to measure the EBL are also shown, along with lower limits from integrating the flux in resolved galaxies (i.e. galaxy counts). The cumulative flux from foreground sources is roughly 100 times the EBL flux.

The brightest and most problematic foreground source is the rapidly varying molecular and atomic line emission, airglow, produced by the Earth's own atmosphere. We have avoided this problem entirely by using HST to measure the total night sky flux from above the atmosphere. The dominant foreground contributing from above the Earth's atmosphere is sunlight scattering off of the large ($\geq 10\mu$m), rough, interplanetary dust grains which concentrate in the ecliptic plane. As the solar system is harder to escape than the Earth's atmosphere,

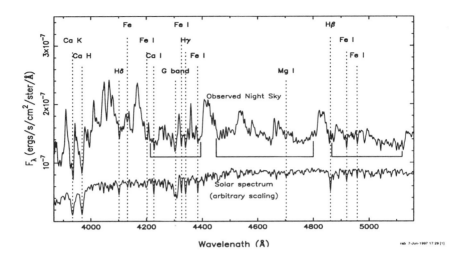

Figure 2. The observed night sky spectrum compared to a scaled solar spectrum. Solar absorption features due to ZL in the night sky spectrum are readily identified.

it is fortunate that the scattering involved is well described by Mie theory and is only weakly wavelength dependent, with scattering becoming more efficient by 5% per 1000Å with increasing wavelength. The mean zodiacal light flux can thus be identified by the strength of solar absorption features (Fraunhofer lines) which are preserved in its spectrum (see Figure 2).

Stars in our own Galaxy can be resolved and subtracted relatively easily with modern optical CCD detectors. However, interstellar dust scatters incident starlight, producing diffuse Galactic light at optical wavelengths. The dust column density and the interstellar radiation field strength, both of which determine the intensity of scattered DGL, are well correlated with the $100\mu m$ thermal emission from the dust along the line of sight. Our field was selected for its very low $100\mu m$ emission and therefore minimal DGL. The remaining low-level optical DGL can be estimated using simple scattering models, which are in good agreement with the empirical observations of the DGL from 2500–9000Å (see Witt et al. 1997 and references therein).

In this work, we isolate the EBL23 using three, simultaneous data sets: absolute surface photometry in 1000Å-wide bands centered at roughly 3000, 5500, and 8000Å from WFPC2; low (~ 300Å) resolution surface spectrophotometry at 4000—7000Å from the FOS; and moderate (2.0Å) resolution surface spectrophotometry from the Boller and Chivens spectrograph on the 2.5m duPont telescope at Las Campanas. The HST data are used to measure the mean flux of the total background, while the LCO data are used to measure the zodiacal light (ZL) by the method described above. We estimate foreground diffuse galactic light (DGL) using scattering theory and empirical correlations

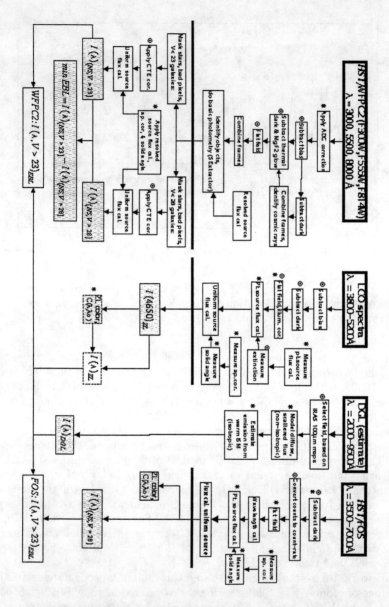

Figure 3. Flow chart of data reduction. Abbreviations are as follows: zodiacal light, ZL; diffuse galactic light, DGL; total night sky flux, NS; surface brightness, $I(\lambda)$; object cut-off at $V=23$, $I(\lambda)_{V>23}$; systematic errors, *; statistical errors, $\oplus$. The thick, horizontal bars divide pre-reduction from analysis. Bold type-face indicates original procedures, as opposed to STScI pipeline. Shaded boxes indicate results of individual data sets; dashed boxes show iterations in the ZL measurement; thick-lined boxes show final EBL23 results.

between the optical DGL and thermal 100μm emission from the same dust. The HST observations were scheduled in three visits of six orbits each, with one month between visits. This allowed us to look for and confirm the expected modulation in the ZL with the Earth's orbital position, to look for possible off-axis scattered light with the satellite oriented at different roll angles, and to safe-guard against unidentified photometric anomalies with the instrument.

Observations from LCO and HST must be simultaneous to ensure that the ZL measured from the ground is exactly the contribution seen by HST. Also, the HST and LCO data sets must be calibrated to the same absolute scale with ∼1% accuracy in order to detect the EBL. The basic data reduction, analysis, and method for combining data sets to detect the EBL23 is shown in the flow chart in Figure 3. When necessary to achieve the required accuracy, original reduction procedures were developed and STScI calibrations were augmented with our own solutions, as indicated in the flow chart. To eliminate stray light, HST observations were made only in the shadow of the Earth, with the Moon greater than 65 degrees from the optical axis of the telescope. We also selected the field to avoid $V > 7$ AB mag stars within 3 degrees.

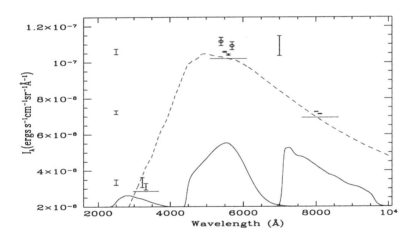

Figure 4. Total sky flux from WFPC2 (dots) and FOS (circles), November and December (offset +100Å) data sets. The 2% difference in flux results from changes in ZL due to Earth's motion. Dashed line shows ZL spectrum. Horizontal bars show ZL convolved with WFPC2 filters. Sky flux at $\lambda > 4000$Å is ∼ 95% ZL. Error bars on all points indicate 1σ RMS errors. Floating error bars at 2500Å show systematic uncertainties for WFPC2 result; the error bar at 7000Å shows the FOS uncertainty. Solid lines show WFPC2 band-passes.

3. Results on the EBL

From each WFPC2 CCD image, we obtain a measurement of the night sky flux. As expected, we find variations in the sky flux between visits due to the varying ZL contribution at different times of year. We find no significant difference between the results from the three WFPC2 chips, and no indication of stray light or photometric anomalies between visits. Final statistical errors are less than 1% in the two redder pass-bands. The statistical error in the measurement at 3000Å is dominated by the accuracy with which we can subtract instrumental backgrounds, which are almost as bright as the total sky flux at this wavelength. Systematic errors are also of order 1%, and are dominated by the aperture correction and point source calibration. The results are shown in Figure 4. The total systematic errors plotted reflect 90% probability limits. FOS results have a much large systematic uncertainty ($\sim 5\%$), due to uncertainty in the aperture solid angle and point spread function, and non–uniform sensitivity across the detector, which affects surface brightness results much more than point source measurements. The WFPC2 and FOS results are discussed in detail in Bernstein et al. 1999a.

We took spectra of blank sky within the HST/WFPC2 field of view using the Boller and Chivens spectrograph at Las Campanas Observatory simultaneously with the November HST observations. We have used those spectra to measure the absolute flux of the ZL at 4650Å with a precision of 0.8%, and a systematic uncertainty of $< 1\%$, using the method outlined in §2. That measurement is discussed in Bernstein et al. 1999b.

The optical DGL was estimated using a simple non–uniform, back–scattering model, $I_\lambda = j_\lambda \omega_\lambda \tau_\lambda S(g, b)$, in which $S(g, b)$ is the scattering phase function in terms of Galactic latitude, b, and the average phase function of the dust, g (Jura 1979). The optical depth, τ_λ, and the surface brightness of the interstellar radiation field, j_λ, are empirically determined. The effective albedo, ω, and phase function, g, are based on extensive laboratory tests and modeling by Draine & Lee (1980) and are in excellent agreement with both optical and IR observations. The predicted DGL for our field (galactic latitude $b \geq 50°$, 150° from the galactic center) is in good agreement with prediction from scaling relations between the optical and UV scattering and the 100μm thermal emission from the galactic dust (see Witt et al. (1997) for a discussion of models and a review of results). The contribution from DGL is small enough that uncertainty in this estimate is an insignificant source of error for our final result.

Combining the HST and LCO results with a model of the DGL, we obtain the detections of the EBL23 shown in Figure 5. The errors are dominated by two systematic effects: unavoidable limitations in the flux calibration of independent data sets to the same absolute scale; and uncertainty in the color of the ZL, which we combine with our absolute ZL flux measurement at 4650Å to obtain the ZL flux at each WFPC2 band.

In the spirit of integrating galaxy counts to estimate the EBL from detected galaxies, we have identified the minimum EBL23 by a simplified aperture photometry method. In brief, we measure a single "sky" value using the whole CCD frame and measure source fluxes within apertures that are 8 times larger than those used for source photometry in, for example, the HDF catalog (see Williams et al. 1996). We find that for galaxies within two magnitudes of our

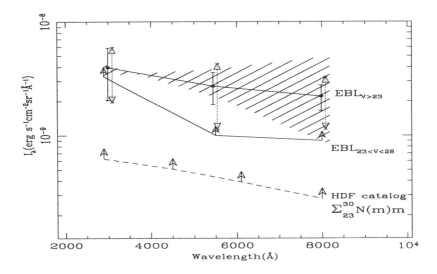

Figure 5. The detected EBL23 — EBL from galaxies with $V > 23$ AB mag. Solid error bars show 1σ RMS errors; dotted error bars show systematic uncertainties. The hatch-marked region shows uncertainty from ZL subtraction. The dashed line connects integrated galaxy counts from the HDF. The solid line connects the minimum EBL23 values we identify from galaxies $23 < V \leq 28$ AB mag.

detection limit, roughly 50% of each galaxy's flux lies outside of the standard-sized apertures used by FOCAS or similar photometry packages. In addition, we estimate that an extragalactic sky pedestal is created by the overlapping wings of *resolved* galaxies and contributes roughly 2×10^{-10} ergs s^{-1} cm^{-2} sr^{-1} Å^{-1} to the sky level. This is a significant component and is undetectable except by absolute surface photometry. We therefore identify a minimum EBL23 from *detected* galaxies as shown in Figure 5. Previous estimates of the flux from resolved galaxies fainter than $V = 23$ AB mag are at least 50% too low as a result of the above effects (e.g. Pozzetti et al. 1998).

4. Discussion

The EBL23 is more than four times the flux recovered in standard galaxy photometry of galaxies with $23 < V < 30$ AB mag (e.g. HDF catalog, Williams et al. 1996). However, photometry errors clearly play a significant role in this difference, as the flux we recover from simplified aperture photometry of galaxies with $23 < V < 28$ accounts for $\sim 50\%$ of the EBL23 we detected. Noting that luminosity functions of LSBs at low redshift contribute an additional 30–50% to the local luminosity density, and noting that even galaxies with *normal* central surface brightnesses (i.e. $\mu_V \sim 21.5$) are undetectable in the HDF at redshifts $z \gtrsim 0.5$, we conclude that the EBL23 we detect can be explained by galaxies which are well observed in the local universe. No exotic explanations are required. The total EBL, including the flux from galaxies *brighter* than $V=23$, is shown in Figure 6.

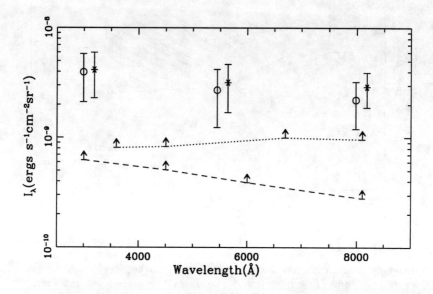

Figure 6. Open circles show the EBL23 from galaxies with $V > 23$ AB mag. Error bars show 2σ statistical errors, encompassing nearly the same range as the 90% confidence interval due to systematic errors. Lower limit arrows connected by a dashed line mark the integrated flux in the HDF catalog. Lower limit arrows connected by a dotted line mark the flux in HDF counts plus ground-based counts for $V < 23$ AB mag. Asterisks show the total EBL: EBL23 from sources with $V > 23$ AB mag, plus the integrated ground-based counts at $V < 23$ AB mag.

References

Bernstein, R.A., Freedman, W.L., Madore, B.F. 1999a, in prep.
Bernstein, R.A., Freedman, W.L., Madore, B.F. 1999b, in prep.
Draine, B.T. & Lee, H.M. 1984, ApJ, 285, 89
Pozzetti, L., Madau, P., Zamorani, G., Ferguson, H.C., & Bruzual, G.A. 1998, MNRAS, 298, 1133
Williams, R.E. et al. 1996, AJ, 112, 1335
Witt, A.N., Friedmann, B.C. & Sasseen, T.P. 1997, ApJ, 481, 809

The Low Surface Brightness Universe, IAU Col. 171
ASP Conference Series, Vol. 170, 1999
J. I. Davies, C. Impey and S. Phillipps, eds.

Optical Diffuse Light in Clusters of Galaxies

Rosendo Vílchez–Gómez

Space Telescope Science Institute, 3700 San Martin Drive, Baltimore, MD 21218, USA

Abstract.
    I present here a review of the observed characteristics of the optical diffuse light in clusters, the possible sources of this light and some of the theories that try to explain the existence of big envelopes around the brightest cluster galaxies.

1. Introduction

The first reference that we can find in the literature about the diffuse light in clusters of galaxies was given by Zwicky (1951): "One of the most interesting discoveries made in the course of this investigation [in the Coma cluster] is the observation of an extended mass of luminous intergalactic matter of very low surface brightness. The objects which constitute this matter must be considered as the faintest individual members of the cluster. [We report] the discovery of luminous intergalactic matter concentrated generally and differentially around the center of the cluster and the brightest (most massive) galaxies, respectively". This is a perfect characterization of the optical diffuse light in clusters: extended, low surface brightness and around the center of the cluster.
    Zwicky was trying to settle three of the problems of the extragalactic astronomy at that moment: (1) this luminous intergalactic matter can account for the dark matter needed in Coma if this cluster were virialized; (2) the shape of the luminosity function (a Gaussian, according to Hubble) is monotonely increasing with decreasing brightness; and (3) the galaxies extend notably far away from their centers[1].
    The characteristics of this diffuse matter published by Zwicky (1951, 1957, 1959) were qualitative: it has an extension of around 150 kpc, the color index is rather blue and produces a local absorption of light of the order of six tenth of a magnitude.
    The first published attempt to obtain a value for the surface brightness of the faint intergalactic matter in Coma corresponds to de Vaucouleurs (1960). He reported an upper limit of B $>$ 29.5 mag arcsec^{-2} at $\langle r \rangle \simeq 0°.9$. With this value, de Vaucouleurs reasons that "a stellar population composed exclusively of extreme red dwarfs of mass M$<$ 0.1 M$_\odot$ and absolute magnitudes $M(pg) > +15$

---

[1] Baum (1955) claims that "galaxies blend into one another with no vacant intergalactic gaps in between".

would, in principle, give an M/L ratio of the order measured in Coma. While such stars are known to exist in the neighborhood of the Sun, it seems very difficult to admit that they could populate intergalactic space with the required density and to the exclusion of all other stars of slightly greater mass"[2]. Thus, de Vaucouleurs concludes that the mass of the intergalactic matter is not enough to account for the mass value estimated through the virial theorem.

The next step in the first studies of the diffuse light in clusters corresponds to Matthews, Morgan & Schmidt (1964). During the analysis of radio sources, they found near the center of a number of Abell's rich clusters, supergiant D galaxies with diameters 3–4 times as great as the ordinary lenticulars in the same clusters. They gave the prefix "c" to these very large D galaxies, "in a manner similar to the notation for supergiant stars in stellar spectroscopy". The reason for this remark is that I believe that there is not a real difference between the detection of intracluster light or the halo of a cD. Whether this diffuse light is called the cD envelope or diffuse intergalactic light is a matter of semantics. In fact, Oemler (1973) in his study of Abell 2670 where he traced a diffuse envelope to almost 1 Mpc says: "An important question is the relation between this diffuse component and the central elliptical galaxy, the combination of which seems to produce the cD galaxy". Nevertheless, there are clusters without a cD in the center where a diffuse light has been detected in its central part, as it is the case in Cl 1613+31 (Vílchez–Gómez, Pelló & Sanahuja 1994a,b).

Before CCD detectors were widely used, most of the observations and study of the diffuse light in clusters was carried out in the Coma cluster: Abell (1965); Gunn (1969); de Vaucouleurs & de Vaucouleurs (1970); Welch & Sastry (1971, 1972); Gunn & Melnick (1975); Mattila (1977); Melnick, White & Hoessel (1977); Thuan & Kormendy (1977). There are also some studies in Virgo: Holmberg (1958); Arp & Bertola (1969); de Vaucouleurs (1969). Finally, there are also studies of the faint envelopes of elliptical and cD galaxies: Arp & Bertola (1971); Baum (1973); Kormendy & Bahcall (1974); Oemler (1973, 1976). I will consider here, basically, the problems associated with the use of CCD's in the study of the diffuse light as well as the results obtained with that kind of detector.

2. Problems and Errors

If we consider that the intracluster light is expected to be extremely faint, about 25–26 mag arcsec^{-2} in a red filter (if it represents 10 to 25% of the total light in the center of an intermediate redshift cluster), it is easy to understand how hard it can be to obtain a reliable detection and analysis of the diffuse light in a cluster. We have to be sure that our detection is not the result of spurious effects, such as instrumental scattering or contamination due to bright stars or faint galaxies. I will comment on some of this error sources:

---

[2] Actually, Boughn & Uson (1997), studying three rich Abell clusters, where they don't detect any anomalous reddening in the intracluster medium, conclude that no more than $2h^{-1}$% of the dark matter can be in the form of low mass (~ 0.1 M$_\odot$) subdwarfs or old disk dwarfs.

Instrumental Scattering. The diffuse light due to the mirrors of the telescope is the first source of parasitic light. If the cleanliness of the telescope optics is not correct, some of the results that we could ascribe to the intracluster light would be masked or spoiled.

Flat Fielding. Our images must be cleaned of any kind of residual ghost image structures as well as free of fringing. A good level of flattening should be lower than 0.5%.

Contamination due to Bright Stars. As we are trying to obtain accurate surface brightness profiles at 25 mag arcsec^{-2} and lower, it is necessary to carry out an accurated removal of the halos of stars and bright clusters members. An unaccurate subtraction can alter the result by more than 0.5 mag arcsec^{-2}. It is also essential to check for the possibility of contamination due to halos of stars located outside but near the field. Some comments about the removal of the halos can be found in Gudehus (1989); Uson, Boughn & Kuhn (1991); Mackie (1992); Vílchez–Gómez et al. (1994a).

Sky Level. If we want to fix what is the real extension of our intracluster light, we need a correct determination of the sky level. Thus, we ought to be sure that we are far enough from the central part of the cluster to reach the end of the diffuse light profile. We can get this either working with a big field (i.e., making a tessellation of different images as in Scheick & Kuhn 1994) or studying a relatively distant cluster in order to be sure that all the cluster is inside our CCD.

Faint Galaxies. We have to correct from the contamination due to the galaxies fainter than the completeness limit in magnitude for our sample. One possibility is to extrapolate a Schechter luminosity function fitted to our data, until the detection limit. But if we use the k-correction, then, we have to assume a Hubble type of this galaxies. If we consider that they are E/S0 galaxies we tend to overcorrect the diffuse light in the red filters with respect to the blue ones (Vílchez–Gómez et al. 1994a).

Other Sources of Errors. For example, vignetting in the image, incorrect determination of the galactic absorption, statistical errors associate with the measure, wrong redshift for the cluster, the presence of galactic cirrus as reported by Haikala & Mattila (1995) or Szomoru & Guhathakurta (1998).

3. Characteristics

I will try to summarize some of the most important characteristics associated with the diffuse light in clusters of galaxies:

Luminosity. It shows a wide range. The intracluster light can represent between the 10% and the 50% of the total light of the region where it is detected. Schombert (1988) finds some correlation, but faint, between the luminosity of the cD envelope and that of the underlying galaxy. This correlation can hint that the process of formation of the Brightest Cluster Galaxy (BCG) has some reflection in the origin of its envelope.

Color. Different authors have report various results. Valentijn (1983) in $B-V$ and Scheick & Kuhn (1994) in $V-R$ find blueward gradients that vary between 0.1 to 0.6 mag drop. Schombert (1988) in $B-V$ doesn't find any evidence of strong color gradients or blue envelope colors. Finally, Mackie (1992) in $g-r$ reports a reddening at the end of the envelopes, in one case of the order of 0.15 mag.

Structure. Schombert (1988) and Mackie, Visvanathan & Carter (1990) find an apparent break in the surface brightness profile of the underlying cD galaxies. According to Schombert (1988), this break is found near the $24V$ mag arcsec^{-2} but there are no sharp changes in either eccentricity or orientation between the galaxy and the envelope. However, Uson et al. (1991) and Scheick & Kuhn (1994) don't see such a break in their studies.

Reinforcing the idea of common evolutive processes, Schombert (1988) and Bernstein et al. (1995) find that the diffuse light, globular cluster density and galaxy density profiles seem to have similar radial structure, proportional to $r^{-2.6}$. However, Cl 1613+31 shows a different profile for the diffuse light and the galaxies (Vílchez–Gómez et al. 1994a).

Cluster properties and diffuse light. Schombert (1988), in one of the most comprehensive studies of cD envelopes, finds the following correlations between the luminosity of the envelope ($L_{\rm env}$) and the general properties of the cluster: (1) There is a clear correlation between $L_{\rm env}$ and cluster richness for compact, regular clusters; (2) there is no evident correlation with velocity dispersion; (3) there is a slight correlation with the Bautz–Morgan or Rood–Sastry cluster type; (4) there is an unambiguous correlation with the X-ray luminosity.

Finally, there are no reports of galaxies with envelopes in the field and the cD-like galaxies observed in poor clusters dwell in local density maxima, comparable to the central region of rich clusters (West & Van den Bergh 1991). That is, a cluster or subcluster environment with high local density contrast looks like an unambiguous requirement for the presence of cD envelopes or intracluster light.

4. Sources for the diffuse light

Basically, there are five processes to explain the origin of the intracluster light:

Stars from the outer envelopes of galaxies. Sometimes the extension of the diffuse light is so large (several core radius) that is hard to believe that these stars are gravitationally bound to any galaxy, and probably, they are stripped material after the interaction between galaxies. This could be the case in Cl 1613+31 (Vílchez–Gómez et al. 1994a). Also, it could be that the stars have born directly in the intergalactic medium from a cooling flow, for example (Prestwich & Joy 1991).

Dwarf galaxies and globular clusters. Part of the light in the intergalactic medium in distant clusters, where it is not possible to resolve dwarf galaxies and globular clusters, can have this origin. Nevertheless, Bernstein et al. (1995) have measure in the Coma cluster a diffuse light apart from dwarf galaxies and globular clusters.

Hot intracluster bremsstrahlung radiation in the optical. Woolf (1967), Mattila (1977) and Bernstein et al. (1995) established that, at least for the Coma cluster, this contribution is not significant if we take into account the boundaries imposed by the observations in X-ray and the observed Hβ intensity.

Light scattered by intergalactic dust. The existence of dust in rich clusters of galaxies as established by Zwicky (1959) or Hu (1992) would suggest the production of diffuse scattered light. Mattila (1977) makes an estimation of around 12% of the total surface brightness of the Coma cluster can be due to the surface brightness of the scattered light with origin in the dust.

The radiative decay of particles. Partridge (1990) considers that the radiative decay of low mass particle ($m_\nu \sim 4$ eV) would produce extragalactic light in the visible.

The first source seems to be the most important. Scheick & Kuhn (1994), studying the diffuse light "granularity" in Abell 2670 established that the luminosity of each source is less that 10^4 L$_\odot$. This suggests that the main origin of the diffuse light is light from stars since the luminosity associated with the sources is about a factor 100 smaller than the luminosity of the faintest dwarf galaxies. A similar result is found by Bernstein et al. (1995) for the Coma cluster.

5. Origins of the cD envelopes

There are basically four theories that try to elucidate what is the origin and evolution of the cD envelopes. None of them offers a complete picture of the problem.

Stripping theory. This theory was initially proposed by Gallagher & Ostriker (1972). According with this theory, the origin of the envelope is the debris due to tidal interactions between the cluster galaxies. These stars and gas are then deposited in the potential well of the cluster where the BCG is located. This process begins after the cluster collapse and the envelope grows as the cluster evolves. The fact that different cD envelopes show different color gradients can be explained as the result of different tidal interaction histories: in some clusters the tidal interactions involve mainly spirals, but in others, early type galaxies are the source material (Schombert 1988). The main problem to this hypothesis is the difficulty explaining the observed smoothness of the envelopes as the timescale to dissolve the clumps is on the order of the crossing time of the cluster (Scheick & Kuhn 1994).

Primordial origin theory. This hypothesis, suggested by Merrit (1984), is similar to the previous one but, in this case, the process of removing stars from the halos of the galaxies was carried out by the mean cluster tidal field and took place during the initial collapse of the cluster. The BCG, due to its privileged position in relation with the potential well, gets the residuals that make up its envelope. However, this picture is difficult to reconcile with the fact that there are cD's with significant peculiar velocities (Gebhardt & Beers 1991) as well as with the smoothness of the diffuse light either the envelope is affixed to the cD or the cD is moving through it. Moreover, if the origin of the diffuse light is primordial, how can we explain the observation of blue color gradients in some envelopes, supposed little activity after virialization?

Cooling flows. Fabian & Nulsen (1977) proposed that the radiative cooling of hot X-ray gas can produce an increase of the densities around the BCG until star formation takes place. But this process is confined to the first one hundred kpc from the center of the cluster (Prestwich & Joy 1991), insufficient to explain the big envelopes of several hundred of kpc observed. Moreover, a blue color gradient is expected if recent star formation is taking place.

Mergers. Villumsen (1982, 1983) found that after a merger with the BCG, and under special conditions, it is possible to form a halo similar to that present in cD galaxies since there is a transfer of energy to the outer part of the merger, resulting an extended envelope. Although this theory reproduces the profile observed for the envelopes, it is not possible to account for the luminosities and masses associated with the diffuse light. However, in poor clusters where there are cD-like galaxies without a clear envelope this mechanism can play a more important role (Thuan & Romanishin 1981; Schombert 1986).

6. Conclusions

After this review, it is clear that it is necessary to carry out a more systematic study of the diffuse light in clusters to obtain a better comprehension of the origin and evolution of its properties and its relation with the global characteristics of the cluster.

Acknowledgments. I would like to thank R. Pelló and B. Sanahuja for their help and comments in my study of the diffuse light in clusters. I am grateful to Kalevi Mattila for enabling my participation in this conference. I thank also STScI and IAU for financial support.

References

Abell, G. O. 1965, ARA&A, 3, 1
Arp, H., & Bertola, F. 1969, Astrophys. Letters, 4, 23
Arp, H., & Bertola, F. 1971, ApJ, 163, 195
Baum, W. A. 1955, PASP, 67, 328
Baum, W. A. 1973, PASP, 85, 530
Bernstein, G. M., Nichol, R. C., Tyson, J. A., Ulmer, M. P., & Wittman, D. 1995, AJ, 110, 1507
Boughn, S. P., & Uson, J. M. 1997, ApJ, 488, 44

Fabian, A. C., & Nulsen, P. E. J. 1977, MNRAS, 180, 479
Gallagher, J. S., & Ostriker, J. P. 1972, AJ, 77, 288
Gebhardt, K., & Beers, T. C. 1991, ApJ, 383, 72
Gudehus, D. H. 1989, ApJ, 340, 661
Gunn, J. E. 1969, BAAS, 1, 191
Gunn, J. E., & Melnick, J. 1975, BAAS, 7, 412
Haikala, L. K., & Mattila, K. 1995, ApJ, 443, L33
Holmberg, E. 1958, Medd. Luns. Astron. Obs. II, No. 136, 63
Hu, E. M. 1992, ApJ, 391, 608
Kormendy, J., & Bahcall, J. N. 1974, AJ, 79, 671
Mackie, G. 1992, ApJ, 400, 65
Mackie, G., Visvanathan, N., & Carter, D. 1990, ApJ, 73, 637
Matthews, T. A., Morgan, W. W., & Schmidt, M. 1964, ApJ, 140, 35
Mattila, K. 1977, A&A, 60, 425
Melnick, J., White, S. D. M., & Hoessel, J. 1977, MNRAS, 180, 207
Merrit, D. 1984, ApJ, 276, 26
Oemler, A. 1973, ApJ, 180, 11
Oemler, A. 1976, ApJ, 209, 693
Partridge, R. B. 1990, in The Galactic and Extragalactic Background Radiation, S. Bowyer & C. Leinert, Dordrecht: Kluwer, 283
Prestwich, A. H., & Joy, M. 1991, ApJ, 369, L1
Scheick, X., & Kuhn, J. R. 1994, ApJ, 423, 566
Schombert, J. M. 1986, ApJS, 60, 603
Schombert, J. M. 1988, ApJ, 328, 475
Szomoru, A., & Guhathakurta, P. 1998, ApJ, 494, L93
Thuan, T. X., & Kormendy J. 1977, PASP, 89, 466
Thuan, T. X., & Romanishin, W. 1981, ApJ, 248, 439
Uson, J. M., Boughn, S. P., & Kuhn, J. R. 1991, ApJ, 369, 46
Valentijn, E. A. 1983, A&A, 118, 123
Vaucouleurs, G. de 1960, ApJ, 131, 585
Vaucouleurs, G. de 1969, Astrophys. Letters, 4, 17
Vaucouleurs, G. de, & Vaucouleurs, A. de 1970, Astrophys. Letters, 5, 219
Vílchez–Gómez, R., Pelló, R., & Sanahuja, B. 1994a, A&A, 283, 37
Vílchez–Gómez, R., Pelló, R., & Sanahuja, B. 1994b, A&A, 289, 661
Villumsen, J. V. 1982, MNRAS, 199, 493
Villumsen, J. V. 1983, MNRAS, 204, 219
Welch, G. A., & Sastry, G. N. 1971, ApJ, 169, L3
Welch, G. A., & Sastry, G. N. 1972, ApJ, 171, L81
West, M. J., & Van den Bergh, S. 1991, ApJ, 373, 1
Woolf, N. J. 1967, ApJ, 148, 287
Zwicky, F. 1951, PASP, 63, 61

Zwicky, F. 1957, Morphological Astronomy, Berlin: Springer-Verlag, 48
Zwicky, F. 1959, in Encyclopedia of Physics, S. Flügge, Berlin: Springer-Verlag, 53, 398

Diffuse Ultraviolet Background Radiation

Richard C. Henry[1,2,3]

Henry A. Rowland Department of Physics and Astronomy, The Johns Hopkins University, Baltimore, MD 21218-2686

Abstract. Diffuse ultraviolet background radiation may contain important information concerning the dark matter of the universe. I briefly review new *Voyager* observations of the diffuse background, which give a very low upper limit on the background radiation shortward of Lyman α, and I review the capabilities for detection and characterization of diffuse radiation that will be provided by a proposed new NASA mission. Low-surface-brightness radiation remains largely an unexplored frontier, particularly in the ultraviolet.

1. Introduction

While most research in astronomy has focused on study of celestial point sources, the astronomers of the present IAU Colloquium consider, instead, the low-surface-brightness universe: extended objects (and backgrounds) that most astronomers have by and large neglected. Diffuse backgrounds always contain independent information about the universe, in some cases of considerable significance: witness the famous 3K background radiation.

O'Connell (1987) has pointed out the great importance of the *ultraviolet* spectral region as a hunting ground for low-surface-brightness radiation. He notes that in the visible, the much brighter zodiacal light renders such searchs difficult. This is the same point that was made by Disney (1976), who suggested that astronomers are blinded in the visible by the extraordinarily high sky background.

In Figure 1, I show the spectrum of the diffuse background radiation at high galactic latitudes, from the near infrared to the X-ray, including not only the ultraviolet, but also the extreme ultraviolet, where the interstellar medium is opaque.

The data that are presented in the figure indicate that there is a drop of $2\frac{1}{2}$ orders of magnitude in the diffuse background between the visible diffuse background and the soft X-ray background. The region from 912 Å to the soft X-ray is obscured by interstellar gas photoionization. One key question is,

---

[1] Director, Maryland Space Grant Consortium

[2] Principal Professional Staff, Applied Physics Laboratory, The Johns Hopkins University

[3] rch@pha.jhu.edu

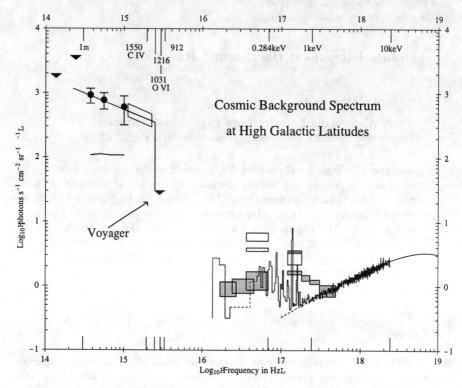

Figure 1. The black triangles are DIRBE upper limits (Hauser et al. 1998), while the three black circles are the HST optical background radiation data of Bernstein (1997); below the latter, the horizontal line shows her measure of the contribution (to that) from galaxies. Vogeley (1997) has shown that the excess is truly diffuse. The line through Bernstein's three points is the extrapolation of the model that was found by Henry & Murthy (1995) to fit the UV background radiation measurements (box). The extraordinary drop to ≤ 30 photons cm^{-2} s^{-1} Å^{-1} ["photon units"] at ~ 1000 Å was discovered by Holberg (1986), using the *Voyager* ultraviolet spectrometer, and confirmed by Murthy, Henry, & Holberg (1991) and Murthy et al. (1998). It has been interpreted by Henry (1991) as indicating that the high-galactic-latitude background, longward of Ly α, is largely redshifted recombination radiation from an ionized intergalactic medium. In the X-ray part of the spectrum, I compare a HUBE-CUBIC (see text) simulated spectrum of the high-galactic-latitude CXRB [based on the rocket spectrum of Mendenhall & Burrows (1998)], with the spectrum of Boldt (1987, solid line); and with the summary (shaded boxes) of McCammon & Sanders (1990), and the early observations (open boxes) of Bowyer, Field, & Mack (1968), Henry et al. (1968, 1971), and Davidsen et al. (1972). ROSAT's imaging has revealed an extremely complex soft X-ray sky, but even so, Wang & McCray (1993) have suggested that radiation from a several-million-degree IGM may have been identified.

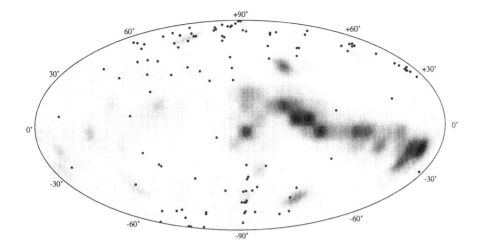

Figure 2. North is at the top, and the galactic center at the center, in this Aitoff map of the sky showing the new (Murthy et al. 1998) Voyager (~ 1000 Å) limits *(filled circles)* on the diffuse ultraviolet background radiation. Of the 431 new observations, 265 (or 62%) are 3σ upper limits. The points plotted in the figure are measurements for which the sum of the measured flux and its error bar are less than 300 photon units. For the points plotted, the average measured flux is 24.5 ± 8.0 photon units. The shaded background is an integration of the light at 1565 Å that is provided by *direct* radiation from TD1 stars. This radiation can scatter from interstellar dust, creating a diffuse background (Bowyer 1991). Note the presence of Gould's Belt, delineated by hot stars. Gould's Belt is a segment of the galaxy that is tipped 19° relative to the galactic plane.

where precisely does the large drop in background occur? I believe it occurs at 1216 Å, and that the longer wavelength radiation that is observed at high galactic latitudes is largely redshifted Lyman α radiation. If that view should turn out to be correct, then longward of 1216 Å we are seeing recombination radiation from an ionized intergalactic medium.

The exciting ideas concerning dark matter in the universe that are suggested by the data of Figure 1 are in part predicated on the reality of the sharp drop in the diffuse ultraviolet background radiation at 1216 Å that the *Voyager* data suggest exists. In Figure 2, I show the distribution on the sky of the new *Voyager* upper limits of Murthy et al. (1998), one example of which is shown in Figure 3 (a). While there are, now, a very large number of such upper limit points, the importance of the result demands a dedicated confirmation investigation.

I have proposed such an investigation, as one part of *Hot Universe Background Explorer* (HUBE), a proposal which was submitted to NASA in 1998 August as a candidate MIDEX mission. HUBE is an improved version of the

HUBE proposal that was accepted by NASA in 1996 as a MIDEX *Alternate* Mission. The original HUBE has been described by Henry (1997, 1998), while the history of my search for the intergalactic medium is detailed in Henry (1995).

The new HUBE has been expanded to include additional instruments that are also dedicated to the elucidation of the nature of the high-energy diffuse background radiation, and that are well matched to the original HUBE instruments: in field of view, optimum target dwell, and science objectives. The new HUBE concept has been presented by Henry, Murthy, Ford, Peacock, Burrows, Smith, & Bloch (1998).

The new HUBE instruments include Imaging Spatial Heterodyne Spectrometer (ISHS), which provides extremely high resolution ($\Delta\lambda = 0.08$ Å) spectroscopy of the interstellar CIV line (1548.20 Å and 1550.77 Å). ISHS will allow study of the dynamics not only of the hot phase of the interstellar medium, but also of the cold phase, as the chosen passband also provides velocity information on two H_2 emission lines (1547.4 Å and 1562.4 Å).

A second new instrument is "Cosmic Unresolved Background Instrument using CCD's" (CUBIC), which has been flown previously, on the ill-fated Argentine SAC-B mission. CUBIC will provide high-sensitivity, excellent-resolution spectroscopy of the diffuse background radiation from 1.24 Å to 250 Å. With CUBIC, too, there is hope of addressing the question of an intergalactic medium (see Figure 1).

2. Neutrino Decay Radiation

The final new instrument in the HUBE complement is a spectrometer that straddles Lyman α, permitting detection of a step at Lyman α with a single spectrometer (as opposed to separate spectrometers above and below Lyman α). A simulation of the result of an observation with this new spectrometer at high galactic latitudes is presented in Figure 3 (b), on the assumption that the observed drop at Lyman α is real. The new spectrometer has been specifically designed to reject terrestrial/solar-sytem Lyman α radiation. The spectrometer is double-pass, and on the first pass, Lyman α is blocked. This means that only *doubly-scattered* Lyman α can reach the wrong part of the spectrum, and with modern gratings, this amount of radiation is totally negligible. Thus, if the break at Lyman α is real, this spectrometer (designated LαS) will establish that fact unequivocally.

If the diffuse ultraviolet background at high galactic latitudes is truly redshifted Lyman α radiation, the intensity is such that to explain the radiation, not only must there be an intergalactic medium containing all of the missing baryonic dark matter, but that medium must be so clumped that it would have already recombined, unless some strong source of ionizing radiation is present beyond that which is provided by quasars and galaxies. Sciama (1998) has suggested just such an ionizing radiation source, namely, massive neutrinos that decay with the emission of an ionizing photon. Our *Voyager* observations can be taken to imply that Sciama's suggestion is correct.

A direct search for the ionizing photons themselves is underway, the EURD detector aboard the Spanish MINISAT 01 satellite, but in Figure 4 we show that this mission is likely to fail, due to the interfering effects of terrestrial airglow.

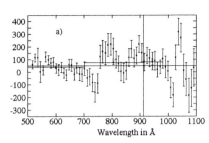 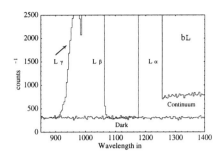

Figure 3. **a)** Example of a *Voyager* spectrum of the diffuse ultraviolet background at high galactic latitude (Murthy et al. 1998). The data are corrected so that no signal appears below the interstellar hydrogen photoionization edge at 912 Å, as is expected. It can be seen that there is no detection of diffuse radiation *longward* of this wavelength either. **b)** Simulation of an observation at high galactic latitudes of the cosmic diffuse ultraviolet background radiation, with our spectrometer which straddles the intense geocoronal/solar-system Lyman α radiation. The spectrometer is double-pass, allowing strong rejection of (blocked) Lyman α. The resolution is 40 Å. Longward of Lyman α, a continuum of extragalactic radiation is seen, but that continuum does not continue below Lyman α, indicating that it is surely made up of red-shifted Lyman α radiation. Strong terrestrial/solar-system Lyman β and Lyman γ radiation are also seen. No continuum is expected below 912 Å in any model of the origin of the high-latitude diffuse UV background. That there actually does exist an *extragalactic* diffuse ultraviolet background (longward of 1216 Å), of 300 ± 80 photon units, is supported by Witt & Pettersohn (1994).

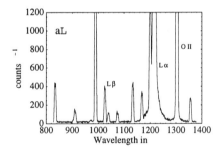 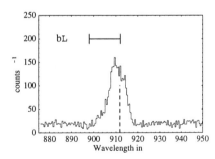

Figure 4. **a)** The day airglow observed by the *Hopkins Ultraviolet Telescope* (HUT) between 800 Å and 1400 Å. **b)** A magnified view, highlighting the critical region, near the Lyman limit *(dashed line)*, where Sciama (1998) has predicted emission from decaying neutrinos (the horizontal error bar shows the range that is permitted by Sciama's theory). The interested reader should consult Feldman et al. (1992) for examples showing HUT spectra of the *night* airglow, which demonstrate the strong time/place variation of the airglow complex around 910 Å.

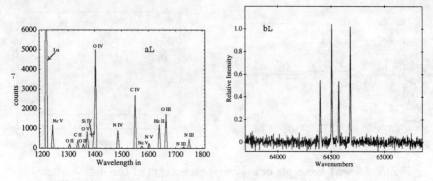

Figure 5. a) A one-hour exposure with HUBE-FUVS of a 1.5-arc-minute by 2.7-arc-minute section of the Cygnus Loop nebula, showing, for example, strong CIV emission. The HUBE instruments will be used to make a detailed map of these emissions over the entire Cygnus Loop nebula. b) Simultaneously, HUBE-ISHS obtains a very-high-resolution spectrum of the CIV emission, not only resolving the two components, but, thanks to Doppler shifts, separating the emission from the side of the shell that is nearest to us, from the emission from the portion of the shell that is farthest from us. (This simulation was created by Jeffrey J. Bloch, Barham W. Smith, and John M. Harlander.)

3. HUBE and the Cygnus Loop

The power of the new version of HUBE in studying diffuse radiation can be shown most clearly by displaying what HUBE might accomplish in the examination of a relatively *bright* source of diffuse ultraviolet and X-radiation, the Cygnus Loop nebula. In Figure 5, I show simulations of observation of the Cygnus Loop with our Far Ultraviolet Spectrometer (HUBE-FUVS), and with our Imaging Spatial Heterodyne Spectrometer (HUBE-ISHS). ISHS co-inventors are HUBE Co-Investigators John Harlander and Fred Roesler (Harlander and Roesler 1990).

In Figure 6, I show observations of the Cygnus Loop (again simulations, of course) with our Extreme Ultraviolet Spectrometer (HUBE-EUVS), and with our X-ray Spectrometer (HUBE-CUBIC). CUBIC is the creation of Dr. David Burrows (Burrows 1996).

So, the HUBE complex of instruments can be seen to provide powerful diagnostics for celestial hot plasmas! In addition, HUBE offers strong synergism with the FUSE, HST, and GALEX missions. FUSE observes OVI in absorption; HUBE observes OVI in emission; together, they determine the filling factor. The same is true for molecular hydrogen fluorescence. GALEX, because it uses a grism, can carry out *no* spectroscopy of the diffuse background radiation, and (like HST) has no capacity of any kind below Lyman Alpha. Only our imager, HUBE-UVI, overlaps with GALEX: but HUBE-UVI, at f/1.7, has 15 times the diffuse-background sensitivity of f/6 GALEX (which is optimized to detect point sources against the diffuse background). It is, of course, our hope that NASA

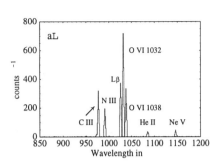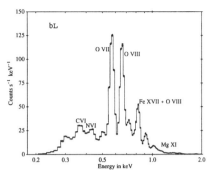

Figure 6. a) A simulated one-hour exposure with HUBE-EUVS of a 1.5-arc-minute by 2.7-arc-minute section of the Cygnus Loop nebula, showing strong OVI emission, with the two components resolved. Note that terrestrial Lyman β emission is small compared with the nebular emission, and that the 3 Å resolution of EUVS is enough to clearly separate OVI from Lyman β. b) Simultaneously, HUBE-CUBIC obtains a soft X-ray spectrum of the Cygnus Loop. (This simulation was created by David N. Burrows.)

will select HUBE for implementation as the next Astrophysics MIDEX mission: in 1998 December, NASA is scheduled to pick two MIDEX finalists, and in mid-1999 to select, from these, a single mission for implementation.

Acknowledgments. I am grateful to the HUBE science team: Richard D. Belian, Luciana Bianchi, Jeffrey J. Bloch, Joel Bregman, David N. Burrows, Renyue Cen, Philip A. Charles, Jane Charlton, Mark Clampin, Eric Conway, Donald P. Cox, Laura Danly, Richard J. Edgar, Eric Feigelson, Holland Ford, Ernest Hammond, Gordon P. Garmire, John M. Harlander, Andrew D. Holland, Peter Jakobsen, Randy A. Kimble, Jeffrey Kruk, Dan McCammon, Jayant Murthy, David A. Neufeld, John H. Nousek, Robert W. O'Connell, Larry J. Paxton, Keith Peacock, William C. Priedhorsky, John C. Raymond, Ronald J. Reynolds, Fred L. Roesler, Diane Roussel-Dupré, Robin Shelton, Mark Skinner, Oswald H. W. Siegmund, Barham W. Smith, Stefan Thonnard, Martin J. Ward, Robert S. Warwick, and Richard Willingale. I am also most grateful for the support of the US National Aeronautics and Space Administration (NASA)!

References

Bernstein, R. A. 1997, PhD Thesis, CalTech
Boldt, E. 1987, Physics Reports, 146, 215
Bowyer, C. S. 1991, ARAA, 29, 59
Bowyer, C. S., Field, G. B., & Mack, J. E. 1968, Nature, 217, 32
Burrows, D. N. 1996, CUBIC Instrument Handbook, PSU (publication available from http://www.astro.psu.edu/xray/cubic/papers/handbook/)

Davidsen, A. F., Shulman, S., Fritz, G., Meekins, J. F., Henry, R. C., & Friedman, H. 1972, ApJ, 177, 629

Disney, M. J. 1976, Nature, 263, 573

Feldman, P. D., Davidsen, A. F., Blair, W. P., Bowers, C. W., Durrance, S. T., Kriss, G. A., Ferguson, H. C., Kimble, R. A., & Long, K. S. 1992, Geophys. Res. Lett., 19, 453

Harlander, J. M., & Roesler, F. L. 1990, Proc. SPIE, 1235, 622

Hauser, M. G., Arendt, R. G., Kelsall, T., Dwek, E., Odegard, N., Weiland, J. L., Freudenreich, H. T., Reach, W. T., Silverberg, R. F., Moselely, S. H., Pei, Y. C., Lubin, P., Mather, J. C., Shafer, R. A., Smoot, G. F., Weiss, R., Wilkinson, D. T., & Wright, W. L. 1998, ApJ (Letters), submitted

Henry, R. C., Murthy, J., Ford, H., Peacock, K., Burrows, D. N., Smith, B. W., & Bloch, J. J. 1998, BAAS, 30, 926

Henry, R. C. 1998, in Multifrequency Behaviour of High Energy Cosmic Sources, F. Giovannelli & L. Sabau-Graziati (eds.), Mem. S.A.It. Vol. 69 (in press)

Henry, R. C. 1997, in Advances in Stellar Evolution, ed. R. T. Rood & A. Renzini, Cambridge Contemporary Astrophysics (Cambridge: CUP), p. 337

Henry, R. C. 1995, in The Physics of the Interstellar Medium and Intergalactic Medium, ASP Conference Proceedings, 80, ed. A. Ferrara, C. Heiles, C. McKee, & P. Shapiro, (San Francisco: ASP), p. 561

Henry, R. C. 1991, ARAA, 29, 89

Henry, R. C., Fritz, G., Meekins, J. F., Friedman, H., & Byram, E. T. 1968, ApJ (Letters), 153, 11

Henry, R. C., Fritz, G., Meekins, J. F., Chubb, T., & Friedman, H. 1971, ApJ (Letters), 163, 73

Henry, R. C., & Murthy, J. 1995, in Extragalactic Background Radiation, ed. D. Calzetti, M. Fall, M. Livio, & P. Madau, (Cambridge: Cambridge University Press), 51

Holberg, J. B. 1986, ApJ, 311, 969

McCammon, D., & Sanders, W. T. 1990, ARAA, 28, 657

Mendenhall, J. A., & Burrows, D. N. 1998, in preparation

Murthy, J., Hall, D., Earl, M., Henry, R. C., & Holberg, J. B. 1998, ApJ, submitted

Murthy, J., Henry, R. C., & Holberg, J. B. 1991, ApJ, 383, 198

O'Connell, R. W. 1987, AJ, 94, 876

Sciama, D. W. 1998, A&A, 335, 12

Vogeley, M. S. 1997, BAAS, 29, 1207; and ApJ, submitted

Wang, Q. D., & McCray, R. 1993, ApJ (Letters), 409, 37

Witt, A. N., & Petersohn, J. K. 1994, in ASP Conference Series 58, The First Symposium on the Infrared Cirrus and Diffuse Interstellar Clouds, ed. R. M. Cutri & W. B. Latter (San Francisco: ASP), 91

Detecting the Low Surface Brightness Universe: the Extragalactic Background Light and LSB Galaxies

P. Väisänen

Center for Astrophysics, 60 Garden St, Cambridge, MA 02138 and Observatory, P.O.B. 14, 00014 University of Helsinki, Finland

E. V. Tollestrup

Center for Astrophysics, 60 Garden St, Cambridge, MA 02138

Abstract.
Individual sources are detected ideally to a confusion limit at a given wavelength, but there is still much information beyond that. Absolute background brightness measurements provide a crucial constraint to models describing the undetected population of sources in the distant – and/or low surface brightness – universe. We calculate how much low surface brightness galaxies (LSB) would contribute to the the overall extragalactic background light (EBL) and review the status of EBL measurements in the optical and infrared wavelengths. To be able to push deeper the detection limits on very faint sources a fluctuation analysis method is introduced. The use of these different approaches together is essential when studying the very faint and low surface brightness universe.

1. Introduction

During the past few years the Hubble Space Telescope and the new generation of 8-meter class ground based telescopes have dramatically quickened the pace of detecting ever fainter and more distant sources in the universe. Consequently the study of galaxy evolution has all but entered into a renassaince era. However, when analyzing these faintest sources, the question still remains: are we truly detecting everything that there is to detect? For example, are the dark areas in the Hubble Deep Field (HDF) really empty? Are we underestimating the light which we do see?

One possibility is that some of the light is absorbed by dust along the line of sight and reradiates in another part of the spectrum. This problem is best studied in the far infrared and submillimeter wavelengths and, indeed, has recently produced some exciting results with SCUBA and ISO (Hughes et al. 1998, Barger et al. 1998, Elbaz et al. 1998). On the other hand the light might still be in the observed waveband, but either severely underestimated or just not detected as objects. This can well arise because of the *low surface brightness* nature of objects.

However, even if the light from individual sources is underestimated for some reason, it is counted in the *total* light coming from that region of sky.

The study of the integrated background has been very fruitful (in the X-ray, for example) and is now beginning to bear results in the far infrared using the COBE data (eg. Guiderdoni et al. 1997, Hauser et al. 1998). In the optical and near infrared few results have so far been obtained due to the relative faintness of the extragalactic background sky (eg. Mattila 1990, Bernstein 1998). Figure 1 summarizes the current observational status from UV to mid-IR.

In the following sections we discuss the detection of individual galaxies including LSB galaxies, the LSB contribution to the EBL, and fluctuation analysis methods to analyze confusion limited images.

2. Detecting galaxies

Counting galaxies is a tricky business. Regardless of the source extraction method used, incompleteness due to noise characteristics and overlapping galaxies must be corrected, as well as compensating the photometry for aperture size and isophotal limit effects. All of these effects become all the more complicated in the presence of LSB's (Dalcanton 1998).

To study the completeness effects of detecting faint galaxies we produced four different simulated images (seeing FWHM = 2.5 pixels) that include the same input source counts, but have different faint end cut-offs. What information can be extracted from these ideal (infinite integration time) images? Fig. 2 shows the differential extracted number counts (BEST magnitudes, SExtractor v.2.0.8; Bertin & Arnouts 1996). It is evident that the images are confusion limited, because the same number of sources are detected in all four images, even though there are three orders of magnitude more input sources in image d than there are in image a.

The dotted vertical lines show the flux levels where the areal density of objects equals 50, 25, and 1 beams per source. Incompleteness sets in at about 50 beams per source and at 25 the observed counts are already 70 % incomplete. The exact behaviour of the completeness limit is complex and depends on the slope of the intrinsic source counts in addition to sky noise and the extraction algorithm.

Realistic selection effects on galaxy surveys including LSB effects have been investigated by eg. Davies et al. (1994), Ferguson & McGaugh (1995), and Dalcanton (1998). Ferguson & McGaugh present models A and B where the LSB population is assumed to have either of the following properties: model A) central surface brightness μ_0 is not correlated with luminosity L, or model B) μ_0 decreases with L (constant size relation). The first indicates that there are large, luminous LSB galaxies in the local universe, which would go undetected in galaxy surveys. This assumption increases modestly the normalization of the luminosity function (LF) of galaxies. With the second assumption one can easily hide large *numbers* of faint LSB's beyond detection limits – hence the result is a steep faint end slope for the LF. For case B, Ferguson & McGaugh find that even though the intrinsic normalization of LF more than doubles and the faint end slope steepens to -1.8, the observed number counts would still be consistent with current observations.

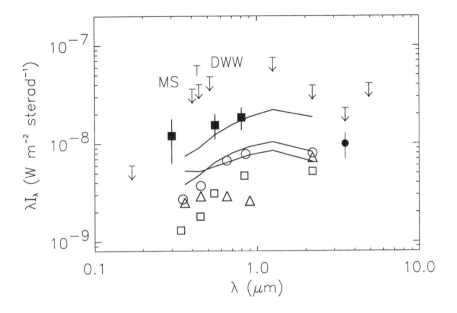

Figure 1. EBL in the UV to mid-IR wavelengths. Lower limits to EBL can be obtained by integrating observed number counts of galaxies: Tyson (1995, triangles), Cowie (1994, open squares), and the deepest counts to date from the HDF (Pozzetti 1998, circles). Models of different galaxy populations are overplotted as solid lines. The two lower curves have extra populations of dwarfs or some extra luminosity evolution and the highest curve has LSB's included; see text for details. Direct measurements of the optical EBL by Bernstein (1998, and these proceedings) are shown with filled squares, and various upper limits as arrows: Mattila (1990, MS), Toller (1983, T), Dube et al. (1979, DWW), and Armand et al. (1994) in the UV. The recent tentative detection of 3.5 μm background by Dwek & Arendt (1998) using DIRBE data is shown as a filled circle. This measurement is dependent on an assumed K-band background: if the K-band EBL would be at the highest model curve, the 3.5 μm measurement would rise by a factor of 1.4. The arrows in the IR are DIRBE data from Hauser et al. (1998). There have also been attempts to estimate the EBL using indirect methods via fluctuations of the background: Vogeley (1998) arrived at basically the same value as the integrated HDF point at 0.8 μm, and Kashlinsky et al. (1996) NIR estimations from DIRBE-data lie just above the direct DIRBE upper limits.

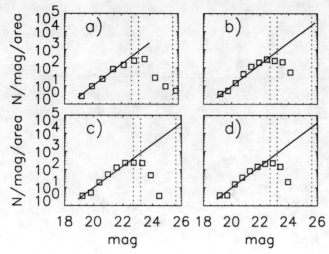

Figure 2. Observed differential number counts (squares) for four different input source counts (solid line). The intrinsic slopes are all euclidian, but the faint end cut-offs are different – in c and d the cut-offs are at 29 and 30.5 mag, respectively. The dotted lines refer to surface densities of objects; see text.

To show the effects of LSB sources in galaxy counts, we have constructed two simulated deep images of a populaton of disk galaxies. The model is simple: galaxy counts with a slope of $d(\log(N))/d(\mathrm{mag}) = 0.4$ have been turned into an image using a relation between angular size and apparent magnitude appropriate for pure disks

$$r_0 \propto 10^{(-0.12 \times m)}.$$

The relation is normalized using data in Roche et al. (1995). Another image is made using the same relation, but multiplying the scale size r_0 of *each* galaxy by three. This produces LSB galaxies which have central surface brightnesses 2.2 magnitudes lower than the corresponding galaxies in the first image. Thus the model mimics the LSB galaxies of model B of Ferguson & McGaugh.

Figure 3 shows the extracted number counts from each case, compared to the input source counts. There is a 2 magnitude difference in the completeness level between the simulated 'normal' disks and LSB disks. In terms of total number of galaxies, 75% of the galaxies detected in the first simulation were *not* detected after they were turned into LSB galaxies. This fraction naturally depends on the slope of the intrinsic counts, but nevertheless the incompleteness is very significant. Without careful consideration of selection effects due to surface brightness any galaxy catalog will remain suspect.

Note that these are purely confusion limited images with no sky or system noise added. We find an excess brightening of sources near the confusion limit, contrary to the usual expectation of underestimating fluxes at faintest detection levels in noise-limited images.

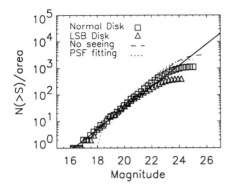

Figure 3. Cumulative number counts extracted from simulated images with normal disk galaxies only (squares) and with LSB's (triangles). The images have been convolved with a $1''$(2.5 pixel) FWHM Gaussian to include a seeing effect in the images. For comparison, the counts from a non-convolved image are shown as a dashed line, and the dotted line represents the normal disks extracted using a PSF-fitting method like DAOPHOT (intended for unresolved objects).

3. Contribution of LSB galaxies to EBL

A completeness limit will always be reached for detecting galaxies individually unless there are actual blank areas on the sky. Therefore, it would be very informative to measure the flux, ie. the EBL, coming out of these 'blank areas'. Originally the measurement of EBL was thought of as a powerful cosmological test, since the total EBL varies as a function of cosmological parameters. Later it was realized that evolutionary effects of galaxies actually affect the EBL more than those changes due to cosmology (eg. Tyson 1995, Davies et al. 1997).

The lower limit for the total EBL can be easily calculated by integrating together all the flux coming from individually detected galaxies. Figure 1 shows some of these integrations: Cowie (1994, open squares), Tyson (1995, triangles), and the HDF (compiled in Pozzetti et al. 1998, circles).

We have also plotted in Fig. 1 the EBL originating from three models of galaxy populations. The models are constructed so that they fit the available number counts (see Väisänen 1996 for details). The lowest line, dubbed 'EDP' in Väisänen (1996), includes pure luminosity evolution and some extra brightening of late type spirals looking back to $z \sim 1$, as well as a population of blue dwarf galaxies. The middle curve ('BBG') is an extreme dwarf-dominated case where all of the 'blue excess' in the faint counts is explained by blue dwarf galaxies.

It is interesting to note that the EBL from both of these galaxy models start to converge to about the level of the total light from the HDF, approximately $\nu I_\nu = 4 - 8$ nWm^{-2}str^{-1} between B and I bands. This is because the slopes of the number counts at the faintest end at all wavelengths (except, perhaps, U-band) are shallow enough that fainter sources would not contribute significantly to the EBL, even if the number counts continue to rise below the detection limits.

So, is all the light in the universe seen in the HDF? The result changes significantly only if we somehow underestimate the light coming from the galaxies. Dust is one contributing factor, but will not be discussed here. The two models introduced above are ideal in the sense that every galaxy that is brighter than a given total magnitude limit would be detected. The upper-most curve in Fig. 1 is a calculation based on model B of Ferguson & McGaugh (1995; 'FMB-LE' in Väisänen 1996). As mentioned above, this takes into account actual detection effects and results in an increase of the intrinsic normalization of the LF and steepening of the faint end of the LF.

In this extreme, but not impossible, case one finds a doubling of the EBL in the UV to near IR wavelengths compared to 'standard' ultra deep, number count fitting, galaxy models. Intriguingly this level is close to the observational results of Bernstein (1998). New galaxy population modeling by Jimenez & Kashlinsky (1998) also predict total fluxes at this level, higher than most previous models. The tentative measurement at 3.5 μm (Dwek & Arendt 1998) is also consistent given the dependancy of this detection on the K-band EBL value. Nevertheless, the highest curve in Fig. 1 can be considered as the upper limit of LSB contribution to the EBL. Generally, even a very large population of LSB's would not be expected to contribute very much to EBL because most of them are also expected to be faint. They contribute more to space density of galaxies, and if their M/L ratios are high as some studies suggest (Impey & Bothun 1997), they could contribute more significantly to the baryonic mass of the universe.

As a final example we estimate the 'lost' amount of EBL due to disk galaxy and LSB selection effects, for the models shown in Fig. 3. The integrated EBL from the differential number counts corresponding to these curves can be calculated using

$$I_{\text{EBL}} \propto \int N(m) 10^{-0.4m} dm.$$

If one would take the turnover of detected sources to be intrinsic, and not a selection effect, one would estimate the EBL to be 72% and 25% of the true EBL in the simulated disk-galaxy and LSB-disk dominated universes, respectively. Of course these models are extreme since *only* LSB's (or only disks) were present, but on the other hand, many LSB's are expected to be of much lower surface brightness than the 2 mag/"difference of this simulation.

4. Fluctuation analysis

Until a consensus value for EBL is reached at a given wavelenght, there is motivation to try to push the source counts deeper than the detection limits. This is possible by using fluctuation analysis, that is, using the statistical properties of the background sky produced by galaxies that are not detected individually.

One such fluctuation analysis method is the $P(D)$-analysis, which has been succesfully used in radio and x-ray wavelengths (Scheuer 1974, Condon 1974). It is most sensitive to sources at about one beam per souce level. We have investigated a different technique using the variances of pixels in an image. As shown in Väisänen & Tollestrup (1998), it can be a useful tool in deriving number counts from the one beam per source level to flux levels brighter than the confusion

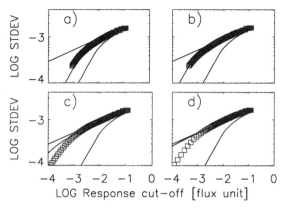

Figure 4. 'Confusion curves' for the same models as in Fig. 2. The standard deviation of all pixels in an image below a cut-off pixel value (or response) is plotted against that cut-off value. By modeling this curve, either analytically or by a Monte Carlo simulation, one is able to fit the observed confusion curve and find models which are consistent with the data. See text for description of the solid curves.

and completeness limits, ie. bridging the number counts between $P(D)$ analysis and direct source counts.

For the same models as in Fig. 2, Fig. 4 shows a plot (squares) of the standard deviation of all pixels in an image below a cut-off flux density. The middle solid curve represents the analytically calculated relation using the same input counts (the differences are due to some specific simplifications, eg. the pixel size was not taken into account in this analysis). The upper-most curve shows how the 'confusion curve' behaves if the model source count slope would continue to infinity, and the lowest curve is the case if it would roll over at the magnitude where the extracted counts turn sharply down (see Fig. 2).

Without going into details, it should be clear that in its simplest form one can generate a set of simulated images with differing faint flux cut-offs and find the best fitting model to the data. The 'observed' curves in panels c and d cannot be separated from each other, but since case b still has a unique curve, we can predict the source count cut-off at least down to the 1 source per beam level. Obviously the slope of the counts affects the confusion curve, so that can be fitted as well. Our simulations suggest that slope variations of about 0.03 (log N - mag slope) can be easily differentiated with this method down to one source per beam flux levels.

5. Summary

Detecting individual galaxies is practical to a surface density of 25-50 beams per source. However, photometry above this limit can still be severely biased due to surface brigtness and confusion effects, especially in the presence of LSB's. On the other hand, an accurate EBL measurement would yield a powerful constraint on the nature of faint extragalactic sources. The maximum LSB contribution to the EBL is estimated to be about the same amount as from normal galaxies. Until a definite value for EBL in optical and near infrared wavelengths is measured, fluctuation analysis methods are very useful in trying to constrain the properties of the faintest universe.

Acknowledgments. PV acknowledges support from the Finnish Academy and the Smithsonian Institution.

References

Armand, C., Milliard, B., & Deharveng, J. M. 1994, A&A, 284, 12
Barger, A. J., Cowie, L. L., Sanders, D. B., et al. 1998, Nature, 394, 248
Bernstein, R. 1998, PhD thesis, California Institute of Technology
Bertin, E. & Arnouts, S. 1996, A&A Suppl., 117, 393
Condon, J. J. 1974, ApJ, 188, 279
Cowie, L. L., Gardner, J. P., Hu, E. M., et al. 1994, ApJ, 434 114
Dalcanton, J. J. 1998, ApJ, 495, 251
Davies, J. I., Phillipps, S., Trewhella, M., & Alton, P. 1997, MNRAS, 291, 59
Davies, J. I., Phillipps, S., Disney, M., Boyce, P., & Evans, Rh. 1994, MNRAS, 268, 984
Dwek, E. & Arendt, R. G. 1998, astro-ph 9809239
Dube, R. R., Wickes, W. C., & Wilkinson, D. T. 1979, ApJ, 232, 333
Elbaz, D., Aussel, H., Desert, F. X., et al. 1998, astro-ph 9807209
Ferguson, H. C. & McGaugh, S. S. 1995, ApJ, 440, 470
Guiderdoni, B., Bouchet, F. R., Puget, J.-L., Lagache, G., & Hivon, E. 1997 Nature, 390, 257
Hauser, M. G., Arendt, R. G., Kelsall, T., et al. 1998, ApJ, in press
Hughes, D. H., Serjeant, S., Dunlop, J., et al. 1998, Nature, 394, 241
Impey, C. & Bothun, G. 1997, ARA&A, 35, 267
Jimenez, R. & Kashlinsky, A. 1998, astro-ph 9802337
Kashlinsky, A., Mather, J. C., & Odenwald, S. 1996, ApJ, 473, L9
Mattila, K. 1990, in Bowyer S., Leinert, Ch. (eds.) Galactic and Extragalactic Background Radiation, IAU Symp. 139, Kluwer, Dordrecht, p. 257
Pozzetti, L., Madau, P., Zamorani, G., Ferguson, H. C., & Bruzual, G. A. 1998, MNRAS, 298, 1133
Roche, N., Ratnatunga, K., Griffiths, R. E., Im, M., & Neuschaefer, L. 1996, MNRAS, 282, 820
Scheuer, P. A. G. 1974, MNRAS, 166, 329
Toller, G. N. 1983, ApJ, 266, L79
Tyson, J. A. 1995, in D. Calzetti, M. Livio, P. Madau (eds.), Extragalactic Background Radiation, Cambridge University Press, p. 103
Vogeley, M. 1998, these proceedings
Väisänen, P. 1996, A&A, 315, 21
Väisänen, P. & Tollestrup, E. V. in preparation

Searching for LSB - VII

Please say you are joking, it cannot be across the sea.

Cosmic Baryon Density from Primordial Nucleosynthesis and Other Evidence

B.E.J. Pagel

Astronomy Centre, CPES, Sussex University, Falmer, Brighton BN1 9QJ, UK

Abstract. I comment on primordial abundances, their implications for cosmic baryon density (all are consistent with $4 \leq \eta_{10} \leq 5$), the nature of baryonic dark matter (probably mainly hot intergalactic gas), its metallicity and implications for the overall yield. Recent estimates of the cosmic star formation rate as a function of red-shift are discussed in the light of the present-day densities of stars and heavy elements.

1. Introduction

It has been recognised for a long time that a 'low' primordial deuterium abundance of a few $\times 10^{-5}$ implies the existence of both baryonic and non-baryonic dark matter, since $\Omega_{vis} < \Omega_{bbns} < \Omega_M$ (see Table). Some of this baryonic dark matter (BDM) could be in the form of low surface-brightness galaxies (Bristow & Phillipps 1994), but from the results reported at this and other conferences it seems that these are likely to make only a minor contribution over and above what has been taken into account already (cf. Fukugita, Hogan & Peebles 1998). In this talk I give an update on estimates of primordial abundances, which seem to be converging to well-defined limits on the baryon density, and compare that density with estimates based on other arguments. The existence of substantial quantities of BDM, now thought to reside mainly in the form of intergalactic hot gas, raises interesting questions about its metallicity and the likely existence of 'dark metals', which in turn have implications for the cosmic history of star formation which has recently undergone intensive study on the basis of optical red-shift surveys, UV drop-out galaxies in the Hubble Deep Field and elsewhere, and infra-red and sub-mm observations. The upshot is that there may well be a substantial amount of 'dark' metals residing in intergalactic gas, and comparison of this with the density of visible stars leads to estimates of the yield that are substantially higher than what one normally assumes for the solar neighbourhood.

2. Primordial abundances

2.1. Deuterium and ^{3}He

The primordial D/H ratio is the chief test of baryonic density because of its steep negative dependence thereon. Extrapolations of the well-determined value of 1.6×10^{-5} in the local interstellar medium (Linsky et al. 1993) to a primordial value give only rather vague limits: Hata et al. (1996), updating an argument by Yang et al. (1984) based on solar D + ^{3}He, give 95% confidence limits 10^5 (D/H)$_P$ = $3.5^{+2.7}_{-1.8}$. A recent detection of the D I hyperfine structure line in the Galactic anti-centre gives a D/H ratio $(3.9 \pm 1) \times 10^{-5}$ (Chengalur, Braun & Burton 1997), which is an alternative lower limit in a region that has undergone less astration than the solar neighbourhood. More precise estimates of the primordial value are available in principle from direct measurements of D/H in Lyman-limit absorption-line systems in front of quasars, which can be expected to have essentially primordial chemical composition, but only a few systems are suitable from the spectroscopic point of view and there has been some controversy with both high ($\sim 2 \times 10^{-4}$) and low values ($\sim 2 \times 10^{-5}$) reported. The best estimates for two high red-shift systems ($z = 2.5$ and 3.6) observed with the Keck Telescope give D/H = $(3.4 \pm 0.3) \times 10^{-5}$ (Burles & Tytler 1998ab), but Webb et al. (1997) have found a high value in a system with $z = 0.7$ observed with IUE and HST, as well as a low one in another system at $z \simeq 0.5$. These results have encouraged speculations that there might be cosmic variations in η (Jedamzik & Fuller 1995; Copi, Olive & Schramm 1996) or in neutrino degeneracy (Dolgov & Pagel 1997), but doubts have been raised as to whether the differences are real: Levshakov, Kegel & Takahara (1998ab) have argued that the conventional microturbulent approximation used in simulating line profiles is invalid in these systems and have found solutions involving mesoturbulence with lopsided velocity distributions that give the same D/H ratio, 4.1×10^{-5}, in all three of the prominent cases. It is possible, therefore, that the data are homing in on a universal primordial deuterium abundance somewhere between 3 and 5×10^{-5}, which implies that the baryonic density parameter η_{10} is between 4 and 6, or $0.015 \leq \Omega_b h^2 \leq 0.022$.

This result implies that D in the interstellar medium has been destroyed by a factor between 2 and 3, which is reasonable from the point of view of Galactic chemical evolution. One might then expect a corresponding increase in the abundance of ^{3}He, resulting from D destruction and some fresh production. However, the latest results on ^{3}He in Galactic H II regions (Balser et al. 1998) give a 'plateau' with abundances equal within errors to the proto-solar value ^{3}He/H $\simeq 1.5 \times 10^{-5}$ over a wide range of Galactocentric distances and metallicities indicating an apparent balance between production and destruction. Should this value be identified with the primordial one, they estimate a corresponding value of $\eta_{10} = 3.2^{+4.4}_{-1.9}$, consistent with the value derived from deuterium. The existence of 'extra mixing' in giant stars leading to significant ^{3}He destruction has substantial support from carbon isotope ratios in red giants (Hogan 1995; Charbonnel & do Nascimento 1998), and it seems that production and destruction may be nearly in balance (Sackmann & Boothroyd 1998).

2.2. Helium-4

The primordial helium abundance is not a good measure of baryon density, to which it is comparatively insensitive, but it is an important test of consistency and of N_ν, the number of relativistic neutrino types present at BBNS, which there are good grounds for believing to be 3. The most accurate estimates come from measurements of recombination emission lines in extragalactic H II regions as a function of oxygen (and nitrogen) abundance representing the heavy-element fraction Z and extrapolating linearly to $Z = 0$ (Peimbert & Torres-Peimbert 1974). Over the years, such studies have often led to a rather low value $Y_P \simeq 0.23$ (e.g. Lequeux et al. 1979; Pagel et al. 1992; Skillman & Kennicutt 1993; Skillman et al. 1994), which raised doubts in some quarters as to the viability of conventional BBNS theory (Hata et al. 1995), and to rather high values of 3 or more for $\Delta Y/\Delta Z$.

Pagel et al. (1992) noted that there could be systematic errors of various sorts and suggested a 95% confidence upper limit of 0.242; Olive, Skillman & Steigman (1997) and Hogan, Olive & Scully (1997) suggested a similar limits of 0.242 to 0.244. The assumption of such a limit has been criticised on various grounds which were mostly spurious (e.g. Sasselov & Goldwirth 1995), but the critics' conclusions now seem to have been correct if for the wrong reasons. The trouble centres in large part on the blue compact galaxy I Zw 18, where it has been shown by Izotov & Thuan (1998) and by Vílchez & Iglesias-Páramo (1998) that there was a greater effect of underlying stellar absorption lines than we had allowed for. There are also other difficulties with the lowest-metallicity H II regions arising from their high electron temperatures and the resulting sensitivity of collisional-excitation corrections to the electron density. We used densities based on [S II] (the only method available in 1992), whereas Izotov, Thuan & Lipovetsky (1997), Izotov & Thuan (1998) and Izotov et al. (1999) have produced a more homogeneous set of data than was previously available and use ratios of helium lines with different sensitivities to collisional and radiative transfer effects, resulting in $Y_P = 0.2452 \pm 0.0009$, slightly above the upper limits that were previously suggested. This is probably a better value than ours, but the error bar looks very optimistic. I should prefer to conclude that the primordial helium abundance is between 0.24 and 0.25, corresponding to η_{10} between 2.5 and 8.5, and thus quite compatible with the the deuterium result within the standard BBNS framework.

2.3. Lithium-7

The final piece of the BBNS jig-saw is ^{7}Li. Bonifacio & Molaro (1997), using improved model atmospheres, especially with regard to the effective-temperature determination from the infra-red flux method of Blackwell et al. (1990), find the Spite plateau to be remarkably well defined for [Fe/H] < -1 and $T_{\text{eff}} > 5700$ K, with $12 + \log \text{Li/H} = 2.24 \pm 0.05$ (syst.). The same authors have measured a subordinate Li I line in HD 140283, checking its concordance with the resonance doublet and thereby ruling out significant non-LTE effects (Bonifacio & Molaro 1998). Destruction factors in the stellar atmospheres are severely constrained by this tightness and by the presence of ^{6}Li in HD 84937 (Smith, Lambert & Nissen 1993), and they are estimated in any case to be less than 0.2 dex (Vauclair & Charbonnel 1998), so that it now seems reasonable to take this as the primordial lithium abundance. Owing to the bimodal dependence of lithium on η, this leads to two solutions corresponding to low η and high D/H and to high η and low D/H respectively. The low-η solution now seems unlikely in view of the results for D and ^{4}He reported above, while the high-η solution gives $3 \lesssim \eta_{10} \lesssim 5$. Combining this with the the deduction from D/H, we end up with $4 \lesssim \eta_{10} \lesssim 5$, or $0.015 \lesssim \Omega_B h^2 \lesssim 0.018$.

3. Abundances of different forms of matter

The attached Table gives estimates of the smoothed-out cosmic densities of different forms of (mostly baryonic) matter deduced from estimates of mass:light ratios combined with luminosity densities from red-shift surveys and other evidence and arguments. They are given as Ω, normalised to a Hubble parameter of 70 km s^{-1} Mpc^{-1}, which seems to be the fashionable value of the month. Figures on the left of the two columns are taken from Persic & Salucci (1992), while those on the right are from the more recent work of Fukugita, Hogan & Peebles (1998).

The two estimates of the amount of mass in the form of stars are not very different, amounting to only a tenth of the total from BBNS, and in the Persic & Salucci estimates allowing for gas does not make a great difference. At high red-shifts, a substantial part of the stellar population is accounted for by the damped Ly-α population, whereas all the baryonic matter is accounted for by ionized gas associated with the Lyman forest (Rauch et al. 1998). The subsequent fate of this (presumably largely intergalactic) ionized gas has been uncertain, but it has now been suggested that it still exists in the form of hot, rarefied intergalactic gas today (Cen & Ostriker 1998). The density of this hot intergalactic gas was previously derived on the assumption that it bears the same relation to field and group elliptical galaxies as does hot gas to elliptical galaxies in clusters (Mushotzky & Loewenstein 1997; Fukugita, Hogan & Peebles 1998), and the resulting overall density of stars plus gas is then $0.021\, h_{70}^{-1.5}$, which is within striking distance of the BBNS-derived total density of $(0.03 \text{ to } 0.04) h_{70}^{-2}$.

There is thus the possibility that the dark baryonic matter can all be accounted for, but there is another issue associated with its chemical composition. If it is primordial, then most of the 'metals' in the universe are in stars and can be accounted for on the basis of conventional galactic chemical evolution theory with overall yields comparable to solar abundances, although larger yields may be needed in cluster ellipticals to account for the metallicity of intra-cluster gas (e.g. Pagel 1997). Mushotzky & Loewenstein argue, on the other hand, that the metallicity of diffuse intergalactic gas should be the same as in the intra-cluster gas, i.e. about 1/3 solar, and in this case the yield averaged over the whole universe has to be 2.5 times larger than solar abundance.

4. Implications for overall star formation history

Cosmic star formation rates as a function of red-shift have been intensively studied in the last few years on the basis of Hα surveys, red-shift surveys and UV drop-out galaxies in the Hubble Deep Field and elsewhere. According to Madau, Pozzetti & Dickinson (1998), optical and near IR data are well accounted for by a co-moving star formation rate (SFR) in units of $h_{50} M_\odot$ yr^{-1} Mpc^{-3} rising from 0.01 at $z = 0$ to a peak of 0.1 at $z \simeq 1.5$ and then declining exponentially to reach 0.02 at $z = 5$, assuming a Salpeter IMF with a lower limit of 0.1 $M_\odot$ and SMC-type dust in a foreground screen with $E_{B-V} = 0.1$. These figures are likely underestimates by a factor of 2 or so, according to extinction corrections based on supplementary near-IR and radio observations of Canada-France red-shift survey fields by Hammer & Flores (1998) and on G−R colours of HDF UV

Inventory of cosmic baryons and 'metals'

Densities expressed as Ω, in units of $\rho_{\rm crit} = 1.54 \times 10^{11} h_{70}^2 \, M_\odot \, {\rm Mpc}^{-3}$

All baryons from BBNS ($D/H = 3.4 \times 10^{-5}$ [a])		$0.035 \, h_{70}^{-2}$
Stars in spheroids	$.0015$ [b]	$.0026 \, h_{70}^{-1}$ [c]
Stars in disks	$.0007$ [b]	$.0009 \, h_{70}^{-1}$ [c]
Total stars	$.0022$ [b]	$.0035 \, h_{70}^{-1}$ [c]
Cluster hot gas	$.0006 \, h_{70}^{-1.3}$ [b]	$.0026 \, h_{70}^{-1.5}$ [c]
Group/field hot gas	$.0002 \, h_{70}^{-1.3}$ [b]	$.014 \, h_{70}^{-1.5}$ [c]
Total stars + gas	$.003 \, h_{70}^{-0.4}$ [b]	$.021 \, h_{70}^{-1.5}$ [c]
Machos		?? [c]
Ω_Z (stars, $Z = 0.02$ [d])	4.4×10^{-5} [b]	$7 \times 10^{-5} \, h_{70}^{-1}$ [c]
Ω_Z (hot gas, $Z = .006$)	$5.0 \times 10^{-6} \, h_{70}^{-1.3}$ [b]	$1.0 \times 10^{-4} \, h_{70}^{-1.5}$ [c]
		$1.2 \times 10^{-4} \, h_{70}^{-1.3}$ [e]
Yield ρ_Z / ρ_*	$.022 \, h_{70}^{-0.2}$ [b]	$.051 \, h_{70}^{-0.3}$ [c]
Damped Ly-α		$.0015 \, h_{70}^{-1}$ [c,f]
Ly-α forest		$.04 \, h_{70}^{-1.5}$ [c,g]
Gals + DM halos ($M/L = 210 \, h_{70}$)		0.25 [c,h]
All matter ($f_B = .056 \, h^{-1.5}$)		$0.37 \, h_{70}^{-0.5}$ [c,i]

[a] Burles & Tytler 1998
[b] Persic & Salucci 1992
[c] Fukugita, Hogan & Peebles 1998
[d] Edmunds & Phillipps 1997
[e] Mushotzky & Loewenstein 1997
[f] Storrie-Lombardi, Irwin & MacMahon 1996
[g] Rauch et al. 1998
[h] Bahcall, Lubin & Dorman 1995
[i] White & Fabian 1995

drop-out galaxies by Pettini et al. (1997). Similar numbers up to $z = 1$ result from a UV-selected galaxy red-shift survey by Treyer et al. (1998).

The resulting model is one among many discussed by Blain et al. (1998), who refer to it as Peak-G (peaking at $z \simeq 2$). According to their computations, it ends up with a present-day stellar density parameter $\Omega_* h_{50}^2 = 0.01$ or $\Omega_* h_{70}^2 = 0.005$, slightly higher than the figure of 0.0035 in the Table, but not disastrously so. With a typical heavy-element yield of 0.02, this gives $\Omega_Z h_{70}^2 \simeq 10^{-4}$, which would account for 1/2 of the Ω_Z given in the Table, including diffuse intergalactic gas. There is thus a choice between (a) assuming somewhat lower abundances in the diffuse intergalactic gas, which is quite reasonable in view of the existence of relatively young dwarf galaxies with low metallicity; (b) assuming a higher stellar density than suggested in the Table (LSB galaxies, intergalactic stars?); and (c) assuming a larger yield, e.g. from a top-heavy IMF. This higher yield is probably needed anyway for elliptical galaxies in clusters, but the new evidence suggests that it could be more universal.

Far infra-red and submm observations, notably with the SCUBA detector on JCMT, together with new COBE data on the submm diffuse background, are very important as probes of high red-shift star formation in galaxies hidden by dust because of the negative K-corrections. These have so far given rise to widely divergent interpretations, however. Lilly et al. (1998) find from their studies of CFRS fields that about 1/2 of star formation and metal production occurs in obscured souces with a similar red-shift distribution to the one derived from optical data, which agrees with the peaking model discussed above. Blain et al., on the other hand, present 'anvil' models based on the submm observations which imply substantially higher star formation rates, at least for $z > 1$, resulting in predicted present-day star densities that are nearly an order of magnitude higher ($\Omega_* \simeq 0.03 h_{70}^{-2} \simeq \Omega_{bbns}$!) and in significant disagreement with counts of observable stars. Blain et al. discuss possible reasons for this disagreement, which include a contribution from AGNs and a top-heavy IMF. In the latter case, the cosmic metal density is more closely related to the measured luminosity density than is either of them to the total SFR by mass, so that the predicted metal density at the present time provides a more specific test of the models. The submm-based models discussed by Blain et al. typically give $\Omega_Z h_{70}^2 \simeq 5 \times 10^{-4}$, which is 2.5 times too much for the estimates in the Table. Putting this another way, this figure would require all baryonic matter (most of which is presumably diffuse gas) to have a mean abundance of 0.7 solar, and that does not look very likely. It is not clear whether there could be a sufficiently large AGN contribution to fill the gap.

References

Bahcall, N.A., Lubin, L. & Dorman, V. 1995, ApJ, 447, L81

Balser, D.S., Bania, T.M., Rood, R.T. & Wilson, T.L. 1998, ApJ, in press

Blackwell, D.E., Petford, A.D., Arribas, S., Haddock, D.J. & Selby, M.J. 1990, A & A, 232, 396

Blain, A.W., Smail, I., Ivison, R.J. & Kneib, J.-P. 1998, MNRAS, subm., astro-ph 9806062

Bonifacio, P. & Molaro, P. 1997, MNRAS, 285, 847
Bonifacio, P. & Molaro, P. 1998, ApJ, 500, L175
Bristow, P.D. & Phillipps, S. 1994, MNRAS, 267, 13
Burles, S. & Tytler, G, 1998a, ApJ, 499, 699
Burles, S. & Tytler, D. 1998b, astro-ph 9803071
Cen, R. & Ostriker, J.P. 1998, Science, subm., astro-ph 9806281
Charbonnel, C. & do Nascimento, J.D. Jr 1998, A & A, 336, 915
Chengalur, J.N., Braun, R. & Burton, W. B. 1997, A & A, 318, L35
Copi, C.J., Olive, K.A. & Schramm, D. 1996, astro-ph 9606156
Dolgov, A. & Pagel, B.E.J. 1997, astro-ph 9711202
Edmunds, M.G. & Phillipps, S. 1997, MNRAS, 292, 733
Fukugita, M., Hogan, C.J. & Peebles, P.J.E. 1998, ApJ, 503, 518
Hammer, F. & Flores, H. 1998, in *Dwarf Galaxies and Cosmology*, Moriond Conference, Ed. Frontières, Paris, astro-ph 9806184
Hata, N., Scherrer, R.J., Steigman, G., Thomas, D., Walker, T.P., Bludman, S. & Langacker, P. 1995, Phys.Rev.Lett, 75, 3977
Hata, N., Scherrer, D., Steigman, G., Thomas, D. & Walker, T.P. 1996, ApJ, 458, 637
Hogan, C.J. 1995, ApJ, 441, 17
Hogan, C.J., Olive, K.A. & Scully, S.T. 1997, ApJ, 489, L119
Izotov, Y.I., Chaffee, F.H., Foltz, C.B., Green, R.F. & Guseva, N.G. 1999, Poster at this meeting
Izotov, Y.I., Thuan, T.X. & Lipovetsky, V.A. 1997, ApJS, 108, 1
Izotov, Y.I. & Thuan, T.X. 1998, ApJ, 497, 227
Jedamzik, K. & Fuller, G. 1995, ApJ, 452, 33
Lequeux, J., Peimbert, M., Rayo, J.F., Serrano, A. & Torres-Peimbert, S. 1979, A & A, 80, 155
Levshakov, S.A., Kegel, W.H. & Takahara, F. 1998a, A & A Lett., 336, L29
Levshakov, S.A., Kegel, W.H. & Takahara, F. 1998b, ApJ, 499, L1
Lilly, S.J., Eales, S.A., Gear, W.K., Bond, J.R., Dunne, L., Hammer, F., Le Fèvre, O. & Crampton, D. 1998, in 34th Liège Astrophysics Coll.:*NGST: Science and Technological Challenges*, ESA Publ., astro-ph 9807261
Linsky, J.L., Brown, A., Gayley, K. et al. 1993, ApJ, 402, 694
Madau, P., Pozzetti, L. & Dickinson, M. 1998, ApJ, 498, 106
Mushotzky, R.F. & Loewenstein, M. 1997, ApJ, 481, L63
Olive, K.A., Skillman & Steigman, G. 1997, ApJ, 483, 788
Pagel, B.E.J., Simonson, E.A., Terlevich, R.J. & Edmunds, M.G. 1992, MNRAS, 255, 325
Peimbert, M. & Torres-Peimbert, S. 1974, ApJ, 193, 327
Persic, M. & Salucci, P. 1992, MNRAS, 258, 14P

Pettini, M., Kellogg, M., Steidel, C.C., Dickinson, M., Adelberger, K.L. & Giavalisco, M. 1997, in J.M. Shull & C.E. Woodward (eds.), *Origins*, ASP Conf. Ser., astro-ph 9708117

Rauch, M., Miralda-Escudé, J., Sargent, W.L.W. et al. 1998, ApJ, 489, 1

Sackmann, I.-J. & Boothroyd, A.I. 1998, ApJ, in press, astro-ph 9512122

Sasselov, D. & Goldwirth, D. 1995, ApJ, 444, L5

Skillman, E.D. & Kennicutt, R.C. Jr 1994, ApJ, 411, 655

Skillman, E.D., Terlevich, R.J., Kennicutt, R.C., Jr, Garnett, D. & Terlevich, E. 1994, ApJ, 431, 172

Smith, V.V., Lambert, D.L. & Nissen, P.E. 1993, ApJ, 408, 262

Storrie-Lombardi, L.J., Irwin, M.J. & MacMahon, M.J. 1996, MNRAS, 283, L79

Treyer, M.A., Ellis, R.S., Milliard, B., Donas, J. & Bridges, T.A. 1998, MNRAS, subm., astro-ph 9806056

Vauclair, S. & Charbonnel, C. 1998, ApJ, in press

Vílchez, J.M. & Iglesias-Páramo, J. 1998, in *Abundance Profiles: Diagnostic Tools for Galactic History*, D. Friedli, M. Edmunds, C. Robert & L. Drissen (eds.), ASP Conf. Series no. 147, p. 120

Webb, J.K., Carswell, R.F., Lanzetta, J.M. et al. 1997, Nature, 388, 250

White, S.D.M. & Fabian, A.C. 1995, MNRAS, 273, 72

Yang, J., Turner, M.S., Steigman, G., Schramm, D.N. & Olive, K.A. 1984, ApJ, 281, 493

Chemical Constraints, Baryonic Mass and the Chemical Evolution of Low Surface Brightness Galaxies

M.G. Edmunds
Department of Physics and Astronomy, Cardiff University, P.O. Box 913, Cardiff, CF2 3YB, U.K.

Abstract. A brief review of primordial helium and deuterium abundances suggests a baryonic mass density of $\Omega_B \approx 0.04 - 0.045$ (for $H_o = 70$). This mass may be dominated by intergalactic gas in clusters and groups of galaxies. The observed low chemical abundances in evolved dwarf galaxies might suggest that outflow was the origin for such gas, and we make general suggestions for the interpretation of the data from the next generation of X-ray spectroscopic satellites. The effects of both outflow and inflow on the chemical evolution of galaxies is discussed, particularly in the context of low surface brightness galaxies, and we comment on their dust content.

1. Baryonic Mass

Pagel *et al* (1992) attempted to derive an accurate abundance for primordial helium, but neglected one systematic effect which may well have a significant influence. As pointed out by Izotov *et al* (1997) and Vilchez & Iglesias-Páramo (1998) the neglect of underlying helium absorption lines of the stellar population in HII regions can lead to a slight underestimate of the strength of the helium emission lines, thus underestimating the helium abundance. A reasonable correction suggests a value of the primordial helium fraction closer to 0.24 than the Pagel *et al* value of 0.228. The higher implied baryonic density is indicated in Figure 1. I think everyone aknowledges that Y must be less than 0.25, and so the helium abundances come into better agreement with the deuterium data. For the latter, Tosi *et al* (1998) give a lower limit on astration destruction from detailed numerical chemical evolution of the Galaxy which agrees quite well with analytical limits (Edmunds 1994), and the observed Galactic deuterium abundance provides an upper limit. Tytler's qso absorption-line measurements lie nicely in between (Tytler, Fan & Burles 1996, scaled to $H_o = 70$; a slightly lower value of 0.039 may be indicated by more recent work - Burles & Tytler 1998). Looking at Figure 1 suggests that a value of Ω_B between 0.04 and 0.05 would fit the He and D data - I hesitate to include lithium because of the difficulties associated with its interpretation (but see Pagel in this volume or his book 1997). We are assuming $H_o = 70$, but the value of say 0.040 - 0.045 for Ω_B would be multiplied by $(H_o/70)^{-2}$ for other values of H_o.

In the lower part of Figure 1, we give some estimates of "observed" baryonic mass. It would be better to say "inferred" rather than "observed", since the argument is rather indirect and must proceed through various assumptions about

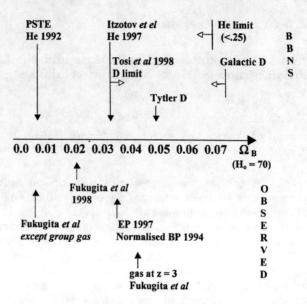

Figure 1.   Baryonic Mass Estimates

mass-to-light ratios and so on. The lowest point here is from a useful compilation by Fukugita, Hogan & Peebles (1998). This low point includes intergalactic gas in clusters of galaxies (where it dominates the baryonic mass, comprising perhaps three times the total baryonic mass actually *in* galaxies) but neglecting any intergalactic gas in groups. What Fukugita *et al* point out is that if groups had an intergalactic gas fraction similar to that in clusters it would dominate the baryonic mass, and push the point to the right as shown. Now by stretching mass-to-light ratios etc, it is possible to push the inferred galaxies' mass further to the right, as indicated by the point EP 1997 - which represents using the galaxy numbers etc detailed in Edmunds & Phillipps (1997) with the extreme mass-to-light ratios of Bristow & Phillipps (1994). This is probably a bit of an overestimate for the galaxy mass, and it includes cluster gas, but not group gas. However, it is evident that it is almost reaching the Ω_B implied by the big-bang nucleosynthesis results, and could certainly do so with judicious addition of gas in groups. A final point is to note the (*very* uncertain) baryonic mass of gas at redshift 3, as implied from qso absorption clouds - and presumably representing what will subsequently become galaxies or remain as intergalactic gas. So we see that there is a sort of consistent picture with $\Omega_B \approx 0.04 - 0.045$, but that it probably implies the existence of considerable intergalactic gas, much of which could be in groups. This group gas, being in a smaller potential well and rather cooler than cluster gas, might well have escaped significant detection so far - but be a prime target for the next generation of X-ray telescopes. We now speculate on its origin and composition.

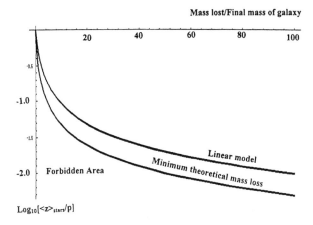

Figure 2.   Inevitable Mass Loss for Low-Metallicity Galaxies

2. Galactic Mass Loss and Intergalactic Gas

The possibility of global mass loss influencing the chemical evolution of galaxies, and enriching the intergalactic medium - particularly in clusters - has been discussed for many years. What motivates my comments is the realisation that it is easy to give (Edmunds 1999) a *lower* limit on the gas mass flowing out of galaxies, under certain (fairly standard and reasonable) conditions. The conditions are the assumptions usually made in analytic chemical evolution modelling - instantaneous recycling, that the gas in the galactic system is well-mixed (i.e. any outflow is not preferentially metal-enhances by, for example, direct outflow of supernova products), and that the intrinsic chemical yield (mainly constrained by the initial mass function for star formation) does not vary. Then a system which has lost mass, so that only a fraction M of its initial mass remains when all the gas has been converted into stars, must have a mass-weighted stellar metallicity equal to or *greater* than $p[1-(1-1/M)\text{Ln}(1-M)]$. Thus if we see a gas-exhausted system with a low mean stellar metallicity $<z>_{stars}$ it *must* have undergone considerable mass loss - unless one or more of the assumptions is invalid. The implied gas loss is shown in Figure 2, which is a plot of the above equation. It also shows the even higher mass loss implied for the popular "linear" outflow model in which the outflow rate is simply proportional to the star formation rate. Now suppose (following references in Phillipps & Edmunds 1996) that we assume a metallicity-luminosity relation for galaxies - for which there is reasonable observational evidence - of the form $z(L) \alpha L^k$ where $k\sim0.5$ (or perhaps a bit shallower: van Zee, Haynes & Salzer's 1997 Figure 1 suggests $z \sim L^{0.36}$ for gas-rich, lsb dwarfs. Here we are really talking about gas-poor, evolved systems). Also assume a luminosity function of the form $N(L) \alpha L^\alpha$, at least for moderate and low mass galaxies, and a constant mass-to-light ratio. The metallicity-luminosity relation implies (via the mass-loss equation) the minimum mass that a galaxy of present luminosity L must have lost, and one can integrate over the luminosity function to find the ratio (minimum mass lost from galaxies)/(mass now in galaxies). A little numerical integration, or approxima-

Metals in groups and clusters
(Detailed X-ray analysis awaited!)

	High metals	Low metals
Little gas	*Metal-enhanced outflow*	*Outflow not universally important*
Much gas	*Current yields very wrong and outflow very important*	*Either (i) outflow important- if z ~ as expected or (ii) if z very low, outflow not important*

Figure 3.   "Outflow" Interpretations of Intergalactic Gas

tion, soon shows that this ratio can be of order three, if we take an L_* galaxy to have lost no mass (and hence have $<z>/p = 1$) and integrate down to, say, $0.001 L_*$. So again a picture can arise in which the intra-cluster and intra-group gas has been through the galaxies, but now dominates the baryonic mass by the right amount to give consistency with the baryonic density implied by big-bang nucleosynthesis. Rough estimates can be made of the metallicity of the gas (for details see Edmunds 1999), and it must be low-ish, perhaps 1/5 to 1/4 solar, if we stick to the standard assumption of the yield p being around solar. Of course, this and our other assumptions could be wrong - in particular, metal-enriched outflow might be occuring. The outflowing gas may mix with unenriched gas that never went through galaxies, although we might then start predicting too *much* intergalactic gas if there is significant outflow! Here abundances would be very low. We can make a rough table (Figure 3) of what we would conclude from future X-ray observations if they show little gas or lots, and its metallicity.

3.   Chemical Evolution of LSB Galaxies

Is the chemical evolution of Low Surface Brightness (LSB) galaxies much the same as for high surface brightness (HSB) galaxies? As an initial try at answering this question, Figure 4 is a plot of gas metallicity versus gas fraction for

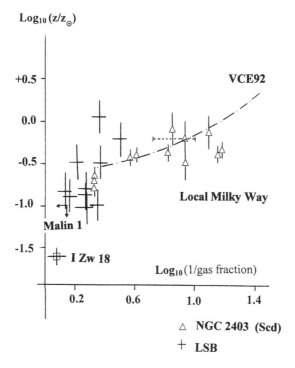

Figure 4.   Gas Abundances versus Gas Fraction

both LSBs and HSBs. On such a diagram the simple "closed box" model for chemical evolution would have z/p = Ln(1/gas fraction), and it is possible to show (Köppen & Edmunds 1999, Edmunds 1999) that nearly any type of inflow - particularly time-decreasing inflow - does not cause a major displacement from this line. To move away significantly requires either outflow or rather sudden, rapid accretion onto a well-evolved system. It is quite easy (within the errors) to pass a single simple model curve through most of the points on Figure 3 - so *there is no evidence here that LSB galaxies are intrinsically different in their chemical evolution from HSB galaxies.* Van Zee, Haynes & Salzer (1997) reach a similar conclusion. On the diagram the crosses are LSB data mainly taken from McGaugh (1994), McGaugh & de Blok (1997). The Malin I point is *not* the nucleus whose spectrum does not look particularly metal poor, the I Zw 18 abundance is from Pagel *et al* 1992 with gas fraction from data in van Zee *et al* 1998 with star mass from luminosity rather than dynamics. The triangles are the Scd spiral NGC 2403 from Garnett *et al* (1997). The dash-dot line is a mean reation for spiral disks from Vila-Costas & Edmunds (1992), and the local Milky Way (i.e. solar neighbourhood) is marked as a short horizontal dashed line.

Since LSB galaxies apparently show less obvious spiral structure than HSB galaxies, one might wonder if they might *not* show similar radial abundance gradients - since the organisation of star formation by spiral structure is a good candidate mechanism for gradient generation (e.g. Wyse & Silk 1989). Both spiral structure and abundance gradients seem to die in galaxies fainter than $M_V \sim -17$

(Edmunds & Roy 1993), and it would be fascinating to know if LSB galaxies with absolute magnitudes brighter than this also show a lack of both attributes. Thijs van der Hulst and his colleagues have started looking at this problem (e.g. de Blok & van der Hulst 1998), but more data is needed before any conclusions can be drawn.

There is, of course, the apparent observational correlation of abundance with surface brightness (e.g. Ryder 1995, and Skillman in this volume) - but its true origin is still unclear, although self-regulation of star formation is a possibility. It is important to make sure that any observed correlation is not simply due to gas fraction (i.e. a lower star formation rate and consequently less heavy elements) rather that some special coupling to other local conditions. I do wonder if some of the "second parameter" behaviour in the luminosity/metallicity relation mentioned in Evan Skillman's paper (this volume) might simply arise from different gas fractions.

4. Dust

I hesitate to say much about dust in LSB galaxies, except to emphasise that the formation/destruction cycle for dust is probably not yet understood in *any* type of galaxy - although there are indications of progress (Dwek 1998, Tielens 1998). Steve Eales (Edmunds & Eales 1998) and I have tried to set some very elementary limits on dust masses in galaxies from simple abundance constraints, assuming that an approximately fixed fraction of interstellar metals condense into/onto dust. If metal abundance is really the controlling factor, then LSBs should be no different in dust content (for a given gas fraction) than HSBs. If it is other mechanisms that dominate - e.g. destruction rate through supernova explosions, growth rate of dust grain mantles (which is where the bulk of mass lies) in dense clouds - then LSBs might well have different dust content, if the supernova rate or interstellar cloud mass/number/density spectrum is significantly different. We are beginning work to try and understand the dust creation-destruction cycle better, and any (even qualitative) predictions lie only in the future - but comparisons between LSB and HSB galaxies will provide a useful testbed for ideas.

5. Final Thought

One of the participants at this meeting rather nicely described a low surface brightness galaxies as "a big baryonic gas bag" - I fear that after this talk, he'll apply the description to me....

References

Bristow, P.D. & Phillipps, S. 1994, MNRAS, 267, 13
Burles, S. & Tytler, D. 1998, ApJ, 507, 732
de Blok, W.J.G. & van der Hulst, J.M. 1998, A&A, 335, 421
Dwek, E. 1998, ApJ, 501, 643

Edmunds, M.G. 1994, MNRAS, 270, L37
Edmunds, M.G. 1999, MNRAS *in preparation*
Edmunds & Eales, S.A. 1998, MNRAS, 299, L29
Edmunds, M.G. & Phillipps, S. 1997, MNRAS, 292, 733
Edmunds, M.G. & Roy, J-R., 1993, MNRAS, 261. L17
Fukugita, M., Hogan, C.J. & Peebles, P.J.E. 1998, ApJ, 503, 518
Garnett, D.R., Shields, G.A., Skillman, E.D., Sagan, S.P. & Dufour, R.J. 1997, ApJ, 489, 63
Izotov, Y.I., Thuan, T.X. & Lipovetski, V.A. 1997, ApJS, 108, 1
Köppen, J. & Edmunds, M.G. 1999, MNRAS, *in press*
McGaugh, S.S. 1994, ApJ, 426, 135
McGaugh, S.S. & de Blok, W.J.G. 1997, 481, 689
Pagel, B.E.J. 1997 *Nucleosynthesis and the Chemical Evolution of Galaxies*, Cambridge Univesity Press.
Pagel, B.E.J., Edmunds, M.G., Simonson, E.A. & Terlevich, R.J. 1992, MNRAS, 255, 325
Phillipps, S. & Edmunds, M.G. 1996, MNRAS, 281, 362
Ryder, S.D. 1995, ApJ, 444, 610
Tielens, A.G.G.M. 1998, ApJ, 499, 267
Tosi, M., Steigman, G., Matteucci, F. & Chiappini, C. 1998, ApJ, 498, 226
Tytler, D., Fan, X.M. & Burles, S. 1996, Nature, 381, 207
van Zee, L., Haynes, M.P. & Salzer, J.J. 1997, AJ, 114, 2497
van Zee, L., Westpfahl, D., Haynes, M.P. & Salzer, J.J. 1998, AJ, 115, 1000
Vila-Costas, M.B. & Edmunds, M.G. 1982, MNRAS 259, 121
Vilchez, J.M. & Ingesias-Páramo, J. 1998, in *Abundance Profiles: Diagnostic Tools for Galaxy History, ASP Conference Series, Vol 147 eds D. Friedli, M.G. Edmunds, C. Robert & L. Drissen* p120
Wyse, R. & Silk, J. 1989, ApJ, 339, 700

The Low Surface Brightness Universe, IAU Col. 171
ASP Conference Series, Vol. 170, 1999
J. I. Davies, C. Impey and S. Phillipps, eds.

Helium Abundance in the Most Metal-Deficient Dwarf Galaxies

Y. I. Izotov

Main Astronomical Observatory, Goloseevo, Kiev-22, 252650 Ukraine

Abstract. The high-quality long-exposure spectroscopic observations of the two most-metal deficient blue compact galaxies I Zw 18 and SBS 0335–052 are discussed. We confirm previous findings that underlying stellar absorption strongly influences the observed intensities of He I emission lines in the brightest NW component of I Zw 18, and hence this component should *not be used* for primordial He abundance determination. The effect of underlying stellar absorption, though present, is much smaller in the SE component. The extremely high signal-to-noise ratio spectrum (≥ 100 in the continuum) of the BCG SBS 0335–052 allows us to measure the helium mass fraction with precision better than 2% in nine different regions along the slit. The weighted mean of helium mass fraction in two most metal-deficient BCGs I Zw 18 and SBS 0335–052, $Y=0.2462\pm0.0009$, after correction for the He production in massive stars results in primordial He mass fraction $Y_p = 0.2452\pm0.0009$.

1. Introduction

Blue compact galaxies (BCGs) are ideal objects for the determination of primordial helium abundance and hence for determination of one of the fundamental cosmological parameters – baryon mass fraction in the Universe.

One of the important questions is how robust are measurements of He abundance in BCGs. We discuss this problem using new highest signal-to-noise ratio observations of two most metal-deficient BCGs known, I Zw 18 and SBS 0335–052. Due to the very low oxygen abundances ($Z_\odot/50$ and $Z_\odot/40$ in I Zw 18 and SBS 0335–052 respectively) the helium mass fraction in these galaxies is very close to the primordial value Y_p which we derive in this paper as mean value in two galaxies.

2. Observations

Spectrophotometric observations of I Zw 18 were obtained with the *Multiple Mirror Telescope (MMT)* on the nights of 1997 April 29 and 30. The total exposure time was 180 minutes and was broken up into six sub-exposures, 30 minutes each. The slit was oriented in the direction with position angle P.A. $= -41°$ to permit observations of both NW and SE components. The Keck II telescope optical spectra of SBS 0335–052 were obtained on 1998 February 24 with Low Resolution Imaging Spectrometer (Izotov et al. 1998). The slit

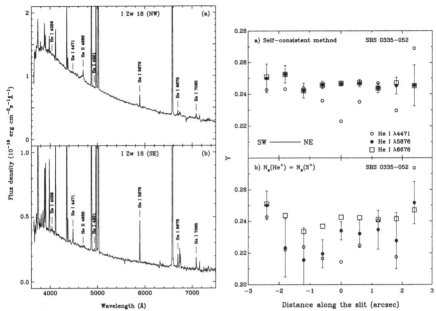

Figure 1. **Left panel:** The MMT spectra of brightest parts of the NW and the SE components of the I Zw 18. Note that all marked He I lines in the spectrum of the SE component are in emission while two He I λ4026 and λ4921 lines are in absorption in the spectrum of the NW component. **Right panel:** The spatial distributions of the helium mass fractions in the SBS 0335−052. The helium mass fractions (upper plot) are derived self-consistently from the observed He I λ3889, λ4471, λ5876, λ6678 and λ7065 emission line intensities. The helium mass fractions in lower plot are derived from the He I line intensities corrected only for collisional enhancement with electron number density N_e(S II).

was oriented in direction with position angle P.A. = 60° perpendicular to the SBS 0335−052 major axis. The total exposure time was 40 min, broken to two exposures of 30 min and 10 min.

In Figure 1a (left panel) we show one-dimensional spectrum of the NW component of I Zw 18 in its brightest part with aperture $0.6'' \times 1.5''$ which shows broad WR bumps at λ4650 and λ5808 and which have been discussed by Izotov et al. (1997). In Figure 1b (left panel) the spectrum of SE component in aperture $0.6'' \times 1.5''$ is shown at the angular distance $5.4''$ from the NW component. All He I lines in the spectrum of the SE component are in emission while two He I λ4026 and λ4921 lines are in absorption and He I λ4471 emission line is barely seen in the spectrum of the NW component. Other three He I lines marked in the spectrum of the NW component are in emission, although their intensities are reduced due to the presence of underlying stellar absorption.

Thanks to the very highest signal-to-noise ratio Keck II spectrum of SBS 0335–052 with SNR $\geq$ 100 for the continuum in its brightest part, several apertures have been applied for extraction of one-dimensional spectra. We report here the discovery of the weak WR bump, which is detected only in the brightest part of SBS 0335–052. Hence Wolf-Rayet stars are present in two most metal-deficient galaxies I Zw 18 and SBS 0335–052.

3. Helium Abundance

To derive element abundances, we have followed the procedure detailed by Izotov, Thuan & Lipovetsky (1994, 1997). He emission-line strengths are converted to singly ionized helium $y^+ \equiv \mathrm{He^+/H^+}$ and doubly ionized helium $y^{++} \equiv \mathrm{He^{++}/H^+}$. The main mechanisms deviating He I emission line intensities from the recombination values are collisional and fluorescent enhancements. In order to correct for these effects, we have adopted the following procedure: we have evaluated the electron number density N_e(He II) and the optical depth $\tau(\lambda 3889)$ in the He I $\lambda 3889$ line in a self-consistent way, so that the He I $\lambda 3889/\lambda 4471$, $\lambda 5876/\lambda 4471$, $\lambda 6678/\lambda 4471$ and $\lambda 7065/\lambda 4471$ line ratios have their recombination values, after correction for collisional and fluorescent enhancements. The very high signal-to-noise ratio Keck II observations of SBS 0335–052 allow us to derive with great precision the helium mass fraction in nine different regions in this BCG. We find that both collisional and fluorescent enhancements of He I emission lines are important in SBS 0335–052 and should be taken into account properly. It is shown (Figure 1, right panel) that electron number density derived from [S II] emission lines cannot be used due to the overcorrection of the He I emission line intensities for said effects by 5 – 10% and consequently to underestimation of He mass fraction in this BCG. When the self-consistent method is used the helium mass fractions Y derived from He I $\lambda 5876$ and $\lambda 6678$ emission line intensities are in perfect agreement in each of 9 regions while the He mass fraction derived from He I $\lambda 4471$ emission line is systematically lower due to the presence of underlying stellar absorption. The weighted mean of He mass fraction for all 9 regions of SBS 0335–052 is $Y = 0.2463\pm 0.0009$ if He I $\lambda 5876$ and $\lambda 6678$ emission lines are used. This Y value leads to primordial value $Y_p = 0.2453\pm 0.0009$, after correction for He enrichment by massive stars, and corresponds to baryon-to-photon number ratio $\eta = (4.7\pm 0.4)\times 10^{-10}$ which translates to baryon mass fraction $\Omega_b h_{50}^2 = 0.068\pm 0.006$ ($h_{50} = H_0/50$ km s^{-1}Mpc^{-1}).

References

Izotov, Y. I., Chaffee, F. H., Foltz, C. B., Green, R. F., & Guseva, N. G. 1998, in preparation

Izotov, Y. I., Foltz, C. B., Green, R. F., Guseva, N. G., & Thuan, T. X. 1997, ApJ, 487, L37

Izotov, Y. I., & Thuan, T. X. 1998, ApJ, 497, 227

Izotov, Y. I., Thuan, T. X., & Lipovetsky, V. A. 1994, ApJ, 435, 647

Izotov, Y. I., Thuan, T. X., & Lipovetsky, V. A. 1997, ApJS, 108, 1

What We Don't Know About the Universe

Chris Impey

Steward Observatory, University of Arizona, Tucson, AZ 85721, USA

Abstract.
Despite progress on many fronts in cosmology, outstanding questions remain. What is the nature of the dark matter? Is the inflationary big bang model viable? Must we accept a non-zero cosmological constant? Do we know the true population of galaxies? What is the range of star formation histories in the universe? Can gravity alone explain the large scale structure we observe? Studies of the low surface brightness universe may provide the answers to many of these questions.

1. Introduction

It was the best of times, it was the worst of times. In many ways, this is a golden age in cosmology. Large new telescopes and efficient detectors have enabled surveys of unprecedented depth and scope. There are nearly a hundred thousand published galaxy redshifts; soon there will be several million. Quasars and other AGN are being successfully used to measure diffuse baryons (via absorption lines) and the distribution of dark matter (via gravitational lensing) over 90% of the lookback time. There is good evidence that our universe is adequately described by the hot big bang model – with understandable uncertainty over the first billion years or so when gravitational collapse was nonlinear and many stellar systems formed (e.g. Peebles 1993). Supercomputer models successfully reproduce the observed basic characteristics of large scale structure.

On the other hand, we remain ignorant of important aspects of cosmology (for an overview, see Bahcall and Ostriker 1997). The need for dark matter is inescapable, but the most viable candidates for a cold dark matter particle involve new or unknown physics. We believe that gravity can explain the large scale motions and clustering of galaxies, but neither supercomputer simulations nor pure theory can account for all of the existing observations. The quest to measure the parameters of the standard model continues. This was the goal of the Palomar 200 inch telescope when it was built in the 1930s, it was a goal of the Hubble Space Telescope when it was launched nearly ten years ago, and it is a goal of current and prospective 8-10 meter telescopes and microwave satellites. Only those who fall prey to the ancient Greek sin of hubris would claim that the end of cosmology is in sight. Astronomy is an observational science, and the universe has shown an impressive ability to surprise us.

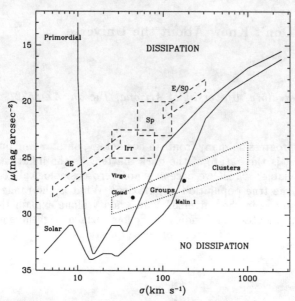

Figure 1. Collapsed structures in the universe, assuming dissipation by intercloud collisions, after Efstathiou & Rees (1983).

2. What We Don't Know About Stars

The "bread and butter" of the universe is the conversion of gas into stars. We know a lot about the modes of star formation in local galaxies all along the Hubble sequence (Kennicutt 1998). We are also getting a basic idea from deep surveys of the history of star formation in the universe (Madau, Pozzetti, & Dickinson 1998). However, it is also clear that there are dim and unrecognized repositories of baryons. At low redshift, the serendipitous discovery of Malin 1 – still that largest and most gas-rich galaxy known – is a reminder that star formation in disks can be slow and very inefficient (Impey & Bothun 1989). Figure 1 shows that the lowest surface brightness galaxies have properties quite distinct from normal stellar systems on the Hubble sequence. At high redshifts, galaxies can escape detection due to the severe effects of surface brightness dimming or due to dust obscuration. We still do not know the full range of star formation histories in the universe. The luminosity-weighted integral $\int \phi(L) L dL$ appears to converge not far below L_*, but there are indications of a steepening fainter than $M_B = -16$ (see the discussion in Impey & Bothun 1997). The most abundant stellar system in the universe is a gas-rich dwarf galaxy, examples of which have rarely been studied outside the Local Supercluster. The motivation to look for additional baryons is strong; Ω_{lum} from the integral of the galaxy luminosity function is a factor of 2-3 below the lower bound on Ω_{baryon} from nucleosynthesis arguments (Copi, Schramm, & Turner 1995; Persic & Salucci 1992; Bristow & Phillipps 1994).

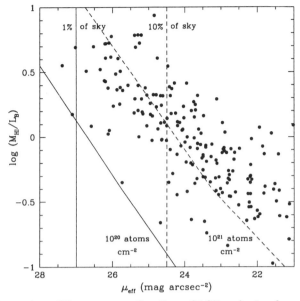

Figure 2. The gas mass fraction of LSB galaxies from the APM survey increases with decreasing surface brightness. There are selection effects against finding objects that are diffuse in either gas or stars.

3. What We Don't Know About Gas

Astronomers typically use galaxies as markers of space in many of the standard cosmological tests. Galaxies are easy to detect. Notably, the upcoming Sloan Digital Sky survey will produce digital information for about 100 million of them. All optical surveys will undercount galaxies that are diffuse or small or unevolved (in the sense of having converted a small fraction of the gas mass into stars). While cold gas can be detected effectively by the 21 cm line of neutral hydrogen, the limited bandwidth and sensitivity of "blind" radio surveys mean that they probe relatively small volumes (Schneider, Spitzak, & Rosenberg 1998). Figure 2 shows that galaxies from the APM survey (Impey et al. 1996) have a strong trend of increasing gas richness with lower surface brightness (McGaugh & de Blok 1997). In this diagram, there is strong surface brightness selection — galaxies fainter than 10% of the sky level are under-represented and galaxies fainter than 1% of the sky level do not make it into most optical catalogs. The high surface brightness galaxies have have HI column densities of a few times 10^{21} atoms cm^{-2} and the low surface brightness galaxies have HI columns ten times lower. At any particular surface brightness, radio surveys only detect the most gas-rich galaxies. Below a few times 10^{19} atoms cm^{-2}, disk star formation is inhibited (Kennicutt 1989), and we know very little about the universe at these low column densities.

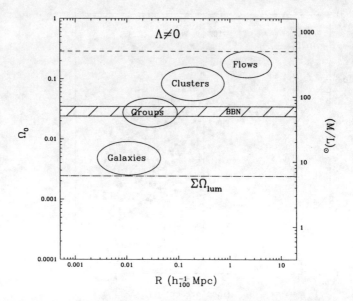

Figure 3. The density of matter on different scales. The dot-dashed line indicates the integral over luminous matter in galaxies and the striped region shows the baryon bounds from big bang nucleosynthesis.

4. What We Don't Know About Dark Matter

The nature of dark matter is the critical unresolved issue in cosmology. On the scale of individual galaxies, the evidence for dark matter is incontrovertible (for a review, see Ashman 1992). From kinematic studies on cosmological scale, two conclusions emerge, as illustrated by Figure 3. Clusters show dark matter in excess of the plausible upper bound from nucleosynthesis, indicating that most of the dark matter is nonbaryonic (Bahcall, Lubin, & Dorman 1995). Structure formation theories would require this dark matter to be non-relativistic at the epoch of recombination, or "cold." This in turns implies a particle physics solution to a cosmology problem, with no experimental data to guide us at this time. Second, cluster evolution and mass-to-light ratios point to a value of Ω_0 well below one (setting aside for a moment the ambiguous results from large scale bulk flows). If this is true, then the spatially flat universe that is a natural consequence of the inflationary big bang model can only be recovered with the addition of a non-zero cosmological constant. Unfortunately, we have no theoretical expectation for why Λ should have a cosmologically interesting value. Few people are willing to contemplate a deeper level of ignorance concerning our theory of gravity, yet there are puzzling problems in the dynamics of galaxies that deserve continued attention (Milgrom 1989; McGaugh & de Blok 1998).

5. Bright Icebergs

Following earlier speculation by Zwicky (1957), Arp (1965), and Peebles (1971), Disney (1976) proposed the existence of galaxies that lurked mostly below the level of the sky brightness. The distribution of surface brightness shows no sign of a turnover down to the limit of current surveys (McGaugh 1996; Dalcanton et al. 1997). This raises the issues of whether there is a limit to the diffuseness of a stellar system and what the mode of star formation is in a disk with only a few $M_\odot$ pc^{-2}. Even if Malin-type disks do not contribute much to the luminosity-weighted integral of galaxies, they are important laboratories for studying the distribution of dark matter in large halos (Navarro, Frenk & White 1997). Also, the large size of the gas disks can result in a significant cross section to quasar absorption, perhaps accounting for some fraction of the damped Lyman-α systems.

6. Dim Icebergs

Supercomputer simulations that incorporate hydrodynamics have been able to reproduce many of the large scale features of the universe. Cold dark matter modulates the formation of a filamentary web of low column density gas (e.g. Ostriker & Cen 1996). Quasar absorption can be used to trace the distribution and metallicity of this gas over nearly ten orders of magnitude in column density. Figure 4 shows the different scales probed by hydrogen absorbers of different column densities. The highest column densities correspond to damped Lyman-α lines due to galaxy disks or their progenitors at high redshift. More modest column densities trace metal-enriched gas in the extended halos of bright galaxies. At $z \sim 2$ most of the baryons in the universe are in diffuse structures of about 10^{14} atoms cm^{-2} (Hernquist et al. 1996; Miralda-Escude et al. 1996). Most of this gas has probably been heated to a high temperature by the present epoch (Cen & Ostriker 1998). Galaxies are the whitecaps that float on this churning sea of diffuse baryons and dark matter.

7. Dark Icebergs

It is possible to find galaxies that are dark matter dominated across the size spectrum — from giant low surface brightness disks to gas-rich dwarfs. Perhaps there are dark halos where the baryonic component is negligible. There have been claims of "dark galaxies" based on the incidence of close quasar pairs where no lens is visible, even though a massive galaxy would be required to cause the observed image splitting (Hawkins 1997). Counter to this claim is the fact that radio and optical surveys do not show the same incidence of close quasars with no visible lensing galaxy (Kochanek, Falco, & Munoz 1998). The alternative explanation is that the quasars are bound pairs, with interaction fuelling the activity in both AGN. At the other end of the mass scale, there is a clear expectation that dark halos exist, since the mass function in hierarchical clustering models is much steeper than the faint end of the galaxy luminosity function. Small halos may not be able to make or retain substantial numbers of stars.

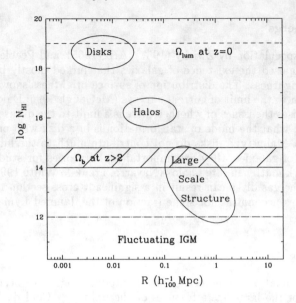

Figure 4. The column density and characteristic size of hydrogen quasar absorbers over ten decades in column density. The dot-dashed line is the level at which individual absorbers blend into a continuously fluctuating Gunn-Peterson effect. The dashed line is the mass density in luminous material measured locally.

8. What We Don't Know About the Universe

Cosmology has moved from its pioneering phase into its middle years, but the end game is not in sight yet. Figure 5 is a schematic view of the basic cosmological parameters, shown in terms of their level of uncertainty. The right hand side of the bar or arrow indicates the level of knowledge that we have now or can anticipate in the near future (for a recent overview, see Turok 1997). Only the Hubble constant is approaching a level of 10% precision, although there is still discordance between results from the conventional chain of of distance indicators and direct techniques (that depend on supernovae, gravitational lensing, and the Sunyaev-Zeldovich effect). Direct measurement of space curvature remains elusive, but the matter density is its proxy in the absence of substantial vacuum energy. The best evidence indicates that the universe cannot be closed by matter, whatever its form. After a series of adjustments to the stellar models, the derived ages of globular clusters have diminished to the point that they can accommodate a flat matter-dominated universe, provided that $H_0 < 67$ km s^{-1} Mpc^{-1} (Chaboyer et al. 1998). The issue of the cosmological constant is still unresolved. Finally, we are gradually defining two key attributes of the universe after it begins to form structures — the star formation rate over a Hubble time and the power spectrum of density fluctuations.

The successful measurement of cosmological parameters is leaving us with some deep mysteries. If the cosmological constant can be ruled out, then we need a natural explanation for an open universe without sacrificing the virtues of the

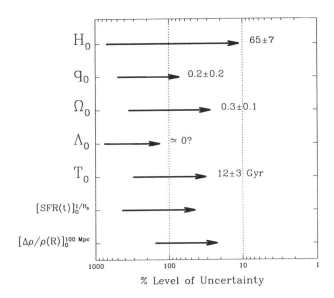

Figure 5. Current estimates of cosmological parameters. The end of the arrows indicate the precision of recent and prospective measurements. The diagram does not account for systematic errors that might affect each of these quantities.

inflationary big bang model. If standard inflation is saved with a vacuum energy term, we must explain why its contribution to the density parameter (based on unknown physics) is similar in magnitude to the matter density. Either way, we have to understand why baryons and dark matter particles — two fundamentally distinct microscopic phenomena — occur with roughly similar cosmic density. The invocation of anthropic arguments is merely a distraction as we try to refine our cosmological models. Ignorance can be a virtue; we should embrace it if it leads us to think of new and creative ways to understand the universe.

Acknowledgments. Thanks go to Mike Disney for illuminating much of our ignorance concerning galaxies, and for coming up with creative ways to counter it. Thanks to Jon Davies and the Local Organizing Committee for arranging a lively and convivial meeting. I acknowledge grant support from the National Science Foundation under AST-9003158 and AST-9617826, and from the International Programs Office of the University of Arizona.

References

Arp, H.C. 1965, ApJ, 145, 402
Ashman, K.M. 1992, PASP, 104, 1109
Bahcall, N.A., Lubin, L.M., & Dorman, V. 1995, ApJ, 447, 81

Bahcall, J.N. & Ostriker, J.P. eds. 1997, Unsolved Problems in Astrophysics, Princton: Princeton University Press
Bristow, P.D., & Phillipps, S. 1994, MNRAS, 267, 13
Cen, R., & Ostriker, J.P. 1998, ApJ, in press
Chaboyer, B., Demarque, P., Kernan, P.J., & Krause, L.M. 1998, ApJ, 494, 96
Copi, C.J., Schramm, D.N., & Turner, M.S. 1995, Science, 267, 192
Dalcanton, J.J., Spergel, D.N., Gunn, J.E., Schmidt, M., & Schneider, D.P. 1997, AJ, 114, 635
Disney, M.J. 1976, Nature, 263, 573
Hawkins, M.R.S. 1997, A&A, 328, L25
Hernquist, L., Katz, N., Weinberg, D.H., and Miralda-Escude, J. 1996, ApJ, 457, 51
Impey, C.D. & Bothun, G.D. 1989, ApJ, 341, 89
Impey, C.D. & Bothun, G.D. 1997, ARA&A, 35, 267
Impey, C.D., Sprayberry, D., Irwin, M.J., & Bothun, G.D. 1996, ApJS, 105, 209
Kennicutt, R.C. 1989, ApJ, 344, 685
Kennicutt, R.C. 1998, ARA&A, 36, 189
Kochanek, C.S., Falco, E.E., & Munoz, J.A. 1998, ApJ, in press
Madau, P., Pozzetti, L., & Dickinson, M. 1998, ApJ, 498, 10
McGaugh, S.S. 1996, MNRAS, 280, 337
McGaugh, S.S., & de Blok, W.J.G. 1997, ApJ, 481, 689
McGaugh, S.S., & de Blok, W.J.G. 1998, ApJ, 499, 41
Milgrom, M. 1989, ApJ, 338, 121
Miralda-Escude, J., Cen, R., Ostriker, J.P., & Rauch, M. 1996, ApJ, 471, 582
Ostriker, J.P., & Cen, R. 1996, ApJ, 464, 27
Navarro, J., Frenk, C.S., & White, S.D.M. 1997, ApJ, 490, 493
Peebles, P.J.E. 1971, Physical Cosmology, Princeton: Princeton University Press
Peebles, P.J.E. 1993, Principles of Physical Cosmology, Princeton: Princeton University Press
Persic, M., & Salucci, P. 1992, MNRAS, 258, 14P
Schneider, S.E., Spitzak, J.G., & Rosenberg, J.L. 1998, ApJ, 507, L9
Turok, N. ed. 1997, Critical Dialogs in Cosmology, Singapore: World Scientific Publishing
Zwicky, F. 1957, Morphological Astronomy, New York: Springer Verlag

Searching for LSB - VIII

At last, Large Supplies of Beer !